Nonlinear Physics of Ecosystems

Ehud Meron
Ben-Gurion University
of the Negev, Israel

CRC Press
Taylor & Francis Group
Boca Raton London New York

CRC Press is an imprint of the
Taylor & Francis Group, an **informa** business

Cover image: Vegetation patterns in an arid landscape. Courtesy of Kevin Sanders.

CRC Press
Taylor & Francis Group
6000 Broken Sound Parkway NW, Suite 300
Boca Raton, FL 33487-2742

First issued in paperback 2019

ISBN-13: 978-1-4398-2631-7 (hbk)
ISBN-13: 978-0-367-37771-7 (pbk)

Visit the Taylor & Francis Web site at
http://www.taylorandfrancis.com

and the CRC Press Web site at
http://www.crcpress.com

For Product Safety Concerns and Information please contact our EU representative: GPSR@taylorand⊠francis.com. Taylor & Francis Verlag GmbH, Kaufingerstraße 24, 80331 München, Germany.

To my wife, Liora, and my children, Elia and Omri,

and

to the memory of my parents,

my father, Yehoshua, who fervently believed in human progress through science and continued to support my scientific inclination even after witnessing the results of igniting hydrogen in my home lab,
and
my mother, Shoshana, who was a wholehearted artist, but her skepticism and analytical thinking would have well suited the soul of a scientist.

Patterns of shapes and colors (Shoshana Meron).

Contents

Preface

A significant part of this book is based on a graduate course on pattern formation which I have been teaching during the past two decades in the Physics Department at Ben-Gurion University. Another significant part of the book reflects the activities of my research group and the interactions with a few close colleagues, throughout this period on pattern formation problems and their applications to ecology. The book has also benefited from numerous joint meetings with Prof. Moshe Shachak's ecology group. These meetings helped to establish a common language, understandable to both physicists and ecologists, which I have used in the book.

The book is primarily intended for graduate students and researchers in nonlinear and interdisciplinary physics, geophysics, biomathematics, mathematical ecology, and ecohydrology. However, a broader readership, including ecologists in general and physical geographers, may also benefit from the book. The book chapters are intentionally divided into three parts: an overview of pattern formation and spatial ecology as disparate research fields that are yet strongly related to one another (Part I), an advanced account of pattern formation theory (Part II), and applications of pattern formation theory to ecological problems (Part III). Readers who are not mathematically oriented may skip Part II, which is pretty technical, and use the basic introduction to pattern formation theory and modeling in Part I to follow the ecological applications of pattern formation theory described in Part III. A fairly good understanding of model results can be achieved without dwelling on their mathematical derivation.

Many studies that are related to the topics addressed in the book are not cited, and I apologize for that. The book is not intended to provide a review of the proliferating studies at the interface between pattern formation and spatial ecology; it is rather intended to introduce the concepts and tools of pattern formation theory and demonstrate their utility in ecological research using selected problems in spatial ecology. It therefore includes representative references rather than citations to all relevant studies.

The content of the book is a result of many collaborations with colleagues and students. I am indebted to Christian Elphick, from whom I learnt more about asymptotic expansions and perturbation theory than from any course or textbook. I am also indebted to Moshe Shachak, who introduced me to the field of spatial ecology and kept updating me with new relevant studies; I particularly benefited from his integrative approach to the field. Special thanks go to Aric Hagberg and Jost von Hardenberg for most enjoyable and fruitful long-term collaborations, and to Golan Bel for the very productive recent collaboration. The outcomes of these interactions fill up many pages in this book. Special thanks go also to Harry Swinney, Anna Lin and other members of the experimental Austin group. The long and fruitful collaboration with this top group helped me realize the importance of confronting theory with experiment and has provided many joyful moments, especially when matching between the two has been achieved. Last but not least, I would like to thank Yagil Osem, Antonello Provenzale, Hezi Yizhaq, and Yair Zarmi for the most productive interactions, the outcomes of which constitute important parts of the book.

First and foremost, however, this book describes the work of graduate students I have been advising and co-advising throughout the years: Arik Yochelis, Erez Gilad, Efrat Sheffer, Assaf Kletter, Rotem Manor, Adam Lampert, Jonathan Nathan, Yair Mau, Lev Haim, Shai Kinast, Paris Kyriazopoulos, and Yuval Zelnik. Their capacity to study new topics and methods, their commitment to hard work, and the often surprising ideas and results they came up with made this book possible.

I would also like to thank all people who helped me in preparing the manuscript through helpful suggestions and corrections or figure preparation: Yair Zarmi, Moshe Shachak, Isaak Rubinstein, Ruhama Lipow, Michele Herman, Yuval Zelnik, Yair Mau, Lev Haim, and Marco Cusmai. Finally, I would like to thank Luna Han, a senior editor of the Taylor & Francis publishing group, for her helpful comments and patience, and my wife, Liora, for her help in designing the cover page and for her enthusiastic and continuous encouragement.

The support of the Israel Science Foundation, the US - Israel Binational Science Foundation, the Ministry of Science, Technology, and Space, and the James S. McDonnell Foundation is gratefully acknowledged.

About the author

Ehud Meron is a professor of physics in the Blaustein Institutes for Desert Research and the Physics Department at Ben-Gurion University of the Negev. His research interests include nonlinear dynamics and pattern formation theory with applications to fluid dynamics and chemical reactions, modeling complex systems, and spatial ecology with a focus on desertification and biodiversity dynamics. Professor Meron has been collaborating with ecologists since the year 2000 in an effort to assimilate the concepts of pattern-formation theory into ecological research, and this book is part of that effort.

Chapter 1

Introduction

1.1 An emerging new scientific discipline

Scientific research is generally conducted within well established disciplines of "normal science" [162] with few cross-disciplinary interactions. Periods of time in which two disparate scientific disciplines begin to interface with one another are rather the exception. Such is the case with spatial ecology and pattern formation, a research field that centers on the nonlinear dynamics of spatially extended systems and the self-appearance of spatial patterns. Field observations in arid and semi-arid regions during the past decade [317, 309, 63, 101] have revealed nearly periodic vegetation patterns that are familiar from a variety of other pattern-formation contexts, including fluid dynamics, chemical reactions and nonlinear optics [56]. A few examples of such patterns are shown in Figure 1.1. They consist of vegetation spots in an otherwise bare area devoid of vegetation (panel (a)), vegetation stripes (panel (b)), or barren gaps in vegetated areas (panel (c)). The understanding that vegetation patchiness is not merely dictated by environmental heterogeneities, but may also be a result of self-organization driven by pattern-forming instabilities of uniform states, has led to a surge of empirical and theoretical studies using the conceptual framework of pattern-formation theory [169, 155, 325, 128, 255, 256, 29, 212].

The relevance of pattern-formation theory to spatial ecology has been pointed out earlier [275, 173, 172], and has motivated modeling studies in various ecological contexts [259, 307, 209, 40, 293, 15, 190, 341, 220]. However, of all contexts, self-organized vegetation patchiness in water-limited systems stands out in providing the best case study, so far, for applying pattern-formation theory to spatial ecology. One reason for that is the wide scope of observed vegetation patterns and the good correspondence to model predictions. These observations not only include nearly periodic spot, stripe and gap patterns, but also a wide variety of non-periodic patterns (see Figures 10.12, 10.17 and 10.19). Another reason is the wide scope of ecological problems that

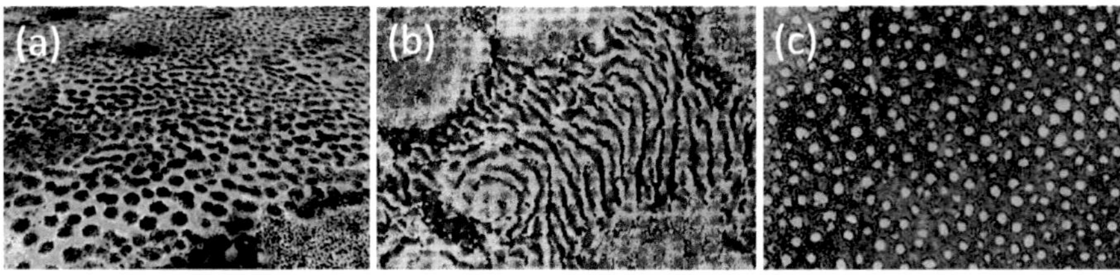

FIGURE 1.1: Aerial photographs of nearly periodic vegetation patterns in nature: (a) a spot pattern in Zambia (from [29]), (b) a stripe pattern in Niger (from [317]), (c) a gap ("fairy circle") pattern in Namibia (courtesy of S. Getzin).

can be addressed, including outstanding questions such as desertification, biodiversity loss and their implication for ecosystem function.

The increasing interest in dryland vegetation has motivated pattern formation studies in wetland vegetation too [85, 46], and in a few other marine ecosystems, such as mussels beds [178]. In all these cases pattern formation results from non-uniform instabilities of uniform states in which the growth of spatially structured modes leads to patterned states. However, pattern formation may also result from uniform instabilities that give rise to a multiplicity of stable uniform states, as patterns consisting of spatial domains occupied by different states are then possible. As will be shown in Part II of this book, such systems can show a wide variety of persistent patterns, including stationary labyrinthine patterns, rotating spiral waves, and spatiotemporal chaos. A multiplicity of stable states has been found in studies of tidal marshes [197], plankton systems [262, 261] and coastal vegetation [140], and is likely to be found in many more marine or marine-related ecosystems. These systems all lend themselves to pattern formation studies.

The main thesis we pursue here is that inasmuch as concepts of nonlinear dynamics, such as multi-stability of steady states, tipping points, oscillations and chaos, have already been integrated into ecological research, pattern formation concepts should be integrated too. The latter include the concepts of a non-uniform instability, periodic stripe and hexagonal patterns, traveling waves, front dynamics, spatial resonances and others. The need to integrate these concepts should be expected on general grounds, as ecosystems are nonlinear spatially extended systems, like all other pattern forming systems in nature, but can also be motivated using concrete examples, as we discuss in the next section.

1.2 Pattern formation—a missing link in ecological research

Much effort is focused in ecology on understanding the reciprocal relationships between the abiotic environment, biodiversity, and ecosystem function [180]. We argue that these relationships are very often mediated by pattern-formation processes as Figure 1.2 schematically illustrates. Pattern formation is directly linked to any of the three components (small dotted arrows). It is linked to the abiotic environment because environmental stresses often induce spatially patterned states. It is linked to biodiversity because it may induce self-organized heterogeneity of biomass and resources that affect inter-specific interactions[1]. It is also linked to ecosystem function since pattern formation can affect water-use efficiency and biomass production, or imply different rates and pathways of nutrient change.

These links form indirect causal relationships between the abiotic environment, biodiversity and ecosystem function through various pattern formation processes (solid arrows in Figure 1.2). The impact of climate change on species diversity through pattern transitions that change inter-specific interactions is an example of an indirect relation between the abiotic environment and biodiversity. A possible example of an indirect relation between biodiversity and ecosystem function is spatial self-reorganization of a community in an alternative stable state of different productivity, and an example of an indirect relation between the abiotic environment and ecosystem function is gradual regime shifts involving cascades across different pattern states.

The pattern formation links depicted in Figure 1.2 involve processes occurring on different length scales and across different organization levels. Figure 1.3 illustrates an example of a series of such processes in the context of dryland vegetation. The processes described are motivated by model studies to be presented in detail in Part III. Local biomass-water feedbacks, involving water transport toward vegetation growth points, can induce spatial instabilities that lead to vegetation pattern formation at the landscape scale. Environmental changes at the landscape scale, such as drought or spate, can induce transitions to a variety of other alternative stable patterns. Associated with these transitions are changes in the spatial soil-water distributions, which, in turn, affect inter-specific interactions at local scales. In woody-herbaceous systems these interaction changes can result in transitions from competition to facilitation and, consequently, in community-structure changes.

The scenario described above includes bottom-up processes whereby plant-plant interactions, mediated by the limiting water resource at the local scale, give rise to the emergence of periodic patterns at the landscape scale. It also

[1]The term *inter-specific* interaction refers to the interaction between individuals of different species, in contrast to the term *intra-specific* interaction, which refers to interactions between individuals of the same species.

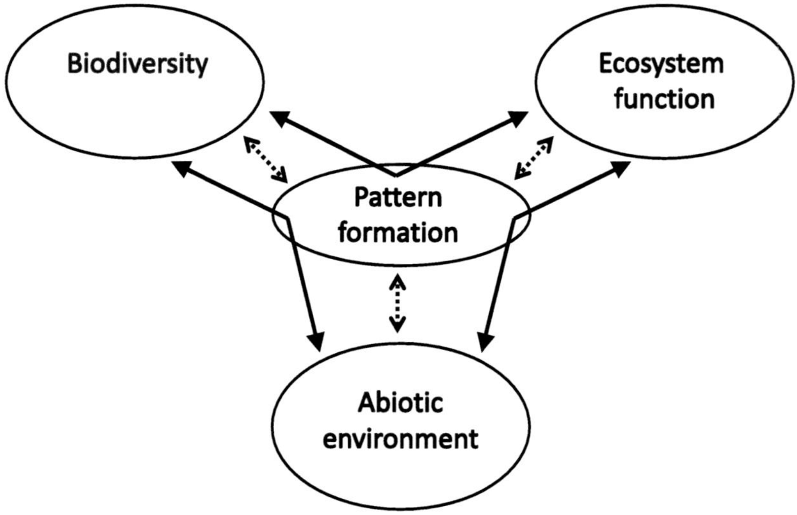

FIGURE 1.2: Pattern formation processes induce indirect causal relationships between the abiotic environment, biodiversity and ecosystem function. The dotted arrows represent different manners by which pattern formation is linked to these three components, and the solid arrows represent compositions of these manners that form indirect relationships between the three components (see the text for examples).

includes top-down processes, in which species-interactions at local scales are affected by pattern changes at the landscape scale. These processes not only involve disparate length scales, but also different levels of organization, starting at the organism level, with species traits that give rise to spatial instabilities, proceeding to the population level, through vegetation pattern formation, and to the community level, through changes in inter-specific interactions.

This example also highlights the integrative role pattern formation can play in ecological research. Ecology, as an empirical science, has branched into different research fields according to the hierarchical levels and spatio-temporal scales that the empirical studies have addressed. As a result, many subdisciplines have emerged, such as population ecology, community ecology, ecosystem ecology[2], and landscape ecology. By bridging over different organization and trophic levels, and over different length and time scales, studies of pattern formation in ecology can contribute to the integration of these subdisciplines.

[2]Ecosystem ecology is a subfield of ecology dealing with the flow of energy and matter through biotic and abiotic ecosystem components. The term is somewhat misleading in that it refers to specific aspects of ecosystems, rather than to all aspects as the term suggests.

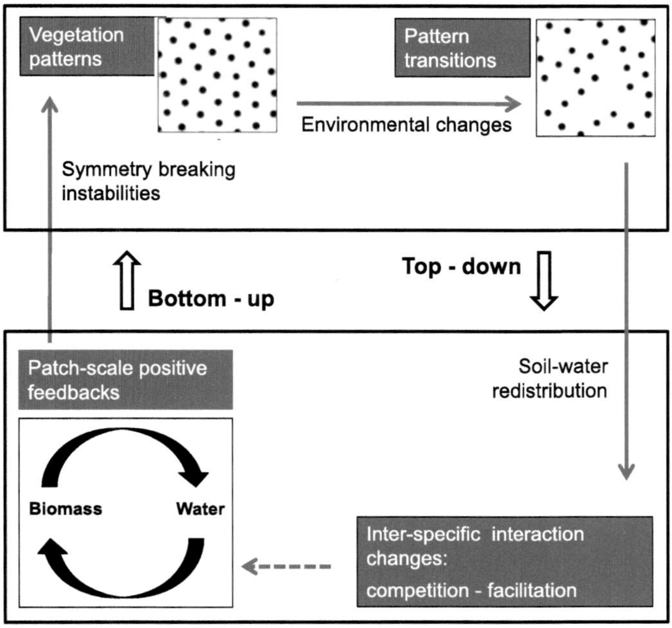

FIGURE 1.3: Pattern formation can link phenomena that occur on different length scales and organization levels - an illustration with a dryland-vegetation example. Local biomass-water feedbacks (lower frame) induce vegetation pattern formation at the landscape scale (upper frame; dark spots represent vegetation patches). Environmental changes at the landscape scale induce transitions to other alternate stable vegetation patterns (upper frame). These pattern transitions change the local soil-water distributions and thereby affect inter-specific interactions (lower frame). In woody-herbaceous systems these interaction changes may induce transitions from competition to facilitation. In other systems they may feed back on vegetation pattern formation (dashed arrow).

1.3 Purpose and scope of the book

The purpose of this book is to assimilate the concepts and methods of pattern formation theory into ecological research and, thereby, to contribute to the development of the newly emerging interdisciplinary research field at the interface between spatial ecology and pattern formation. Much of the book revolves around the diagram shown in Figure 1.2, and the elucidation of various links between pattern formation, on one hand, and the abiotic environment, biodiversity and ecosystem function, on the other hand. We focus on dryland vegetation as a case study, but the general approach is applicable to other ecological contexts, including marine ecosystems. The book is by no means intended to be comprehensive. It is rather intended to demonstrate the utility of pattern formation theory in ecological research and to highlight outstanding open problems that can be handled with this approach, along with the progress that has already been made on selected problems.

Being an interdisciplinary field in its infancy, a significant part of the book is devoted to the introduction of pattern formation theory. The introduction is made at two levels; an elementary level that requires basic knowledge of linear algebra and ordinary differential equations, and a more advanced level which also requires familiarity with partial differential equations. The elementary-level introduction is included in Part I of the book, which is devoted to an overview of pattern formation and spatial ecology as strongly related disparate research fields. The advanced introduction is presented in Part II of the book, and includes descriptions of analytical tools and applications of these tools to the study of general mechanisms of pattern formation and pattern dynamics.

There are several related topics which were left aside. We focus on deterministic dynamics, largely ignoring stochastic aspects, such as demographic noise in small populations [28] and noise-induced patterns [34, 268, 253]. A few pattern formation topics have not been considered, including front propagation into an unstable state [73], which is relevant to species invasion problems [289], and pattern formation in excitable media [211], which has been studied in the context of phytoplankton ecosystems [311, 189]. The presentation of mathematical methods has also been limited to the most common ones. For example, we refer to, but do not describe the derivation of phase equations far from the onset of instabilities [226, 136], nor do we describe free boundary-layer analysis, leading, for example, to kinematic descriptions of curved fronts [20, 119, 121] and spiral waves [124]. Finally, although numerical methods are crucial tools for studying nonlinear spatially extended systems we left a detailed description of these methods outside the scope of this book, as this topic is well covered in the literature.

Part I

Overview

Chapter 2

Spatial self-organization

2.1 Natural and laboratory realizations of pattern formation

The term "pattern formation" refers to processes by which positive feedbacks operating at small spatial scales give rise to self-organization at large scales that results in stationary or time-dependent spatial patterns. Numerous examples of such processes have been found and studied in various fields of science, including fluid dynamics, chemical kinetics, nonlinear optics and geophysics. Much of our understanding of pattern formation phenomena derives from controlled laboratory experiments conducted on simple model systems. Following a brief description of patterns in nature and some of their characteristic features (Section 2.1.1), we introduce two experimental model systems, the Rayleigh–Bénard thermal convection system (Section 2.1.2) and the Belouzov–Zhabotinsky chemical reaction (2.1.3). Both systems show a variety of pattern formation phenomena and have played important roles in uncovering general mechanisms of pattern formation and dynamics.

2.1.1 Patterns in nature

A common feature of spatially extended nonequilibrium systems is the possible emergence of ordered patterns with characteristic length scales. Plenty of natural examples of this phenomenon exist, including cloud streets, sand ripples, stone patterns and vegetation patterns (Figure 2.1). A fascinating aspect of these pattern-formation phenomena is that the order is not imposed by any external factor; it rather results from positive feedbacks operating at small scales, that give rise to self-organization and pattern formation at large scales. We will discuss these phenomena in two specific physical contexts shortly, but some of the underlying principles can already be stated. A uniform force that drives a uniform system out of equilibrium can break the spatial uniformity of the system and induce spatially periodic patterns. The transition to the patterned state is not gradual; spatial variability appears only beyond a critical force strength. We call such a phenomenon a *symmetry breaking instability*. Although we generally cannot prove the emergence of patterns in nature from symmetry breaking instabilities, we often do make this association, relying on experimental studies of model systems and on mathematical analyses of model equations.

Another principle of pattern formation relates to the universal nature of these phenomena, that is, to the observations of similar patterns, such as stripes, hexagons and spiral waves, in completely different physical contexts. Stripe patterns, for example, appear in clouds, in dryland vegetation and in animal coat patterns, although the mechanisms responsible for these patterns are obviously specific to the system in question, and differ from one another. The universality of pattern formation phenomena is tightly related to the symmetry breaking instabilities that induce patterns; different systems that go through the same type of instability behave similarly close to the instability threshold.

Pattern formation is an example of an emergent property [3], that is, a property that appears at the system level, e.g., the level of clouds, sand dunes, or patchy landscapes, and often has no meaning at the level of the system's constituents—the water molecules, sand grains, or plants. At the system level, patterns are affected by global forces, such as temperature, wind, and rainfall, and the resulting pattern dynamics may feed back on small-scale processes (Figure 1.3). Studying pattern formation and pattern dynamics is therefore significant for understanding bottom-up and top-down cross-scale processes in complex natural systems.

2.1.2 The Rayleigh–Bénard system

A classical experimental model for pattern formation is the Rayleigh–Bénard (RB) system of thermal convection [44, 24]. Consider a fluid at rest that is heated from below. If the temperature difference between the bottom and the top of the fluid compartment is smaller than some critical value,

FIGURE 2.1: Patterns in nature. From left to right: cloud stripes, sand ripples on a dune [13], stone patterns [151] and grass patterns [325].

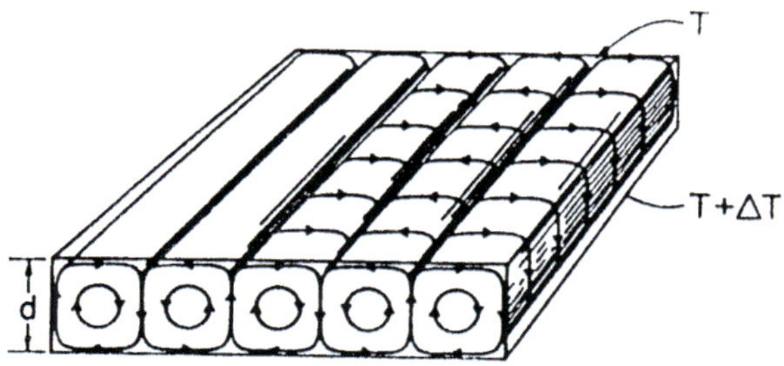

FIGURE 2.2: Schematic representation of convection rolls in the Rayleigh–Bénard system. The fluid compartment is of depth d with temperature at the bottom $T_{\text{bot}} = T + \Delta T$ higher than that at the top $T_{\text{top}} = T$. From [56].

$\Delta T = T_{\text{bot}} - T_{\text{top}} < \Delta T_C$, the fluid remains at rest and the heat transfer upward proceeds by molecular conduction. However, when $\Delta T > \Delta T_C$ a convection sets in, generating ordered parallel fluid rolls as illustrated in Figure 2.2. The number of rolls is approximately determined by the ratio of the length of the fluid compartment to its height, the so-called *aspect ratio.*

Why does the rest state of the fluid become unstable beyond a certain temperature difference? Imagine a fluctuation in which a fluid particle[1] at some height has a temperature which is slightly higher than the surrounding fluid at that height. Because of thermal expansion the fluid particle will have a lower density than that of the surrounding fluid (i.e., will be lighter) and will tend to move upward. As it moves upward the surrounding fluid becomes yet colder and the buoyancy force upwards increases. This is a positive feedback between the height of the fluid particle and the buoyancy force; the higher the

[1]By a "fluid particle" we mean a parcel of fluid which is very small on a macroscopic scale, but still very large on a microscopic scale.

particle's position the stronger the force and the stronger the force the higher the particle's position. Besides the buoyancy force there are also processes that act to stabilize the rest state. Fluid viscosity induces transfer of linear momentum from the upward moving fluid particle to its neighborhood, thus reducing its momentum and speed. In addition, thermal conduction induces diffusion of heat from the fluid particle to its colder neighborhood, thus reducing the buoyancy force that drives the fluid particle upward. The instability therefore sets in at a critical temperature difference, ΔT_C, at which these stabilizing factors just balance the destabilizing buoyancy force.

The instability can be induced by varying other parameters that affect the buoyancy force or the stabilizing factors: the coefficient of thermal expansion, α, the thermal diffusivity, κ, the fluid's kinematic viscosity, ν, the gravitational acceleration, g, and the hight of the fluid layer, d. The instability can be induced, for example, by increasing α which strengthens the buoyancy force, or by decreasing κ and ν, which weakens the stabilizing factors. The effects of all parameters are lumped together in a single dimensionless parameter, the so called Rayleigh number R, given by [44]

$$R = \frac{\alpha g \Delta T d^3}{\kappa \nu} . \tag{2.1}$$

The instability sets in as R exceeds a threshold value R_c. In practice, the Rayleigh number is generally increased by heating the bottom of the fluid compartment, i.e., by increasing ΔT. The roll patterns that form beyond the instability point can be visualized by the shadowgraphy method. This method makes use of the fact that the index of refraction varies weakly with temperature; warmer (colder) fluid regions have a lower (higher) index of refraction. Since the temperature varies periodically across the rolls, so does the refraction index. Passing a beam of light through the fluid layer results in a pattern of alternating bright and dark stripes. The bright stripes correspond to regions of cold fluid flowing downward, that act as converging lenses because of the higher refraction index. Figure 2.3 shows examples of roll patterns observed with the shadowgraphy method.

2.1.3 The Belouzov–Zhabotinsky reaction

The Rayleigh–Bénard system is an example of an experimental pattern-formation model associated with fluid motion. Chemical reactions provide another type of experimental pattern-formation model. A classical example is the oscillatory Belousov–Zhabotinsky (BZ) reaction [90], a catalytic oxidation reaction of malonic acid in an acidic bromate solution. A nice aspect of this reaction is that the oscillations are clearly visible to the bare eye because of the different colors associated with the two oxidation states of the catalyst. The mechanism of this reaction has been worked out by Field et al. [91] and contains many elementary reactions. A reduced model (the Oregonator), consisting of only five reaction steps, captures many qualitative aspects of the

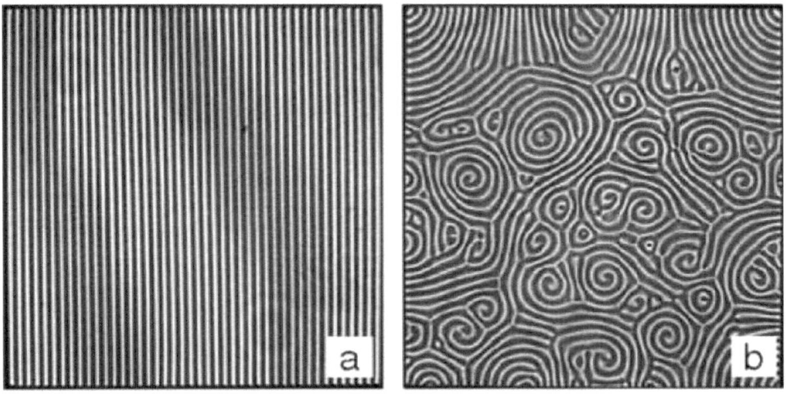

FIGURE 2.3: Regular (a) and chaotic (b) roll patterns in a Rayleigh–Bénard convection experiment. From [24].

BZ reaction dynamics [92]. The five reaction steps are:

$$A + W \rightarrow U + P\,, \tag{2.2a}$$
$$U + W \rightarrow 2P\,, \tag{2.2b}$$
$$A + U \rightarrow 2U + 2V\,, \tag{2.2c}$$
$$2U \rightarrow A + P\,, \tag{2.2d}$$
$$B + V \rightarrow hW\,. \tag{2.2e}$$

where $A = BrO_3^-$ (bromate ions), $W = Br^-$ (bromide ions), $U = HBrO_2$ (bromous acid), $P = HOBr$ (hypobromous acid), V is the oxidized form of the catalyst (e.g., cerium Ce^{4+}), $B = CH_2(COOH)_2$ (malonic acid), and h is a stoichiometric coefficient (note that this simple model is not stoichiometrically balanced). The key steps in this reaction scheme are (2.2b) and (2.2c). Both steps compete for U, but while U is consumed in (2.2b) it is autocatalytically produced in (2.2c). An initial access of W eliminates U in step (2.2b) before step (2.2c) becomes significant. However, as W drops down step (2.2c) takes over. This step involves a positive feedback (U accelerates the growth of itself), which leads to fast exponential production of U. The production of U is accompanied by the production of V, which changes the color of the solution. The growth of U is slowed down in step (2.2d) which, together with step (2.2e), brings the system to the starting point and to the initiation of a new cycle. The chemical composition needed to initiate the reaction consists of an acidic aqueous solution containing bromate ions (e.g., potassium bromate), malonic acid, bromide ions (e.g., potassium bromide), and a metal catalyst in a reduced form (e.g., cerium Ce^{3+}).

The BZ reaction is an example of an *activator-inhibitor* system. In such a system, the activator is a substance that "activates" the growth of itself and of another substance–the inhibitor. The inhibitor inhibits the growth of the

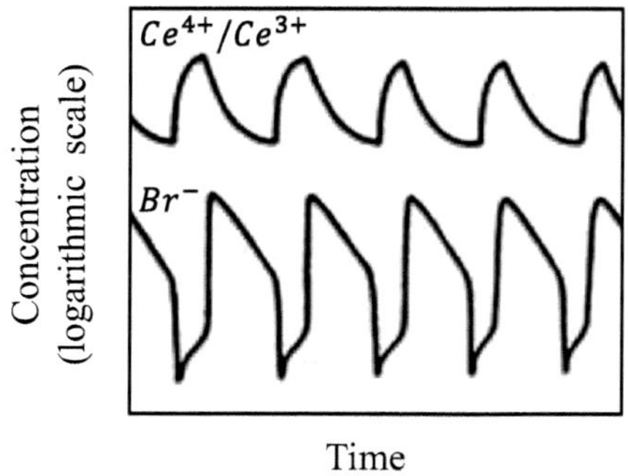

FIGURE 2.4: Relaxation oscillations in the BZ reaction. Shown are time signals of $\log[Ce^{4+}]/[Ce^{3+}]$ (top) and $\log[Br^-]$ (bottom). The oscillation period is of the order of $10^2 s$. Adopted from [91].

activator and often the growth of itself too. In the BZ reaction U plays the role of the activator and V the role of the inhibitor. Quite often the activator changes on a time-scale significantly shorter than that of the inhibitor. The oscillatory dynamics then involve alternate phases of slow and fast concentration changes as Figure 2.4 shows. Oscillations that involve two distinct time scales are often called "relaxation oscillations" [296]. In spatially extended systems with diffusive coupling such oscillations can give rise to traveling-wave phenomena; diffusion of the activator to its neighborhood, before it is damped by the inhibitor, can induce its growth there and therefore its spatial spread. Sufficiently fast inhibitor diffusion, on the other hand, can give rise to stationary patterns; the fast inhibitor diffusion away from an activated domain prevents the local decay of the activator and also the activator's spread to the highly inhibited neighborhood of this domain. In the BZ reaction the activator changes on a time scale much shorter than that of the inhibitor, and the inhibitor diffusion is sufficiently slow to allow for traveling waves.

Early pattern formation experiments in the BZ reaction were made in closed systems (petri-dish experiments) with the inevitable approach to a stationary, uniform equilibrium state. Even in these simple experiments striking traveling-wave phenomena, such as spiral waves, have been observed [331]. More recent experiments have utilized open systems which are continuously fed with fresh chemicals so as to keep the system at a fixed distance from equilibrium [299, 17]. A typical experimental setup consists of a thin reactor layer containing an inert gel or a porous glass that allow diffusion of the reactants but damp convection. The reactor layer is in diffusive contact with one or two stirred reservoirs which are continuously fed with fresh reagents. The reaction dynamics can be controlled by varying the flow rates of chemical reagents, such as bromate or malonic acid, into the reservoirs. The advantage of this type of setup is that it allows conducting long experiments which are needed in studying instability phenomena.

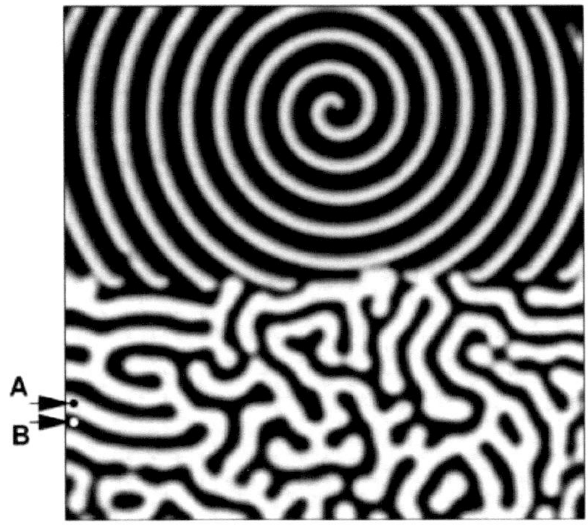

FIGURE 2.5: Patterns in a reactor of the light sensitive BZ reaction whose lower half is subjected to periodic illumination at twice the natural frequency of the oscillatory reaction. The unforced reaction (upper half) shows a rotating spiral wave, while the forced reaction shows a labyrinthine standing-wave pattern. The labels A and B denote points that oscillate out of phase. From [241].

Various modifications of the BZ reaction have been studied. One type of modification is the replacement of the metal catalyst, Ce^{3+}, by other metal ions such as Fe^{2+}, Mn^{2+} and Ru^{2+}. Of particular significance is the use of ruthenium (Ru^{2+}) as the catalyst. This modification makes the chemical kinetics sensitive to light, and allows studying the effects of forcing the chemical oscillations by periodic illumination in time or in space or in both. Experiments on the light-sensitive BZ reaction, subjected to time-periodic, spatially uniform illumination, have shown resonant responses similar to those found in periodically forced oscillators [177]. That is, denoting the oscillation frequency of the unforced system by ω_0, and the forcing frequency by ω_f, resonance bands have been found in which the oscillation frequency of the forced system, ω, locks to a rational fraction of the forcing frequency, $\omega = (n/m)\omega_f$, in a range of ω_f around ω_0 (whose width increases with the forcing amplitude). Various resonances $(\omega_f : \omega) = (m : n)$ have been found in the experiments, revealing part of a Farey tree hierarchy of resonances[2] [74, 108] as Figure 8.1 shows. The forcing, however, can also induce new spatial patterns. A striking example is shown in Figure 2.5, where spiral waves in the unforced reaction destabilize to standing-wave labyrinthine patterns when a sufficiently strong uniform forcing with a frequency $\omega_f \approx 2\omega_0$ is applied. A detailed discussion

[2]In a Farey tree of resonances, between any two resonances $(i : j)$ and $(k : l)$ there is an intermediate resonance $(i + k) : (j + l)$.

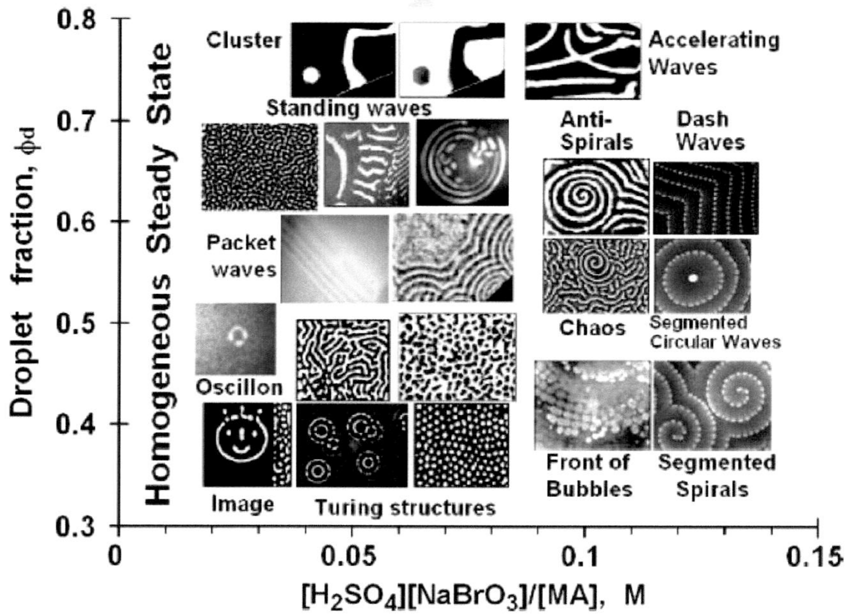

FIGURE 2.6: Patterns observed in the BZ-AOT reaction. From [321].

of periodically forced, spatially extended oscillatory systems is presented in Chapter 8.

Another interesting modification of the BZ reaction is a water-in-oil microemulsion system with nanometer-sized droplets of water surrounded by monolayers of a surfactant (aerosol OT or AOT) and dispersed in oil (octane) [321]. Since most reagents of the BZ reaction are polar, they reside in the water droplets and diffuse at a relatively slow rate characteristic of entire droplets. Some key intermediates, however, are non-polar and therefore escape into the oil and diffuse much faster. This property of the BZ-AOT reaction allows for stationary patterns and a wide variety of traveling wave patterns not observed in the original BZ reaction, as Figure 2.6 shows.

2.2 Pattern-forming systems as dynamical systems

Pattern-forming systems are often described by small sets of fields; the velocity and temperature fields in thermal convection, the concentrations of key species in chemical reactions, biomass and resource fields in patchy landscapes, and so on. Quite often approximate dynamical equations for these fields are known. For relatively simple systems the dynamical equations are derivable from first principles. This is the case with simple fluids that satisfy the Navier–Stokes equations [44], or with electromagnetic radiation in dielectric materials that satisfies the Maxwell equations [229]. More complex systems generally involve some degree of modeling. Pattern-forming chemical reactions, for example, are described by reaction-diffusion models that simplify the complex chemical kinetics of these reactions. Pattern-forming ecosystems require a yet higher degree of modeling (see Chapter 9). We will refer to the set of dynamical equations that describe a pattern forming system as a *dynamical system.*

In what follows we distinguish between small and large pattern forming systems using the concept of *dimension* of a dynamical system (Section 2.2.1). Small pattern forming systems are governed by small numbers of independent degrees of freedom and can be described by small sets of nonlinear ordinary differential equations or "low-dimensional" dynamical systems. The theory of low-dimensional dynamical systems is well developed and is described in many textbooks [296, 165, 163]. We briefly describe it here (Sections 2.2.2, 2.2.3, 2.2.4, and 2.2.5) focusing on concepts that are essential for understanding the presentation of pattern formation theory in Part II and the applications to spatial ecology in Part III.

2.2.1 Dimension and size of a pattern-forming system

The dynamical equations of pattern-forming systems represent infinitely many degrees of freedom; formally, any point in space contributes at least one degree of freedom. However, because of the dissipative nature of these systems, the number of independent degrees of freedom reduces dramatically in the course of time. The asymptotic[3] independent degrees of freedom generally represent slow modes, such as modes that begin to grow at instability points, but may describe faster processes as well, such as transitions between slowly evolving states. We define the *dimension* of a pattern-forming system as the number of independent degrees of freedom that describe the (asymptotic) long-term dynamics of the system.

The dimension of a pattern-forming system strongly depends on its physical size. The size is determined relative to a typical length in the system, such as the wavelength of a periodic pattern. Thus, a small (large) aspect-ratio Rayleigh–Bénard (RB) system that fits in a few (many) pairs of rolls, is an example of a small (large) system. The dynamics of small pattern-forming systems, just above the instability of the equilibrium state, involve a small number of independent degrees of freedom. The tremendous reduction in the number of degrees of freedom in this case is mathematically accounted for by the center manifold theorem [114], but can be intuitively understood using the example of a small RB system. The roll pattern that sets in at the instability

[3]Throughout the book we will use the term "asymptotic" to refer to long times unless otherwise is stated.

point represents the first spatial mode for which the buoyancy force just overcomes all dissipative processes. All other modes describe significantly different spatial structures for which the dissipative processes still dominate the buoyancy force in some range above the instability point. These modes decay to zero unless they are nonlinearly coupled to the growing mode. In that case they do show up but do not represent independent degrees of freedom. As the system becomes larger, more pairs of rolls fit in, and the difference between adjacent modes (i.e., modes describing n and $n+1$ pairs of rolls) becomes smaller. As a result, more modes can grow and the dimension of the system increases.

The independent degrees of freedom of a pattern forming system near an instability point are often represented by the amplitudes of the modes that begin to grow at that point. Consider, for example, a chemical system that goes through an instability to a stationary periodic pattern with a wavelength λ. The concentrations, $c_i(x,t)$ $(i = 1, ..., n)$, of the chemical species that participate in the reaction, can be approximated, near the instability point, by

$$c_i(x,t) \approx \alpha_i u(t) \cos(k_0 x + \phi)\,, \tag{2.3}$$

where the cosine term represents the spatial mode that starts growing at the instability point, with $k_0 = 2\pi/\lambda$ and ϕ being its wavenumber and phase, $u(t)$ is a slowly varying amplitude of that mode, and the factors α_i are constants. The amplitude u represents the independent degree of freedom that describes the long-term dynamics of the system in some parameter range close to the instability point in which k_0 is the only mode to grow[4]. In small systems, i.e., systems whose size L is comparable to λ $(L \sim \lambda)$, this parameter range can be significant and the long-term dynamics are captured by a single degree of freedom—the amplitude of the growing mode k_0. However, in large systems $(L \gg \lambda)$ this range can become diminishingly small, because there are many other modes with wavenumbers close to k_0. At any finite range beyond the instability point the asymptotic dynamics are described by a set of independent degrees of freedom representing a band of modes centered around k_0. We postpone the discussion of large systems to Chapters 5 and 6 and consider in the rest of this section small systems that are describable by low-dimensional dynamical systems.

2.2.2 Basic concepts of low-dimensional dynamical systems

Low-dimensional systems are generally described by small sets of nonlinear ordinary differential equations or ODEs:

$$\dot{\mathbf{u}} = \mathbf{f}(\mathbf{u};\lambda)\,, \tag{2.4}$$

where $\mathbf{u} = (u_1(t), ..., u_n(t))$ is a vector of real valued state variables, representing the independent degrees of freedom, $\lambda = (\lambda_1, ..., \lambda_m)$ is a set of parameters,

[4]Note that the number of independent degrees of freedom is not determined by the number of chemical species n, but rather by the number of modes that grow at the instability point.

$\mathbf{f} = (f_1(\mathbf{u}; \lambda), ..., f_n(\mathbf{u}; \lambda))$ is a non-linear vector function of the state variables, and the dot represents the time derivative ($\dot{u} = du/dt$)[5].

The space spanned by the state variables $\mathbf{u}$ is called *phase space*. The temporal evolution of the system, $\mathbf{u}(t)$, from an initial value, $\mathbf{u}(0)$, traces a trajectory in phase space. Trajectories of this kind can be viewed as "stream lines" of a *flow* in phase space, determined by the specific form of $\mathbf{f}(\mathbf{u})$.

In general, the first objective in studying equations of this kind is identifying steady-state solutions and studying their stability properties. In the phase space of the system such solutions are represented by points and are often referred to as *fixed points*. We say that a steady-state solution, $\mathbf{u_s} = (u_{s1}, ..., u_{sn})$ of (2.4) is *linearly stable* if any infinitesimally small perturbation of $\mathbf{u_s}$ decays in time[6]. Conversely, a steady-state solution $\mathbf{u_s}$ is *linearly unstable* if there exists a small perturbation of $\mathbf{u_s}$ that grows in time. To study the linear stability of $\mathbf{u_s}$ we consider an infinitesimal perturbation of that solution, which we denote by $\delta\mathbf{u} = (\delta u_1, ..., \delta u_n)$, and insert the perturbed form $\mathbf{u}(t) = \mathbf{u_s} + \delta\mathbf{u}(t)$ into (2.4). Linearizing around $\mathbf{u_s}$ we obtain

$$\delta\dot{\mathbf{u}} = \mathbf{J}\delta\mathbf{u}\,, \tag{2.5}$$

where $\mathbf{J}$ is the Jacobian matrix whose (i, j) entry is $\partial f_i/\partial u_j|_{\mathbf{u}=\mathbf{u_s}}$. The solution $\mathbf{u_s}$ is linearly stable when the eigenvalues of $\mathbf{J}$ all have negative real parts, for in that case any perturbation $\delta\mathbf{u}$ decays exponentially in time. It is linearly unstable if the largest real part of all eigenvalues is positive[7].

An instability of a steady-state solution generally takes the system to a new steady-state solution or to a time-periodic solution. We refer to such instabilities as to *stationary instability* and *oscillatory instability*, respectively. In two-dimensional systems steady-state and time-periodic are the only possible asymptotic solutions. In higher dimensional systems chaotic dynamics are possible too [296]. In the following two subsections we analyze a few examples of stationary and oscillatory instabilities and use them to introduce additional concepts of dynamical systems.

2.2.3 Stationary instabilities

Many of the concepts to be introduced here can be explained using the following one-dimensional system and variants thereof:

$$\dot{u} = f(u; \lambda) = \lambda u - u^3\,. \tag{2.6}$$

[5] Readers unfamiliar with ODEs are referred to Ref. [165] or to any other textbook on ODEs

[6] The theory of dynamical systems defines a few forms of stability [107]. The definition given here amounts to *asymptotic stability* in which points near $\mathbf{u_s}$ converge to it directly. Weaker forms of stability include quasi-asymptotic stability in which nearby points eventually converge to $\mathbf{u_s}$, but not necessarily in a direct manner. This weaker form of stability occurs, for example, in excitable systems [211].

[7] The reader is referred to Ref. [165] for a brief introduction to the concepts of matrices and eigenvalues.

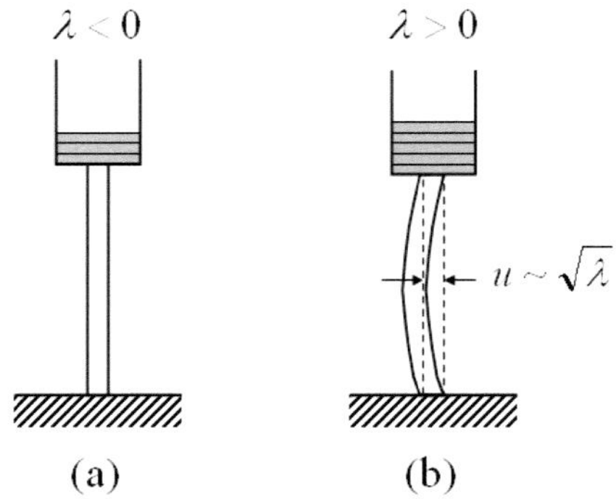

FIGURE 2.7: The buckling instability. (a) Below a critical weight, $W < W_c$ ($\lambda < 0$) the unbuckled state is stable. (b) Beyond the critical weight two symmetric buckled states appear (of which only one is drawn).

An example of a physical process described by such an equation (at least for small λ) is the buckling of a vertical beam that is subjected to a weight W as illustrated in Figure 2.7. Below a critical weight, W_c, the unbuckled state is stable (Figure 2.7(a)), but when W exceeds that critical weight the beam buckles, either to the right or to the left. In this example the variable u measures the degree of buckling and λ denotes the relative deviation from the critical point, that is

$$\lambda = \frac{W - W_c}{W_c}.$$

Equation (2.6) has three steady-state solutions: a zero state, $u_0 = 0$, that exists for any value of λ (the unbuckled state in the above example), and a symmetric pair of solutions, $u_\pm = \pm\sqrt{\lambda}$, that appear at $\lambda = 0$ and exist for any positive value of λ (the two symmetric buckled states). Which of the three different steady-state solutions that exist for $\lambda > 0$ is actually realized? The answer to this question is contained in the stability properties of the various solutions. For a one-dimensional dynamical system Equation (2.5) reduces to $\dot{\delta u} = \sigma\delta u$, where $\sigma = df/du|_{u=u_s} \equiv f'(u_s)$. The stability of the steady state u_s is determined by the sign of the eigenvelue σ; negative (positive) sign implies stability (instability). Since $f'(0) = \lambda$, the zero state is linearly stable in the range $\lambda < 0$ and loses stability at $\lambda = 0$, the value at which the symmetric solution pair $u_\pm = \pm\sqrt{\lambda}$ appears. Since $f'(u_\pm) = -2\lambda$, the solutions $u_\pm$ are both linearly stable in their whole existence range, $\lambda \geq 0$, apart from the instability threshold, $\lambda = 0$, at which they coincide with the zero state. The solution at $\lambda = 0$ cannot be classified as linearly stable or unstable, and we describe it as *marginally stable*. Note, however, that although perturbations about marginally stable states do not grow or decay exponentially they still can grow or decay algebraically. Indeed, writing (2.6) with $\lambda = 0$ as $u^{-3}du = -dt$ and integrating we find that small perturbations decay to zero as $t^{-\frac{1}{2}}$.

We can describe the instability that arises in a system as a control parameter is changed, e.g., λ in (2.6), by a diagram that shows the various steady-state solutions and their stability properties as functions of the control parameter, with the convention that solid (dashed) lines represent linearly stable (unstable) solutions. A diagram of this kind is called a *bifurcation diagram*. The instability appears as a *bifurcation point* where qualitative changes in the solutions take place; solutions may appear or disappear at this point, or may change their stability. The bifurcation diagram for (2.6) is shown in Figure 2.8(b). It describes the *pitchfork bifurcation* in which a zero solution loses stability to a new symmetric pair of stable non-zero solutions. The pitchfork bifurcation arises in many natural contexts, besides beam buckling, including thermal convection and front dynamics (see Section 7.3.1).

Quite often, the new states that appear in a pitchfork bifurcation are not perfectly symmetric. In the example of beam buckling, a non-uniform beam may favor one buckling direction over the other. Situations of this kind are captured by adding to the right side of (2.6) terms that break the inversion symmetry $u \to -u$, i.e., a constant term or nonlinear terms that are even in u. Consider the following variant of (2.6),

$$\dot{u} = f(u; \lambda) = \lambda u - u^3 - au^2 \,, \tag{2.7}$$

where the inversion symmetry is broken by a quadratic term. A straightforward calculation leads to the bifurcation diagrams shown in Figure 2.8(a,c) for positive and negative values of the parameter a, respectively. Two bifurcation points can be identified in these diagrams, $\lambda = 0$ and $\lambda = \lambda_c = -a^2/4$. Consider first the bifurcation at $\lambda = 0$ in which the zero solution loses stability. In the vicinity of this solution we can neglect the cubic term and rewrite (2.7) in terms of $\tilde{u} = au$:

$$\dot{\tilde{u}} = \lambda\tilde{u} - \tilde{u}^2 \,. \tag{2.8}$$

This equation describes the *transcritical bifurcation*, in which a zero solution exchanges stability with a non-zero solution, $\tilde{u} = \sqrt{\lambda}$ at $\lambda = 0$. Consider now the bifurcation point at $\lambda = -a^2/4$, in which the two solutions, $u_\pm = -a/2 \pm \sqrt{a^2/4 + \lambda}$ merge to a common value $u = -a/2$. Expanding $f(u)$ in (2.7) in a Taylor series around $u = -a/2$ and keeping terms to second order only, we obtain

$$\dot{\tilde{u}} = \tilde{\lambda} - \tilde{u}^2 \,, \tag{2.9}$$

where $\tilde{\lambda} = (a^2/4)[\lambda + (a^2/4)]$ and $\tilde{u} = (-a/2)[u + a/2]$. Equation (2.9) describes the *saddle-node bifurcation* or *fold bifurcation*, in which a pair of stable-unstable solutions appears at $\tilde{\lambda} = 0$.

The bifurcation diagram in Figure 2.8(a) for $a > 0$ shows another significant feature, namely, a *bistability* range, $-a^2/4 < \lambda < 0$, in which two distinct solutions, $u = 0$ and $u = -a/2 - \sqrt{a^2/4 + \lambda}$, are linearly stable. A similar bistability exists for the states $u = 0$ and $u = -a/2 + \sqrt{a^2/4 + \lambda}$ for $a < 0$, as

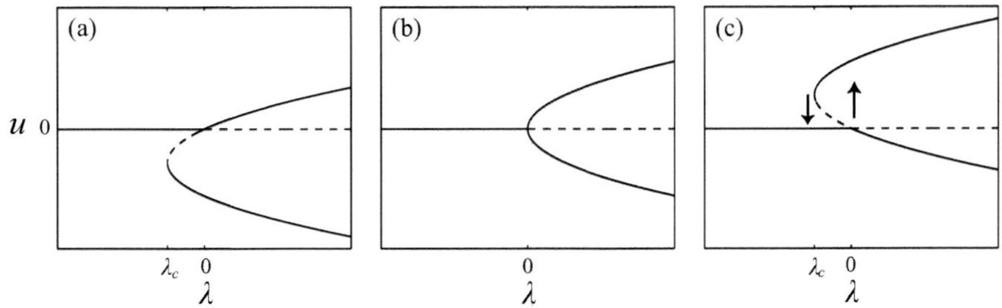

FIGURE 2.8: Bifurcation diagrams in the vicinity of a cusp singularity (see Figure 2.10) showing stable (solid lines) and unstable (dashed lines) steady-state solutions of (2.7). Shown are cuts of the surface $\lambda u - u^3 - au^2 = 0$ at constant a values, $a > 0$ (a), $a = 0$ (b), $a < 0$ (c). The bistability of states in panels (a) and (c) implies hysteresis, as the arrows in panel (c) show.

Figure 2.8(c) shows. Associated with bistability of different states is the phenomenon of *hysteresis*: increasing λ from negative to positive values results in a transition from $u = 0$ to $u = -a/2+\sqrt{a^2/4+\lambda}$ at the threshold $\lambda = 0$, while decreasing λ results in a transition back to the zero state at a lower threshold $\lambda = \lambda_c = -a^2/4$, as the arrows in Figure 2.8(c) illustrate. The phenomenon of hysteresis is familiar from the context of magnetism, but is relevant to a wide variety of other physical, biological and environmental contexts. We will return to this topic when we discuss regime shifts and desertification in Chapter 11.

Note that for $a \neq 0$, a small incremental increase of λ beyond zero results in a large change in u (provided $u > 0$). This is unlike the case $a = 0$ in which the change in u is also small. We call the bifurcation that occurs for $a \neq 0$ a *subcritical bifurcation*, and distinguish it from the *supercritical bifurcation* that occurs for $a = 0$. The new state that the system evolves to slightly beyond the bifurcation point (for $a \neq 0$) exists also below the bifurcation point, hence the term "subcritical." It is important to stress that despite the discontinuity in the stable solution branches in a subcritical bifurcation, there is no discontinuity in the dynamics that take the system from one stable solution branch to another.

Consider now a different variant of (2.6) in which the inversion symmetry is broken by a constant term b:

$$\dot{u} = \lambda u - u^3 + b\,. \tag{2.10}$$

The bifurcation diagrams for positive and negative values of b are shown in Figure 2.9(a,c) and are often referred to as "imperfect pitchfork bifurcation." The single steady-state solution that exists at small λ remains stable for any λ, while a new pair of stable and unstable solutions appears in a saddle-node bifurcation as λ exceeds a threshold $\lambda_c(b)$. Another interesting view of

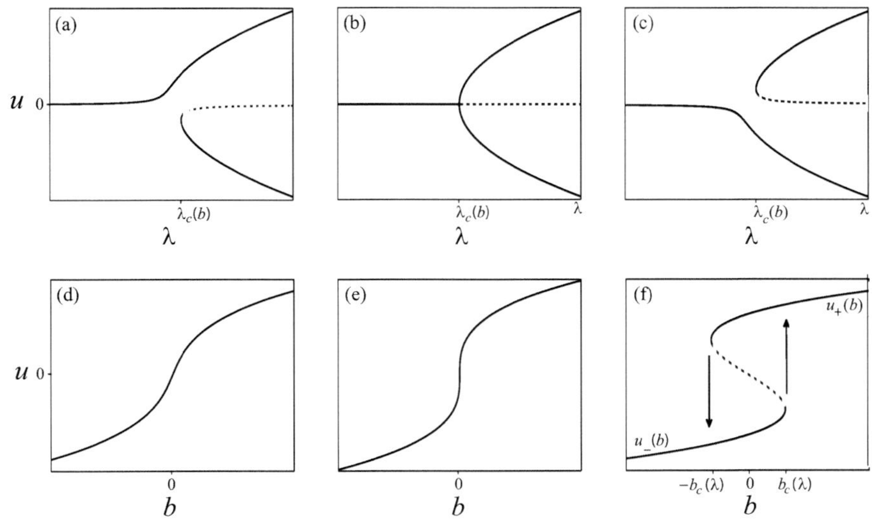

FIGURE 2.9: Bifurcation diagrams in the vicinity of a cusp singularity showing stable (solid lines) and unstable (dashed lines) steady-state solutions of (2.10). Shown are cuts of the surface $\lambda u - u^3 + b = 0$ at constant b values, $b < 0$ (a), $b = 0$ (b), $b > 0$ (c), and at constant λ values, $\lambda < 0$ (d), $\lambda = 0$ (e), $\lambda > 0$ (f). The bistability of states in panel (f) implies hysteresis, as the arrows show.

(2.10) is obtained by regarding b as the control parameter, keeping λ fixed. Figure 2.9(d-f) shows the bifurcation diagrams for negative λ (d), $\lambda = 0$ (e) and positive λ (f). The last diagram (f) shows an existence range, $-b_c(\lambda) < b < b_c(\lambda)$ ($b_c = \frac{2\lambda}{3}\sqrt{\frac{\lambda}{3}}$), of three steady-state solutions. The upper and the lower solution branches are stable while the middle branch is unstable. These solutions appear in a pair of saddle-node bifurcations at $b = \pm b_c(\lambda)$. In the λ, b plane, the saddle-node bifurcation curves meet tangentially in the origin at what is called a *cusp point* as Figure 2.10 shows.

The last variant of (2.6) that we consider in this section is

$$\dot{u} = \lambda u - cu^3 - u^5 \,. \tag{2.11}$$

The case $c > 0$ is qualitatively similar to the pitchfork bifurcation discussed above. In fact, close enough to the bifurcation point $u \approx \sqrt{\lambda/c} \ll 1$, and the quintic term can be disregarded. A qualitative change occurs when c becomes negative, for then, the cubic term accelerates the growth of u and only when u is sufficiently large can the quintic term balance this growth. We leave it as an exercise to the reader to derive the bifurcation diagram shown in Figure 2.11, which implies hysteresis as the arrows indicate. This is another example of a subcritical bifurcation.

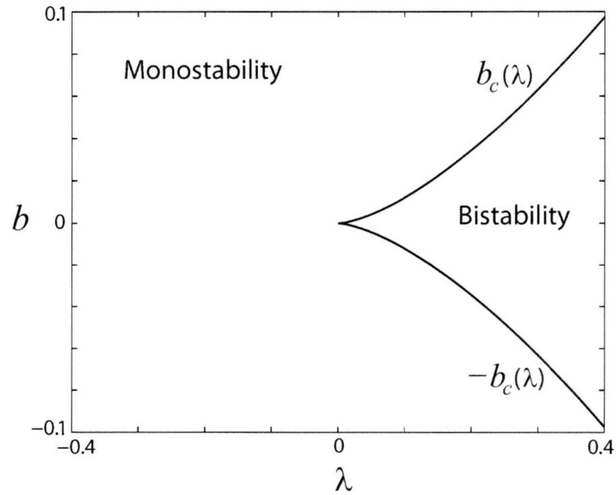

FIGURE 2.10: Phase diagram of steady-state solutions of (2.10) in the λ, b plane showing domains of a single stable solution and of two stable solutions. The boundaries of these domains consist of a pair of saddle-node bifurcation curves, $b = \pm b_c(\lambda)$, that meet at a cusp point at $\lambda = 0$. See also Figure 2.9(f).

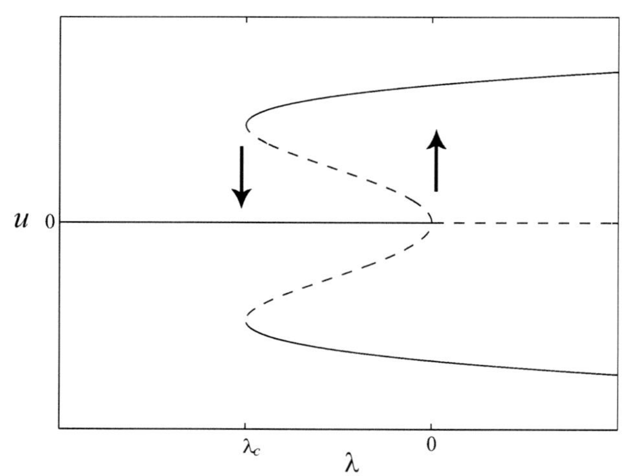

FIGURE 2.11: Bifurcation diagram showing stable (solid lines) and unstable (dashed lines) steady-state solutions of (2.11) for $c < 0$. The arrows indicate state transitions that involve hysteresis.

2.2.4 Oscillatory instabilities

In one-dimensional systems (a single degree of freedom) non-stationary behaviors must be monotonic in time. The onset of non-monotonic dynamics, such as periodic oscillations, requires at least two degrees of freedom. Consider the following two-dimensional system:

$$\dot{u}_1 = \lambda u_1 - \omega u_2 - (u_1^2 + u_2^2)u_1 \quad (2.12a)$$
$$\dot{u}_2 = \omega u_1 + \lambda u_2 - (u_1^2 + u_2^2)u_2 \,, \quad (2.12b)$$

where λ and $\omega \neq 0$ are constant parameters. Rewriting these equations in terms of the polar variables $\rho \geq 0$ and $0 \leq \phi < 2\pi$, which are related to the original variables through $u_1 = \rho\cos\phi$ and $u_2 = \rho\sin\phi$, we obtain

$$\dot{\rho} = \lambda\rho - \rho^3 \quad (2.13a)$$
$$\dot{\phi} = \omega \,. \quad (2.13b)$$

The equation for ρ is identical to (2.6) (except that $\rho \geq 0$). It has a zero solution that loses stability at $\lambda = 0$ to a stable solution $\rho = \sqrt{\lambda}$. While the zero solution describes a steady state of (2.12), the solution $\rho = \sqrt{\lambda}$ corresponds to an oscillatory state. This can be readily seen by integrating the equation for ϕ to obtain $\phi = \omega t + \phi_0$, where ϕ_0 is an integration constant. For then

$$u_1 = \sqrt{\lambda}\cos(\omega t + \phi_0)\,, \qquad u_2 = \sqrt{\lambda}\sin(\omega t + \phi_0)\,. \quad (2.14)$$

That is, u_1 and u_2 oscillate periodically in time. Figure 2.12 shows the dynamics of (2.12) below and above the instability of the zero state. Below the instability point perturbations about the stable zero state decay in an oscillatory manner. The dynamics in the phase space spanned by u_1 and u_2 are described by trajectories that spiral in toward the zero state, as Figure 2.12(a) shows. Above the instability point such perturbations grow in an oscillatory manner and the dynamics converge to periodic oscillations with a constant amplitude. As Figure 2.12(b) shows, the phase-space dynamics involve trajectories that converge to a circle of radius $\sqrt{\lambda}$ centered at the origin $u_1 = u_2 = 0$, which is often called a *limit cycle.* The phase of the oscillation is determined by the constant ϕ_0. Since it is an arbitrary integration constant, there exists a continuous family of oscillatory solutions with phases that span the whole circle $0 \leq \phi < 2\pi$. Note that while the oscillatory solutions are stable to perturbations of the amplitude ρ, they are marginally stable to perturbations of the phase ϕ. This marginal stability is closely related to the phase-shift symmetry of (2.12) or (2.13a), *i.e.*, to the fact that shifting the phase of an oscillatory solution yields another oscillatory solution. The oscillatory instability described above is called the *Hopf bifurcation.* Like the pitchfork bifurcation, the Hopf bifurcation too has various variants, including in particular a subcritical Hopf bifurcation, which implies a bistability range of a stationary zero state and an oscillatory state.

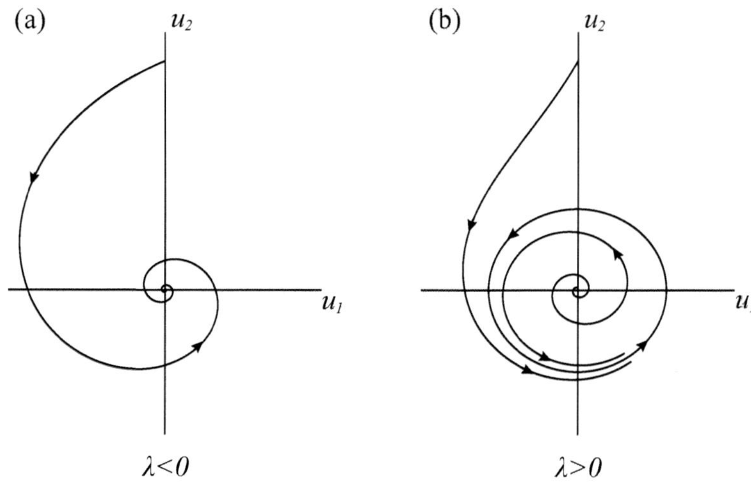

FIGURE 2.12: Solutions of (2.12) demonstrating a Hopf bifurcation of a zero state. (a,c) Oscillatory decay of initial perturbations below the bifurcation point ($\lambda < 0$). (b,d) Oscillatory growth and convergence to periodic oscillations of finite constant amplitude above the bifurcation point ($\lambda > 0$).

A characteristic feature of the Hopf bifurcation is the finite frequency of the oscillations that appear at the bifurcation point. There are other oscillatory bifurcations that do not share this property. Consider the following system[8]:

$$\dot{\rho} = \lambda\rho - \rho^3 \tag{2.15a}$$
$$\dot{\phi} = \omega - a\sin\phi\,, \tag{2.15b}$$

where $\lambda > 0$, and let us first discuss the case $a > \omega$. Equations (2.15) then have two non-zero steady-state solutions, $(\rho_1, \phi_1) = (\sqrt{\lambda}, \arcsin\omega/a)$ and $(\rho_2, \phi_2) = (\rho_1, \pi - \phi_1)$, of which (ρ_1, ϕ_1) is stable and (ρ_2, ϕ_2) is unstable. The two solutions lie on a circle of radius $\sqrt{\lambda}$ in the phase space spanned by $u_1 = \rho\cos\phi$ and $u_2 = \rho\sin\phi$, as Figure 2.13a illustrates. The angular distance between the two solutions, $\Delta\phi = 2\phi_1 - \pi$, diminishes as $a \to \omega$ from above. Consider now the case $a < \omega$. Since $d\phi/dt$ is positive at all times the dynamics are oscillatory, but differ from the case $a = 0$ considered in (2.15) in that the oscillations are no longer uniform along the circle; the dynamics is slow near $\phi \approx \pi/2$, and fast near $\phi \approx 3\pi/2$ as Figure 2.13b illustrates. The oscillation period of this non-uniform dynamics can be obtained by separating variables and integrating:

$$T = \int_0^T dt = \int_0^{2\pi} \frac{d\phi}{\omega - a\sin\phi} = \frac{2\pi}{\sqrt{\omega^2 - a^2}}\,. \tag{2.16}$$

[8]The discussion below follows closely Ref. [296]

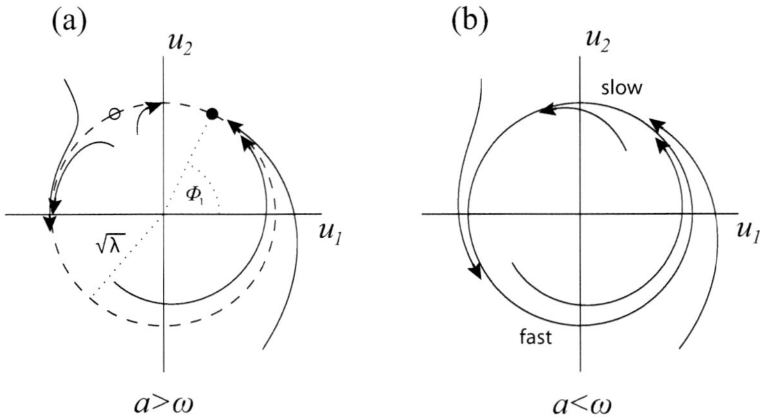

FIGURE 2.13: Illustration of an oscillatory infinite-period bifurcation in (2.15). When $a > \omega$ a stable-unstable pair of states exist on the circle $\rho = \sqrt{\lambda}$ (a) (the stable and unstable states are represented, respectively, by the small full and empty circles on the large dashed circle of radius ρ). While approaching the circle, the dynamics converge to the stable steady state $(\sqrt{\lambda}, \phi_1)$). As a is decreased below ω the pair of steady states approach one another, merge and disappear. The dynamics, once confined to the circle, becomes oscillatory with a slow phase and a fast phase in each oscillation period (b).

This result shows that T diverges as $a \to \omega$ from below. We can discuss now the dynamical behavior of the system (2.15) in the vicinity of the critical point $a = \omega$. Slightly above this point there still exists a stable steady-state solution, (ρ_1, ϕ_1), but as a is decreased below ω it merges with its unstable counterpart (ρ_2, ϕ_2) and disappears. This gives rise to periodic oscillations with a period that scales like $(\omega - a)^{-1/2}$. The instabilty that occurs at $a = \omega$ is called *infinite-period bifurcation*. Like the Hopf bifurcation it renders a steady state unstable and gives rise to oscillations, however, the oscillation period diverges to infinity at the bifurcation point, rather than being finite.

The infinite-period bifurcation is an example of a *global bifurcation*, where the oscillatory solution is not confined to a small neighborhood of the steady-state solution in phase space. Unlike the Hopf bifurcation, in which the control parameter is λ and the radius of the circular trajectory, $\sqrt{\lambda}$, is diminishingly small close to the bifurcation point $\lambda = 0$, the control parameter in the infinite-period bifurcation is a and $\lambda \sim \mathcal{O}(1)$ is fixed. As a result, even close to the bifurcation point, $a = \omega$, the oscillatory solution has a large amplitude, and its circular trajectory explores distant phase-space regions. A different type of an oscillatory instability involving an infinite oscillation period at the bifurcation point is the *saddle-loop* or *homoclinic bifurcation*, which involves the merging of a limit cycle with an unstable (saddle) fixed point [296].

2.2.5 Variational and non-variational systems

The absence of oscillations in a dynamical system is often associated with the existence of a variational principle. According to this principle there exists a real-valued scalar function of the state variables, $L(\mathbf{u};\lambda)$, that cannot increase in the course of time. Such a function is called a *Lyapunov function.* In some cases the equations (2.4) that describe such systems assume the form

$$\dot{\mathbf{u}} = \mathbf{f}(\mathbf{u};\lambda) = -\nabla_{\mathbf{u}} L(\mathbf{u};\lambda)\,, \tag{2.17}$$

where $\nabla_{\mathbf{u}} L = (\partial L/\partial u_1, ..., \partial F/\partial u_n)$ is the gradient of L with respect to the state variables. It is then easy to prove that L in (2.17) is a Lyapunov function, i.e., a function that satisfies $dL/dt \leq 0$. The proof is achieved by applying the chain rule to the time derivative of L and using (2.17):

$$\frac{dL}{dt} = \frac{\partial L}{\partial u_1}\frac{du_1}{dt} + ... + \frac{\partial L}{\partial u_n}\frac{du_n}{dt} = \nabla_{\mathbf{u}} L \cdot \dot{\mathbf{u}} = -(\nabla_{\mathbf{u}} L)^2 \leq 0\,, \tag{2.18}$$

where we assumed that the time dependence of L is due to the time dependence of the state variables, i.e., L does not have an explicit time dependence. The significance of identifying a Lyapunov function of a dynamical system is that it rules out oscillatory dynamics and provides information about the steady-state solutions of the system and their stability properties. Extremum points of L, at which $\partial L/\partial u_1 = \partial L/\partial u_1 = 0$, correspond to steady-state solutions, and minima of L correspond to stable steady-state solutions, since any perturbation must decay in time in order to satisfy the condition $dL/dt \leq 0$. Whatever non-uniform landscape L has in phase space, any initial condition will eventually lead the system to a minimum of L and to a steady-state solution. Note the similarity of (2.17) with the equation of motion of an object in a gravitational potential that is highly damped by a friction force (to the extent that inertia effects can be neglected). In this analogy $\mathbf{u}$ is the position vector of the object and L represents the potential or the height of the object. For this reason the Lyapunov function L is often called a "potential function," and variational systems are called "gradient systems," i.e., systems that move down the gradients of potential functions. Another name in common use is "energy function," in analogy to the free energy of a thermodynamic system near equilibrium.

Consider as an example the dynamical system (2.12) with $\omega = 0$. By integrating the right-hand sides of (2.12a) and (2.12b) with respect to u_1 and u_2, respectively, and equating the outcomes we find

$$\dot{u}_1 = -\partial L/\partial u_1 \tag{2.19a}$$

$$\dot{u}_2 = -\partial L/\partial u_2\,, \tag{2.19b}$$

with

$$L = -\frac{\lambda}{2}(u_1^2 + u_2^2) + \frac{1}{4}(u_1^2 + u_2^2)^2 + L_0\,, \tag{2.20}$$

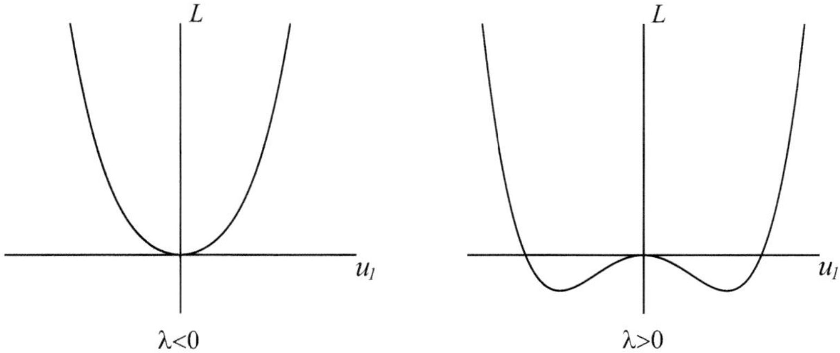

FIGURE 2.14: A graph of the Lyapunov function (2.20) for $\lambda < 0$ (left) and $\lambda > 0$ (right). Identical graphs are obtained along any transect in the (u_1, u_2) plane that goes through the origin.

where L_0 is an arbitrary constant. This shows that (2.12) with $\omega = 0$ represent a variational system. Expressing L in terms of the polar variables $\rho = \sqrt{u_1^2 + u_2^2}$ and $\phi = \arctan(y/x)$ we find that L is independent of ϕ and is given by

$$L = -\frac{\lambda}{2}\rho^2 + \frac{1}{4}\rho^4 + L_0\,. \tag{2.21}$$

Let us use (2.21) to evaluate the steady-state solutions of (2.12) with $\omega = 0$ and their stability properties. Equating the derivative of L with respect to ρ to zero gives the steady-state solution $\rho = 0$ for any λ value, and $\rho = \sqrt{\lambda}$ for $\lambda > 0$. The zero solution corresponds to a minimum (maximum) of L for $\lambda < 0$ ($\lambda > 0$), implying stability (instability) to perturbations with respect to ρ, whereas the solution that is born at $\lambda = 0$ corresponds to a minimum of L and is thus stable. The fact that L does not depend on ϕ implies the existence of a continuous family of steady-state solutions, spanning a circle of radius $\rho = \sqrt{\lambda}$ in phase space. These solutions are stable to amplitude (ρ) perturbations and marginally stable to phase (ϕ) perturbations. These conclusions become evident by drawing the landscape of L in phase space for λ negative and positive, as Figure 2.14 shows. Note that unlike linear stability analysis which provides information about the stability of solutions to infinitesimal perturbations, the Lyapunov function provides information about the basins of attraction of stable solutions and thus about the response to large perturbations. The results of Figure 2.14 imply, for example, that any perturbation of the zero solution, no matter how big, will decay to zero in the range $\lambda < 0$.

Not all variational systems, i.e., systems that have a Lyapunov function, can be cast in the form of (2.19). An example of such a system is [296]:

$$\dot{u}_1 = -u_1 - 2{u_2}^2 \tag{2.22a}$$

$$\dot{u}_2 = u_1 u_2 - {u_2}^3 . \tag{2.22b}$$

It is easy to verify that the function $L = {u_1}^2 + 2{u_2}^2$ satisfies $dL/dt \leq 0$ and is therefore a Lyapunov function.

Returning to (2.12), when ω is not zero the attempt to find a Lyapunov function satisfying (2.19) fails. Indeed, for $\lambda > 0$ equations (2.12) have oscillatory solutions and the dynamics are no longer dictated by the minimization of a Lyapunov function. We say that the dynamical system is *non-variational.* The example of (2.12) shows that changing a system parameter can induce a transition from a variational system to a non-variational system. Such a transition is often accompanied by non-variational effects. In this example it is the appearance of temporal oscillations. In Section 7.3.1 we will encounter a different example of non-variational effect with far reaching pattern formation consequences.

2.3 A glimpse into pattern formation theory

Equipped with basic knowledge of low-dimensional dynamical systems, we now return to the context of spatially extended systems and describe a few fundamental concepts of pattern formation theory—the mathematical theory of spatially extended nonlinear systems. We first address the different types of pattern-forming instabilities that a steady uniform state can go through (Section 2.3.1). Distinguishing between different instability types is significant for understanding the universal character of pattern formation phenomena as different systems that go through the same instability type behave similarly near the instability point (Section 2.3.2). The discussion in this section applies both to small and large systems, although the latter may show richer dynamical behaviors, some of which will be discussed in length in Chapters 6 and 7 where a fuller and deeper account of pattern formation theory is presented.

2.3.1 Instability types and symmetry breaking

The growing modes that destabilize steady uniform states beyond instability points can be uniform or spatially periodic, and the growth can be monotonic or oscillatory. Accordingly, we distinguish among four main types of pattern-forming instabilities:

- *Uniform stationary instability*, in which a monotonically growing uniform mode leads to a stationary uniform state.

- *Non-uniform*[9] *stationary instability*, in which a monotonically growing non-uniform mode leads to a stationary spatial pattern.
- *Uniform oscillatory instability*, in which a uniform mode with oscillatory growth leads to uniform oscillations.
- *Non-uniform oscillatory instability*, in which a non-uniform mode with oscillatory growth leads to traveling waves.

The second and fourth instabilities directly lead to spatially structured states. We say that these instabilities "break" the spatial symmetry of the system. In an RB system, for example (see Section 2.1.2), all points at a given height are equivalent to one another because of the uniform fluid, the uniform heating, the perfect geometry of the compartment and so on. This spatial symmetry, which is preserved by the rest state of the fluid, is broken by the non-uniform roll patterns that appear beyond the convective instability. Mathematically, we can approximate the fluid states by equation (2.3), where the amplitude u satisfies equation (2.6). The physical variables c_i represent in this case the vertical fluid-velocity component and the deviation of the temperature from the rest state value [44]. The bifurcation parameter λ represents the relative deviation of the Rayleigh number R from its critical value R_c: $\lambda = (R - R_c)/R_c$. As λ is increased past zero, the zero state $u = 0$ destabilizes and the amplitude u begins to grow monotonically, approaching the asymptotic values $u_\pm = \pm\sqrt{\lambda}$. This amounts to the destabilization of the stationary uniform rest state of the fluid to one of two symmetric roll patterns, $u_\pm \cos(k_0 x + \phi)$ (differing by the direction of fluid rotation), which break the spatial symmetry of the system.

The first and third instabilities lead to uniform states. In what sense are these instabilities pattern forming? Spatial patterns can result from such instabilities if they involve the creation of multiple stable states, for then patterns consisting of domains of different states can be formed. Consider first a uniform stationary instability of a uniform stationary equilibrium state, such as the pitchfork bifurcation discussed in Section 2.2.3. That is, we consider a system the state variables of which are given by (2.3) with $k_0 = 0$ and $u(t)$ satisfying (2.6). In this bifurcation the zero state, $u = 0$, loses stability to a pair of stable uniform stationary states, $u = u_+ = \sqrt{\lambda}$ and $u = u_- = -\sqrt{\lambda}$ (hereafter the "up state" and "down state"), as the control parameter λ is increased past zero. The coexistence of these two stable states allows for patterns consisting of alternate up-state and down-state domains (although such domain patterns are not necessarily stable as we will see in Section 7.2.1). The appearance of the up and down states is also a consequence of symmetry breaking. Equation (2.6) is symmetric under the transformation $u \to -u$, implying that if u is a solution also $-u$ is a solution. The zero-state solution

[9] Non-uniform instabilities are also called "finite-wavenumber instabilities" since they involve the growth of modes with non-zero wavenumbers.

preserves this symmetry, but any of the two states, $u_\pm$, breaks it and implies the coexistence of its symmetric counterpart.

A uniform oscillatory instability also involves the creation of multiple stable states. Consider the Hopf bifurcation, discussed in Section 2.2.4, in which a uniform stationary equilibrium state loses stability to a continuous family of time-periodic uniform solutions of the form (2.14). These solutions differ from one another by their oscillation phase ϕ_0, and are all stable to amplitude perturbations. They break the phase-shift symmetry, $(u_1, u_2) \to (u_1 \cos\phi - u_2 \sin\phi, u_1 \sin\phi + u_2 \cos\phi)$, which the equilibrium state, $(u_1, u_2) = (0, 0)$, preserves, and allow for long-wavelength traveling-wave phenomena of various forms, some of which will be discussed in Section 7.2.2.

2.3.2 Amplitude equations and universality

The dynamics of a pattern forming system is often described by a set of partial differential equations (PDEs)[10] for some characteristic state variables. These equations are specific to the system in question and vary from one system to another. Yet, equations that describe distinct systems may share the same type of instability. For example, both the fluid equations that describe thermal convection and models of dryland vegetation capture a non-uniform stationary instability of a uniform steady state. A striking result of pattern formation theory is that systems that go through the same type of instability behave similarly near the instability point. Mathematically, starting from the equations that describe each system near the instability point, it is possible to derive new equations that look alike for all systems that share that type of instability. The new equations, often called *normal form equations* or *amplitude equations*, have the same structural form for all systems, but differ in the coefficients of the various terms in the equations, which are expressed in terms of parameters that are specific to the system in question. We will describe how to derive amplitude equations in Chap. 6.1, but their significance can already be highlighted: pattern forming systems near instability points show high degree of *universality*: distinct and unrelated systems, be they fluid, optical, chemical, biological or environmental systems, will generally show the same types of spatial patterns in the vicinities of instabilities that they share [55].

The universal nature of pattern formation is a powerful result. The amplitude equations for the basic instabilities have thoroughly been studied and the identification of such an instability in a particular system makes that information available at once. The identification of a Hopf bifurcation in a chemical reaction, for example, already tells us the possible existence of traveling

[10]PDEs are differential equations for state variables that depend on two or more independent variables, most commonly time and space variables, and involve partial derivatives. The solutions of PDEs are functions of the independent variables, e.g., the steady state temperature, $T(x, z)$, of a roll pattern in thermal convection. Readers unfamiliar with PDEs are referred to textbooks focusing on applications of PDEs in biology, e.g., [78, 165]. See also Ref. [131].

waves, sinks and sources of traveling waves, spiral waves and spatio-temporal chaos [6]. However, not all of these behaviors are necessarily realizable; the parameters of the chemical reaction may restrict the coefficients in the amplitude equation to ranges that exclude some behaviors. It is important to stress that amplitude equations have at least two drawbacks, and therefore are not substitutes for the original equations that describe particular systems. The first drawback is that their validity is limited to the vicinities of instability points, whereas quite often we are interested in behaviors far from the onset of instabilities. Another drawback has to do with circumstances in which the amplitude equations miss significant system behaviors, even close to onset. An example, to be discussed in Section 7.2.1.3, is the pinning of an interface (or front) between a uniform state and a patterned state. Another point to stress is that the appearance of similar patterns and dynamical behaviors in distinct systems *does not* mean that these systems are describable by similar models. The model equations reflect processes and mechanisms that are specific to a particular system. They may show similar patterns in the vicinity of an instability that both models capture, or because of other mechanisms of universal behavior.

Chapter summary

Regular spatial patterns, such as stripes and spots, are often seen in various natural and laboratory contexts. These patterns often result from instabilities of uniform steady states that occur as as some control parameter exceeds a threshold value. The driving forces of such instabilities are positive feedbacks that operate at small spatial scales and give rise to spatial self-organization at large scales throughout the growth of spatially extended modes that determine the nature of the new states the systems evolve to. While the positive feedbacks that operate in different physical systems obviously differ from one another, the instabilities that they induce and, consequently, the modes that grow beyond the instability thresholds are often the same.

Much of our understanding of pattern formation phenomena derives from controlled laboratory experiments conducted on simple model systems. Two model systems of this kind are described. The first is the Rayleigh–Bénard (RB) system in which a layer of fluid at rest is heated from below and destabilizes to an ordered convection state consisting of parallel rolls of hot fluid moving up and cold fluid moving down. The second model system is the Belousov–Zhabotinsky (BZ) chemical reaction in which an equilibrium state with constant concentrations of all reagents loses stability to chemical oscillations as the flow rate of fresh chemicals is changed. The modes that grow beyond the instability points in the RB and BZ systems are different. In the RB system it is a spatially periodic mode that grows monotonically in time and leads to spatially periodic stationary patterns. In the BZ system it is a spatially uniform mode that grows in an oscillatory manner and leads to uniform oscillations or to long-wavelength traveling waves. The RB and BZ systems have served as prototype model systems for studying stationary patterns and oscillating patterns, respectively.

Because of their spatial extent, pattern-forming systems are high-dimensional dynamical systems, that is, systems with large numbers of degrees of freedom. Near instability points, however, the number of independent degrees of freedom dramatically reduces, and in small systems, where the system size is not much larger than the pattern's wavelength, the dynamics may involve just a few independent degrees of freedom. A major part of the chapter is devoted to a brief review of the theory of such low-dimensional dynamical systems. The review is neither comprehensive nor systematic; it rather explains basic concepts, such as phase space, fixed point, limit cycles, bifurcations and linear stability analysis, by working out a few simple examples.

The chapter concludes with a glimpse into pattern formation theory, the topic of Part II. We first describe the main types of pattern-forming instabilities that a steady uniform state can go through, making the point that pattern formation can arise not only from the growth of spatially structured modes; instabilities involving the growth of spatially uniform modes that lead to multiplicity of stable states can also give rise to spatial patterns. We then elaborate on the universal nature of pattern formation phenomena and relate it to the mathematical concept of normal-form or amplitude equations; different systems that go through the same instability type are described by the same amplitude equations near the instability point, and therefore behave similarly.

Chapter 3

Spatial ecology

3.1 The complexity of ecological systems

The term *ecological system* or *ecosystem* is commonly understood as describing a group of organisms that interact among themselves and with their physical or abiotic environment [242]. There are many definitions of this subtle concept and most of them suffer from some degree of vagueness [234]. We will not attempt here an exact definition, but rather emphasize an aspect of this concept which becomes important in modeling complex ecosystems, namely, the distinction between what constitutes the ecosystem and what can be considered as its environment.

We will regard an ecosystem as consisting of biotic and abiotic elements with strong reciprocal coupling, while the environment of an ecosystem will be regarded as consisting of elements having uni-directional influence, strongly affecting the ecosystem's elements but hardly being affected by them. As an example we may consider an arid landscape consisting of patchy vegetation; the vegetation and the limited water resource can be considered as elements of the ecosystem, whereas the rainfall can be regarded as an element of the environment. While the vegetation and water elements are reciprocally related through various positive and negative feedbacks (see Chapter 9), the coupling of the rainfall to these elements is not necessarily reciprocal [271]; the rainfall strongly limits vegetation growth, but may weakly be affected by the low total transpiration of the sparse vegetation and the small albedo change it induces.

The distinction between the ecosystem and its environment makes the study of ecosystem processes manageable by focusing on the ecosystem component. Yet, this component may still be highly complex because of the hierarchy of trophic and organization levels it may contain and the multiple space and time scales ecosystem processes may involve. We address these two aspects in Sections 3.1.1 and 3.1.2.

3.1.1 Hierarchies of trophic and organization levels

The biotic elements of an ecosystem, the organisms, are commonly classified according to their positions in the food chain, which define the trophic levels they belong to. The food chain begins with *primary producers* which form the lowest trophic level. These are organisms that make their own food from inorganic compounds, primarily through the process of photosynthesis. Plants are the most important primary producers in terrestrial ecosystems, but other examples exist, including cyanobacteria, which form soil crusts and play an important role in vegetation pattern formation. An example of primary producers in aquatic ecosystems is the phytoplankton which account for half of all photosynthetic activity on Earth. The next three trophic levels pertain to primary, secondary and tertiary *consumers.* These organisms, mostly animals, achieve their food by consuming other organisms. Primary consumers eat plants and are called herbivores. Secondary consumers eat herbivores and are called carnivores. Tertiary consumers are carnivores which eat other carnivores. The distinction between these trophic levels is not sharp as there are organisms that belong to two or more trophic levels, for example, consumers that are both herbivores and carnivores (omnivores), such as pigs. Another group of organisms, the *decomposers* or detritivores, use dead organisms as their food source. These organisms break down the tissues of dead organisms (plants and animals) and convert them into inorganic compounds that are recycled as nutrients for plants. Examples of decomposers are millipedes, woodlice, and worms. The decomposers can be regarded as another trophic level that closes the food chain to form a loop.

Adding to the complexity of an ecosystem is the existence of a hierarchy of *organization levels*, within a trophic level or across different trophic levels. The hierarchy begins with the *organism level*, the level of a single living entity[1], continues to the *population level*, a group of organisms belonging to the same species, further on to the *community level*, a group of populations of different species, and up to the *ecosystem level* that extends the community level to include the exchange of energy and materials with abiotic elements.

[1]Note that the organism itself consists of yet lower anatomical organization levels: cell, tissue, organ and organ system, and the same holds for the cell itself.

The breakdown of an ecosystem to trophic and organization levels is conceptually helpful, but it should be stressed that quite often these levels are strongly coupled and cannot be considered independently. An important manifestation of this coupling is the emergence of new properties at high organization levels as a result of processes occurring at lower organization levels. A striking example is vegetation patchiness, where biomass-water feedbacks at the organism (plant) level give rise to symmetry breaking instabilities and vegetation pattern formation at the population and community levels.

3.1.2 Multiple space and time scales

Ecosystem processes generally involve multiple space and time scales. The spatial scales are correlated with the levels of organization in the sense that the organism level often sets the smallest scale and the ecosystem level the largest scale. Ecosystems can be small or large and their spatial extent is another factor that affects their complexity. We have already discussed in Section 2.2.1 the relation between the size of a dynamical system and its dimension, i.e., the number of degrees of freedom it contains. The larger the system the higher the dimension and the richer its spatio-temporal dynamics. Sufficiently large ecosystems allow, in particular, the spontaneous emergence of symmetry breaking patterns, such as banded or spotted vegetation in water-limited systems. This form of self-organized heterogeneity contributes to the system's complexity by adding spatial scales and by affecting ecosystem processes across these scales, such as the flow of limiting resources, seed dispersal, and trophic relations, which bear on community structure and ecosystem function.

The time scales of ecosystem dynamics range from long evolutionary scales to short ecological scales, although recent studies suggest that the two ranges of time scales are not necessarily widely separated [42]. Ecological time scales are determined by a variety of factors including rates of organism-level processes, such as growth and mortality, seasonal or inter-annual climate variability, geomorphological processes such as erosion and deposition, and various disturbance forms, including fires, logging, and fragmentation. These and additional factors still span a wide range of time scales.

In general, fast processes tend to equilibrate on the instantaneous states of slower processes, which can result in the elimination of fast degrees of freedom and consequently in simpler dynamics of a reduced dimensionality. The combination of multiple time scales with nonlinearity, however, may lead to richer dynamics involving two or more time scales. This can be the case when the nonlinearity results in a multiplicity of stable states, as slow dynamics along one state may drive the system beyond the existence or stability range of that state and induce a fast transition to an alternative state. The dynamics in this case remain high dimensional, alternating between slow and fast phases that are controlled by different degrees of freedom. Relaxation oscillations and excitability are examples of such dynamics [315, 211]. When the slow dynamics of an ecosystem is induced by environmental elements, state transitions may appear as sudden irreversible changes or regime shifts (see Chapter 11). The impact of regime shifts on ecosystem dynamics and function can be far reaching because of the complex hierarchical structure of ecosystems and the chain processes that can be induced along this hierarchy.

3.2 Outstanding problems

Desertification and the loss of biodiversity, along with climate change, were identified as the greatest challenges to sustainable development during the Earth Summit, a United Nations conference held in Rio de Janeiro in 1992. The Earth Summit has led to two UN conventions, the Convention on Biological Diversity and the Convention to Combat Desertification; both have become international legally binding treaties. Another major effort led by the United Nations is the Millennium Ecosystem Assessment—an international synthesis that analyzes the state of the Earth's ecosystems and provides summaries and guidelines for decision-makers. Two major components of this assessment are the Desertification Synthesis [1] and the Biodiversity Synthesis [71], which cover a broad range of topics and emphasize the role of people as integral parts of ecosystems and their dynamic interactions with other parts. Sections 3.2.1 and 3.2.2 below present brief introductions to both topics. These introductions are by no means comprehensive and mainly serve to motivate the view of desertification and biodiversity change as spatial processes.

3.2.1 Desertification

Desertification is defined as the reduction or loss of biological productivity in drylands[2], as a result of climatic variations or human activities [1]. Quite often it is the combination of these two factors that is detrimental. Desertification often occurs when the demand for ecosystem services, such as food, forage, fuel, and water, is higher than the services that the ecosystem can provide. This imbalance is generally triggered by physical causes, such as droughts, but the failure to adopt appropriate land and water management practices intensifies it because of an unadjusted demand to diminishing services [1].

Drylands, or water limited systems, occupy about two fifths of the Earth's terrestrial area and are home to about a third of the human population. This part of the human population, residing mostly in developing countries, is highly dependent on ecosystem services that are directly related to plant

[2] "Drylands" is a general term describing water-limited regions. Several sub-categories of drylands have been defined according to the degree of water scarcity. The reader is referred to Section 10.1.4 for a description of these sub-categories.

productivity, such as crop, fuel-wood, and construction materials. This makes people in drylands highly susceptible to loss of human well-being as a result of desertification. The adverse impact of desertification, however, is not limited to drylands; non-dryland regions can be affected by dust storms, downstream flooding, impairment of global carbon sequestration capacity, and global climate change, that are induced by desertification processes in drylands. Desertification may also have societal and political impacts that extend to non-dryland areas, e.g., an influx of migrants from dryland areas, following desertification, that brings about social and political strife [1].

The complex entanglement of physical, biological and human elements in desertification processes make these processes hard to study [252]. Contributing to this is the disparity of the disciplines involved, natural sciences vs. social sciences, which lack common research methodologies. Yet, significant tractable problems can still be identified. These problems can be classified as associated with (i) mechanisms and dynamics of desertification, (ii) warning signals of imminent desertification, (iii) measures for prevention or adaptation to imminent desertification, and (iv) restoration of degraded areas. We will mostly be concerned with classes (i), (ii) and (iv) as they involve significant pattern formation aspects that have already been studied to some extent.

Desertification is often conceived as a sudden global transition from a productive stable state to a less productive or unproductive alternative stable state [270]. This view can be illustrated using a bifurcation diagram that shows a bistability range of such states (see Section 2.2.3), as in Figure 3.1. In this diagram the vertical state axis represents a measure of productivity, e.g., biomass, while the horizontal control-parameter axis represents an environmental condition, e.g., precipitation [257]. Desertification can be induced either by decreasing the control parameter below the saddle-node (fold) bifurcation where the productive state no longer exists (solid arrow in Figure 3.1) or by a disturbance that kicks the system out of the attraction basin of the productive state, e.g., by removal of vegetation biomass (dashed arrow).

This conceptual view of desertification has been instrumental in devising warning signals for imminent desertification [269]. Such transitions involve bifurcations at which one eigenvalue or more attains a zero value (see Section 2.2.2). As a result, in the vicinity of a desertification transition the system's response to disturbances slows down, and autocorrelations in time series and temporal variance increase. Thus, monitoring an increase in these quantities may serve as or early indicator for imminent desertification [58]. Another possible indicator is asymmetrical changes in the distribution of time series, quantified by changing skewness [115]. The reader is referred to Section 11.1.2 for a more detailed discussion of early warning signals and for additional references.

The view of desertification presented above overlooks two important spatial aspects. First, desertification seldom involves a transition between two spatially uniform states; most often it is a transition from spotted woody vegetation to ephemeral vegetation or bare soil, as Figure 3.2(a) shows and

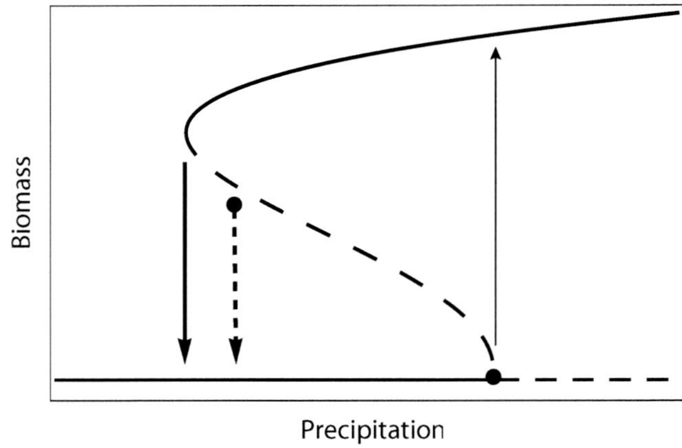

FIGURE 3.1: A bifurcation diagram illustrating two scenarios of desertification in a system that has a bistability range of productive and unproductive states: a precipitation downshift below the saddle-node bifurcation (thick solid arrow), and biomass removal (dashed arrow). The thin solid arrow represents the reverse transition from the unproductive state to the productive state, which occurs at a higher precipitation threshold and thus involves hysteresis.

as discussed in Chapter 10. Second, disturbances are often confined in space, leading to local rather than global transitions to the alternative stable state, as Figure 3.2(b) illustrates. These spatial aspects can significantly affect both the dynamics of desertification and the applicability of the proposed early warning signals, as we will briefly discuss in Section 3.3.2 and, in a more elaborate manner, in Chapter 11.

3.2.2 Biodiversity loss

The concept of biodiversity refers to the full variety and variability of ecological entities within the living world [100]. The ecological entities may be biotic, abiotic or both, and can be grouped into three classes: organism entities, such as genes and species, resource entities, such as various sources of water and nitrogen, and landscape entities, such as soil patches varying in soil properties and vegetation cover [280]. According to the entity class, or the specific entity, different types of diversities can be distinguished, organismal diversity, species diversity, landscape diversity and so on. Of all diversity types species diversity stands out as the most important determinant of ecosystem services. Provision services such as food and timber, regulating services such as wastes and water quality, and supporting services such as photosynthesis and nutrient cycling, are all crucially dependent on the availability of diverse bacterial, plant and animal species. Thus, to a large extent, assessing the

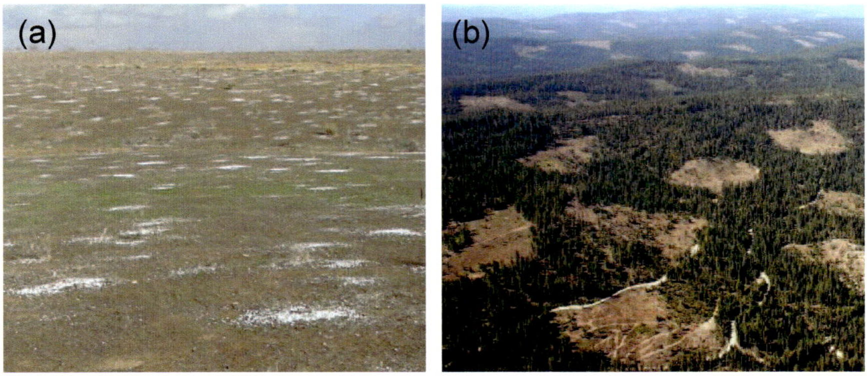

FIGURE 3.2: Spatial aspects of desertification. (a) A vegetation "ghost pattern" in the northern Negev desert, formed by remnant patches of snails at the locations of pre-existing shrubs. The pattern provides an indication of a desertification process involving a transition from non-uniform spotted vegetation to bare soil. Photo courtesy of Moshe Shachak. (b) Clear cutting in Blue Creek Watershed, indicating that disturbances can be strong and yet spatially confined. In bistability ranges such disturbances can form domains of the alternative stable state. Photo courtesy of Bruse Castle–EPFW.

impact of biodiversity change on ecosystem function can be based on measuring species diversity changes.

There are various measures of species diversity. The simplest and most common is species richness, that is, the number of species in a site or a region. This measure does not contain information about species abundance and therefore does not distinguish between a community where all species are evenly abundant and a community with a few or a single dominant species. Obviously the latter community is less diverse in the sense that the chance to randomly choose two different species is smaller, even if the community is as rich as the former one. Various diversity indices have been proposed to integrate richness and degree of evenness. An example is the rank-abundance curve [187] formed by ranking species according to their abundance in a descending order (most abundant species has rank 1) and plotting the abundance relative to the total abundance as a function of the rank. The species richness is given by the end point on the rank axis while the evenness is given by the slope of the curve.

Related to species diversity is the concept of functional diversity, which is attracting an increasing attention [200]. Functional diversity is defined as the diversity of the functional traits of the species in a community. A trait of an organism is said to be "functional" if it affects the organism's role or performance in the ecosystem. Focusing on functional traits and their diversity can help identify the mechanisms that affect community assembly and ecosystem function.

Biodiversity has been reduced during the past fifty years in a rate faster than at any time in human history. The most important drivers of this alarming decline are habitat change, climate change, invasive alien species, over exploitation, and pollution [71]. Many of these drivers involve human activities. Particularly significant in terrestrial ecosystems is the impact of land-use change, such as the clearing of natural vegetation for agriculture or road construction, which leads to habitat fragmentation or destruction. Fragments of a habitat can support smaller species populations which make them more vulnerable to extinction. Moreover, the different climate at the boundary of a habitat may be unfavorable for those species that require an interior habitat. Fragmentation therefore results in smaller areas of favorable conditions, which increase species competition, or in no areas of favorable conditions if the fragmentation is too strong. In marine ecosystems fishing is the major anthropogenic force affecting biodiversity.

Biodiversity loss is a concern because it degrades ecosystem services and may risk ecosystem stability. Biodiversity is related to the stability of an ecosystem to environmental changes by providing a buffer [332]. As different species respond differently to environmental changes, extinction of dominant species does not necessarily impair the function of an ecosystem, for other species, that better fit the new environmental conditions and benefit from a less competitive community, can grow instead. Higher species diversity therefore suggests stability to a wider range of environmental changes.

Understanding of biodiversity-loss mechanisms calls for integrated studies of different diversity types as they are all inherently coupled. One of the earliest expressions of such diversity coupling is the *habitat heterogeneity hypothesis* [303], which relates species, resource and landscape diversities. According to this hypothesis an increase in the number of habitats in a landscape reduces species competition by providing more niches and alternative ways of exploiting the environmental resources, and therefore leads to an increase in species diversity. Relations among landscape, resource and species diversities have been explored in numerous studies [303, 314, 280]. Most studies focused on the response of organisms to landscape and resource heterogeneities, regarding the abiotic elements as "independent variables" that determine biotic elements. Fewer studies have considered the reciprocal relations in which the independent variables are biotic, addressing the role of organisms in modifying their abiotic environment and creating landscape and resource heterogeneities [142]. The actual reality, however, is more complex since both types of elements, biotic and abiotic, are mutually coupled in a highly nonlinear dynamical way through various positive and negative feedbacks. These feedbacks can induce, in particular, pattern-forming instabilities that lead to self-organized heterogeneity, involving both biotic and abiotic elements [255, 105].

The role of self-organized heterogeneity in biodiversity dynamics has hardly been studied [102, 94, 322], and many basic questions have remained unaddressed. Examples of such questions include: (i) What types of spatial patterns a community of species can form along environmental gradients, and how do they differ in species diversity? (ii) Under what biotic and abiotic conditions a multiplicity of stable patterned states can develop and how feasible are these conditions? (iii) What are the controllers of state transitions that involve loss of species diversity? (iv) How is succession in a community affected by pattern formation? (v) How does self-organized heterogeneity mediate the impact of environmental and anthropogenic forces on species diversity? Narrowing down the context, more specific questions can be asked; a few of them are posed in Section 3.3.4 in the context of self-organized vegetation patchiness.

3.3 The relevance of pattern formation theory to spatial ecology

Pattern formation theory is currently the most developed mathematical framework for studying self-organized spatial phenomena in ecology. The theory not only provides the conceptual and mathematical tools for studying the emergence and dynamics of self-organized patchiness, as we discuss in Section 3.3.1, but also many of its ecological derivatives. Three ecological derivatives are discussed. In Section 3.3.2 we address the relevance of pattern formation theory to desertification. The productive vegetation state that is vulnerable to desertification is generally spatially patterned. This fact may have important implications for the dynamics of desertification. In Section 3.3.3 we discuss rehabilitation of degraded areas by water harvesting methods. This is a spatial-resonance problem involving the entrainment of vegetation patterns to periodic ground-modulation patterns. Understanding what patterns better entrain to the ground modulations is important for successful rehabilitation. Finally, in Section 3.3.4 we discuss the possible impacts of vegetation pattern formation, and the resource heterogeneity that it induces, on species coexistence and diversity.

3.3.1 Vegetation patchiness

The similarity of observed vegetation patterns, such as spots, stripes and gaps (Figures 1.1 and 3.3), to universal patterns in other physical contexts, strongly suggests that dryland [63] and wetland [319] landscapes are pattern-forming systems, in which spatial heterogeneity can result from symmetry-breaking instabilities. Such instabilities give rise to spatially-dependent states

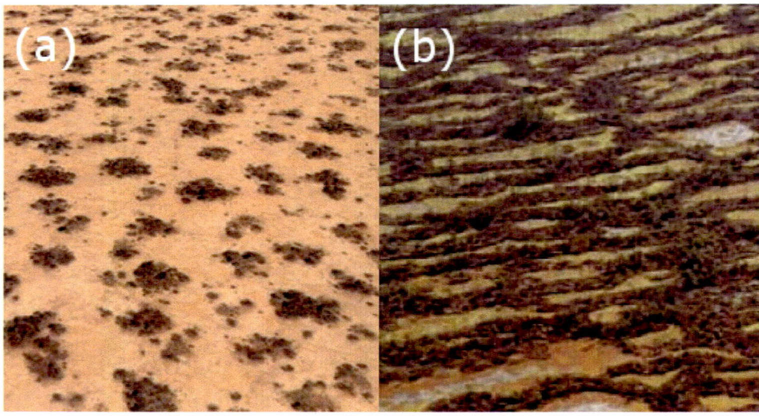

FIGURE 3.3: Vegetation spots in a dryland landscape (a) and vegetation stripes in a wetland landscape (b). From [63] (a) and [319] (b).

even if the systems themselves are homogeneous (see Section 2.3.1). Natural landscapes can seldom be regarded as homogeneous; micro-topography, different types of soils, water-table height and other factors all contribute to the heterogeneity of the landscape. Yet, the deterministic mechanisms that induce symmetry-breaking instabilities often dominate the imposed heterogeneity [283] and give rise to nearly periodic patterns such as those shown in Figure 3.3.

Two methodologies can help corroborate or refute the hypothesis that vegetation patchiness in nature is a self-organization phenomenon involving symmetry breaking. The first is controlled laboratory experiments on systems that are large in the sense discussed in Section 2.2.1, and the second is mathematical modeling. Controlled laboratory experiments on large systems can be carried out in practice only with species that form short-wavelength patterns. Perennial grasses such as *Paspalum vaginatum* (Figure 2.1) and *Poa bulbosa* [285] form patterns with characteristic lengths of the order of 0.1m, and are therefore possible candidates for laboratory-scale experiments. Controlled experiments can be used to test the possible destabilization of uniform vegetation to a pattern state upon decreasing water supply, as well as specific model predictions, including the sequence of vegetation states along a precipitation gradient and bistability of states (see Section 10.1). Such experiments have not been reported yet.

Mathematical modeling can be used to test the self-organization hypothesis by capturing basic processes at the patch scale and testing, by model analysis, whether they induce symmetry-breaking patterns at the landscape scale. In resource-limited vegetation landscapes these processes are generally positive feedbacks involving local biomass growth which by various mechanisms induces the transport of a limiting resource towards the growth area where it is being exploited. The concurrent depletion of the limiting resource in the neighborhood of that area inhibits the biomass growth there and favors the formation of spatial patterns. In Chapters 9 and 10 we will describe a few positive feedbacks of this kind and the progress that has been made in understanding vegetation patchiness using mathematical models that capture these feedbacks. The application of pattern formation theory can provide insight about (i) transient vs. asymptotic patterns, (ii) periodic vs. non-periodic patterns, (iii) scale-free patterns lacking any characteristic length, and (iv) stationary vs. traveling patterns. These insights may help us understand the variety of vegetation patterns observed in nature.

3.3.2 Desertification

Desertification is associated with any state transition that involves loss of productivity. Two transitions of this kind stand out: a transition from uniform vegetation to patchy vegetation, and a transition from patchy vegetation to bare soil (see Figure 10.3). In both cases one of the alternative stable states is spatially patterned. Studies of simple pattern-formation models with bistability ranges of uniform and pattern states have revealed the existence of a multitude of additional stable states [157]. These states, which we call "hybrid states," are spatial mixtures of the basic alternative stable states, consisting of confined domains of one state in an area otherwise occupied by the alternative state. A typical bifurcation diagram that describes this case is shown in Figure 7.13 (see Section 7.2.1.3 for more information).

The significance of these hybrid states for desertification is exemplified in Figure 3.4. Shown in this figure are numerical simulations of a vegetation model [325] in a bistability range of a spot pattern and uniform bare soil. The initial state (left panel) consists of a spot pattern, containing a few defects which locally change the density of spots[3]. This initial state is perturbed by partial biomass removal, uniformly across the pattern, visualized in the figure as a lighter gray shade (left panel). The system responds to the initial perturbation by the decay of spots at the defect cores, as the arrows indicate. The final pattern has recovered from the perturbation, as the darker shades of gray in the right panel imply, but contains "holes" at the locations where spots have decayed. Periodic patterns with holes are among the multitude of hybrid states that exist in the bistability range of uniform and pattern states. Little is known about the dynamics across different hole solutions in response to environmental variability.

As pointed out earlier in Section 3.2.1, ecosystems are generally subjected to spatially confined disturbances, which can induce local state transitions to the alternative stable state. The subsequent dynamics is determined by the

[3]These are "penta-hepta" defects consisting of adjacent spots with five and seven nearest spots, instead of six in a perfect hexagonal spot pattern [249, 243].

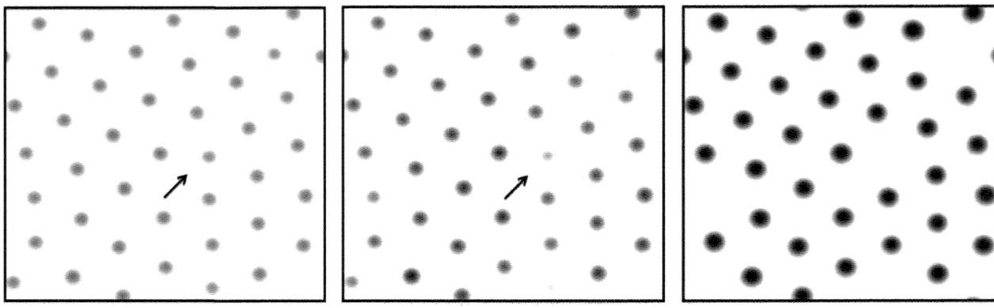

FIGURE 3.4: Transition from a periodic spot pattern to a hybrid pattern consisting of holes in an otherwise periodic pattern. The transition is induced by a uniform biomass-removal disturbance. Time proceeds from left to right, lighter shades of gray denote lower biomass. The left panel shows the initial disturbed periodic pattern, while the right panel shows a nearly asymptotic pattern. The arrows point toward one of the decaying spots. The patterns were computed using the model introduced in [325].

dynamics of the transition zones or fronts that separate the two alternative stable states. Front dynamics is an important driver of pattern formation. We briefly discussed it in Section 2.3.1 and will thoroughly address it in Sections 7.2.1 and 7.3.1. Depending on the control-parameter value, fronts that separate uniform and patterned states can propagate in either direction; the uniform state can expand into the patterned state or vice versa. However, there might be a control-parameter range in which the front is stationary or "pinned" [244][4]. This is unlike bistability of two uniform states in which fronts always propagate in one direction or another except for a particular control-parameter value (the Maxwell point) at which the front is stationary. The existence of pinned fronts, i.e., fronts that are stationary in a *range* of the control parameter, when one of the alternative stable states is spatially patterned is tightly related to the existence of a multitude of stable hybrid states. The implications of these front dynamics are that desertification and recovery are not necessarily abrupt—they can be gradual, proceeding by the expansion and coalescence of degraded areas, or even incipient (see Chapter 11).

Front dynamics play an important role also in shrubland-grassland transitions, such as the invasion of creosote bush (*Larrea tridentata*) and mesquite (*Prosopis glandulosa*) shrublands into black grama (*Bouteloua eriopoda*) grasslands in New Mexico, a process considered to be desertification because of the removal of productive grassland [274]. While shrubland-grassland transitions are a subject of active ecological research [31], important pattern formation aspects of this phenomenon have remained unaddressed. Shrublands in water-limited systems generally form spot patterns, while grasslands form

[4]A front can be depinned, however, when subjected to an additive noise [47].

a continuous land cover. Shrubland-grassland transitions therefore involve the propagation of a pattern state into a uniform state in what appears to be a bistability range of the two, since both plant life forms prevail under the same environmental conditions. The ecological and physical conditions for pinning and the size of the pinning range in actual ecological contexts have never been studied. Identifying biotic and abiotic factors that affect this range, using deterministic and stochastic mathematical models, can shed new light on shrubland-grassland transitions and on means to stop or reverse them.

3.3.3 Rehabilitation of degraded landscapes

Restoring a degraded landscape can be viewed as reversing the state transition that has lead to the degradation in the first place. The reverse transition is expected to occur at a different threshold corresponding to improved environmental conditions relative to the degradation transition as the thin arrow in Figure 3.1 illustrates. This is mainly because of various positive feedbacks that act to stabilize the unproductive state, such as reduced landscape connectivity and consequent resource leakage, enhanced evaporation, soil erosion, etc. Restoration studies focus in part on understanding factors that can lower the threshold of the reverse state transition [297].

Another approach to rehabilitation of degraded areas is runoff harvesting, a term referring to various methods of concentrating rainfall runoff in vegetation growth areas [265]. Runoff harvesting dates back 4000 years to early agriculture in the middle east, where hillside vegetation was cleared to direct spate flow to fields on the plains [88]. Current schemes of runoff harvesting consist of micro-catchments that intercept runoff, and are organized in arrays of different geometries to form linear, trapezoidal or rectangular embankments, as illustrated in Figure 3.5. Figure 3.6 shows an example of a restoration project in the northern Negev, Israel. In the absence of a basic theory, the forms and structures of the micro-catchments and the geometries of their arrays are determined mostly empirically.

Pattern formation theory can be most instrumental in improving existing runoff harvesting practices of vegetation restoration and in developing new practices. The periodic micro-catchment arrays (Figure 3.5) can be regarded as a spatial periodic forcing of a pattern-forming system, and is thus a spatial resonance problem analogous to oscillating systems that are periodically forced in time, except that the time axis is replaced by a space axis. Successful restoration, which corresponds to a vegetation pattern that follows the micro-catchment pattern, amounts to a resonant response. According to pattern formation theory such a resonant response is expected to occur even if there is a mismatch between the forcing pattern of the micro-catchments and the pattern that the unforced system tends to form—the stronger the forcing the bigger the mismatch can be.

The view of vegetation restoration as a resonance phenomenon highlights new aspects of this problem that should be taken into account, such as what

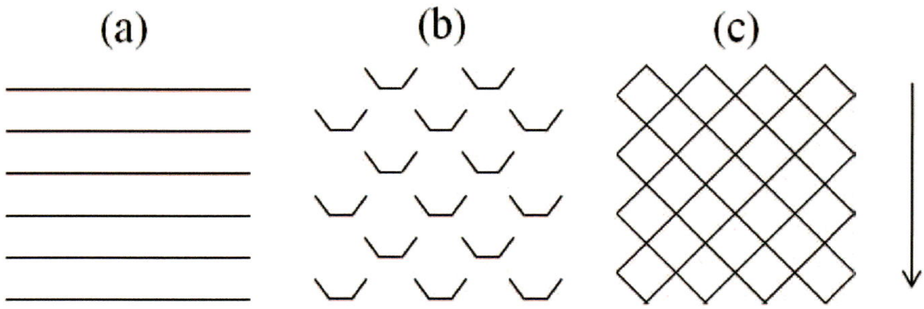

FIGURE 3.5: Schematic illustration of different types of micro-catchment arrays. (a) Parallel bunds, (b) trapezoidal bunds, and (c) diamond bunds ("negarim"). The arrow denotes the down-slope direction.

type of patterns the unforced system tends to form, how big the mismatch with the forcing pattern can be, and what new patterns the forcing can induce. Consider, as an example, the restoration of banded vegetation by means of parallel embankments. Drawing an analogy to temporally forced oscillating systems, resonant response is expected when the forcing wavenumber is sufficiently close to the wavenumber of the pattern that the unforced system tends to form. The maximal wavenumber mismatch that still allows resonant response scales with some power of the forcing strength, a relation that defines a narrow tongue-like domain in the plane spanned by the forcing wavenumber and the forcing strength (see Figure 8.4). Recent studies show that systems with two spatial dimensions have another mode of response—forming resonant two-dimensional patterns; the system adjusts one component of the pattern's wave-vector to the forcing and compensates for the unfavorable wavenumber

FIGURE 3.6: Views at different scales of vegetation restoration in the northern Negev, Israel, by the KKL-JNF. Parallel embankments that intercept runoff provide an increased water supply to the trees that are planted along them.

by forming a wave-vector component in the orthogonal direction [196, 204]. This surprising effect, the induction of a two-dimensional pattern by one-dimensional forcing of a one-dimensional pattern (see Figure 8.15) highlights the importance of understanding pattern formation mechanisms and processes in designing rehabilitation practices.

3.3.4 Mechanisms of species coexistence and diversity change

The habitat heterogeneity hypothesis (see Section 3.2.2) has motivated numerous studies of the relationships between landscape heterogeneity and species diversity [303]. Implicit in these studies is the assumption that the heterogeneity is predetermined and does not respond to environmental changes or to the community dynamics. This assumption considerably limits the scope of phenomena that can be addressed. For example, a question such as what the impact of climate change is on species diversity in water-limited systems, cannot be properly analyzed without taking into account possible transitions between different vegetation patterns, which change the landscape and thus the habitat heterogeneity.

The impact of vegetation pattern formation on species diversity and the possible feedback of diversity change on vegetation patchiness is a hard unstudied problem. Yet, important insights can be gained by considering simplified contexts. Consider, for example, water-limited plant communities, which typically contain two main functional groups, woody and herbaceous. Empirical studies have shown that the interspecific interaction between the two groups changes along a rainfall gradient; at relatively high rainfall rates woody patches compete with herbaceous vegetation and excludes it, whereas at low rainfall rates they can facilitate its growth [133]. Various factors may contribute to this phenomenon, e.g., reduced evaporation in woody patches, nutrient production by litter decomposition, protection against grazers and others. There are, however, additional factors, solely associated with pattern formation effects, that can account for the transition from competition to facilitation as an environmental stress develops. Studies of a two-species model, to be described in greater detail in Sections 12.2.1 and 12.2, indicate that such a transition can result from a change in the relative strength of two biomass-water feedbacks that have opposite effects on the soil-water content in vegetation patches. The relative strength of the two feedbacks can also be affected by a transition from one stable woody pattern to another. Although this is a simplified problem that addresses the interaction between two species only, it provides a conceptual framework for addressing harder questions pertaining to large herbaceous communities, such as the buffering capacity of a pattern-forming woody species to mitigate diversity decline of herbaceous communities in stressed environments.

Another simplified context where pattern formation can affect species coexistence is bistability of uniform and patterned community states. Consider a water-limited plant community in which species compete for two resources, water and light, and assume that there exists a tradeoff in the capacity of a species to capture the two resources: taller species that are better adapted to compete for light have smaller root-to-shoot ratios and are therefore less competitive with regard to water and vice versa. Consider the simplest case of two species that make different tradeoffs. Model studies, to be described in detail in Section 12.3, show the possible existence of two alternative stable states describing pure populations of the two species: a spatially uniform population of one species and a patterned population of the other species. Recall now the discussions in Sections 3.2.1 and 3.3.2 about front pinning and the existence of a multitude of hybrid states in a bistability range of uniform and patterned states. As shown in Section 12.3, similar behavior has been found in this community context too. The significance of front pinning is that none of the two species can invade and displace the other and that many (hybrid) states of species coexistence are possible. This is a pure pattern-formation effect; when the two populations states are spatially uniform one species will always displace the other. In principle, this study can be extended to a whole community of species that are distributed along the tradeoff axis. As discussed in Section 12.4.1, an extension of this kind can be used to study pattern-formation effects on community level properties, such as diversity-productivity relations.

Chapter summary

Ecosystems are complex systems involving hierarchies of trophic and organization levels. Adding to their complexity are spatial patterns and processes occurring across wide ranges of spatial and temporal scales. These aspects of complexity characterize, in particular, two outstanding problems in ecology, desertification and biodiversity loss, and their implications for ecosystem function. Despite the wide attention these problems have received, important spatial aspects have been overlooked in most studies. Desertification is a state transition that generally involves a spatially patterned state and may proceed gradually through the expansion and coalescence of degraded areas. Biodiversity is affected by landscape heterogeneity that can self-organize in response to environmental changes through mechanisms of pattern formation.

Spatial aspects of this kind can be studied using the concepts and mathematical tools of pattern formation theory. Four concrete examples of the relevance of pattern-formation theory to spatial ecology are discussed:

Vegetation patchiness: Pattern-formation theory can be used to test the hypothesis that positive feedback between local vegetation growth and water transport toward the growth location can induce symmetry-breaking instabilities that lead to vegetation patterns.

Desertification: According to pattern-formation theory, bistability of uniform and patterned states often involve a multitude of additional hybrid states. These states, which describe fixed domains of one state in a background of the other, can affect the dynamics of desertification.

Rehabilitation: Vegetation is often restored using runoff-harvesting methods that employ periodic arrays of micro-catchments to intercept runoff and form favorable conditions for vegetation growth. This is a spatial resonance problem—the spatial counterpart of entrained oscillators—that can be studied using the tools of pattern-formation theory.

Species coexistence: Pattern formation processes allow for new mechanisms of species coexistence. Two examples are described, a change in the relative strength of two pattern-forming feedbacks that affects inter-specific interactions, and bistability of uniform and patterned population states of different species, which allows for a multitude of coexistence states.

Chapter 4

Modeling ecosystems

4.1 Why model?

In order to link pattern formation theory to spatial ecology, mathematical models are needed. Such models simplify ecosystem complexity, circumvent empirical limitations, provide inquiry tools for studying ecological mechanisms and processes, and serve as simulation tools for studying various scenarios of ecosystem response to varying environmental conditions and disturbances. Following a brief introduction to non-mathematical descriptive and conceptual models used in ecology we describe and confront two mathematical approaches to modeling the dynamics of a spatially extended ecosystem, computational algorithms vs. partial differential equations (Section 4.1.1), and motivate the need for such dynamic models (4.1.2).

4.1.1 Types of models and the purposes they serve

The term "model" is used in ecology to denote distinct concepts which serve different purposes. Three major types of models can be distinguished, descriptive, conceptual and dynamic [78]. Descriptive models are used to present observed relationships between measured variables by means of graphs or algebraic relations. Examples of such models are hump-shape diversity-productivity graphs, or power-law diversity-area relations. Descriptive models do not contain information about the processes that produce the observed relations, and therefore cannot be used to generalize these relations to other ecological contexts. They are helpful as clear quantitative means to summarize observed relations and motivate theoretical studies. By contrast, conceptual models do seek to unravel basic processes and mechanisms, but remain at the level of qualitative propositions. A conceptual model often consists of a diagram of boxes, representing system elements, and arrows, representing processes that couple different system elements. It may also appear as a graphical presentation of qualitative outcomes predicted by the proposed mechanism. Figure 4.1 shows an example of a conceptual model illustrating the pattern-forming feedback between biomass and water associated with differential infiltration (see Section 9.1.2.1). Conceptual models are instrumental in establishing quantitative dynamic models (see Section 4.2.2).

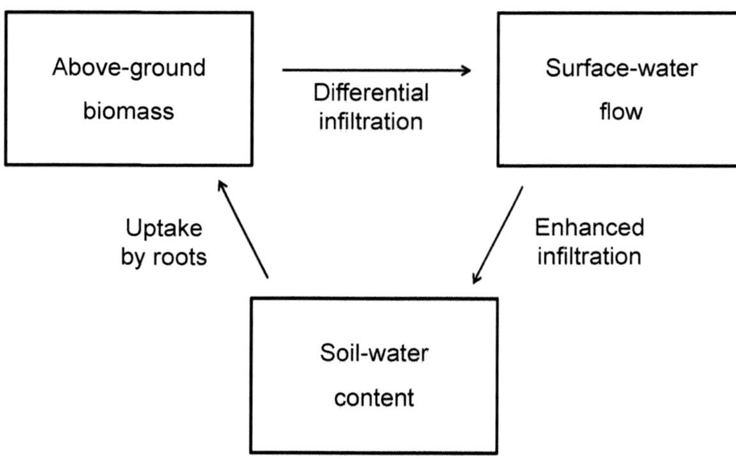

FIGURE 4.1: A conceptual model illustrating the positive feedback induced by differential infiltration that leads to vegetation pattern formation. The boxes represent the relevant biotic and abiotic elements, while the arrows represent processes that couple the elements. Patches of denser vegetation (higher areal density of above-ground biomass) increase the infiltration rate of surface water and create an infiltration contrast with respect to patches of sparser vegetation. The resulting differential infiltration induces overland water flow towards the denser vegetation patches, which enhances the infiltration of surface-water there. The increased content of soil-water in these patches and the uptake by the plants' roots accelerate the vegetation growth, which sharpens even further the differential infiltration. This feedback acts to destabilize uniform vegetation and form patterns because the flow of surface water away from patches of sparser vegetation increases the mortality there.

Dynamic models bridge the gap between conceptual and descriptive models. They provide the mathematical machinery that converts the information contained in conceptual models about processes that couple different system elements, to the measurable outcomes of such processes–the information that is often contained in descriptive models. The former type of information generally pertains to small scales and low organization levels, while the latter pertains to large scales and high organization levels. As their name implies, dynamic models can be used to study how ecosystems change in the course of time given prescribed initial states. Various forms of spatially-explicit dynamic models have been proposed, and an extensive literature reviewing them exists [62, 307, 154, 113, 132, 293, 29, 258]. Of these, two forms stand out, agent-based models, commonly referred to in ecology as individual-based models, and models consisting of partial differential equations (PDE).

Individual-based models are computational algorithms that go down to the level of individual organisms and often describe them in great detail. The individuals in these models have internal states, represented by discrete or continuous variables, and are dispersed in a discrete or continuous space. They interact with one another according to state-dependent rules, which determine the individuals' states in the next time step. PDE models, on the other hand, do not address individual organisms but rather *processes at small spatial scales*. In these models individuals are often described by the space-dependent biomass they form per unit area, which may result from a single or many individuals. While the PDE approach does not capture information related to the life cycle of an individual, it is capable of describing continuous processes, such as resource flow, advection, and erosion-deposition processes, which are often derivable from fundamental theory. From the implementation point of view individual-based models are formulated in terms of algorithms that are executed by numerical computations, whereas PDE models are amenable to mathematical analysis besides numerical computations. This is an important difference between the two modeling approaches; PDE models lend themselves to the powerful tools of pattern formation theory [56, 55], whereas no mathematical theory of this kind is available for individual-based models.

Dynamic models are mainly used for simulating the dynamics of ecological systems ("simulation models") and as inquiry tools for studying ecological mechanisms and processes. Simulation models (mostly individual-based models) are generally very elaborate and often call for high-power computational means. Models used as inquiry tools (mostly PDE) are generally much simpler and are studied using combinations of mathematical analysis and numerical computations.

4.1.2 The need for dynamic models

Empirical studies of ecosystems are inherently limited. The first limitation pertains to the short duration of both field and laboratory experiments. Most experiments are a few years long, but even the longest experiments, which are run in LTER stations[1] and endure a few tens of years, are often too short to monitor slow ecosystem changes, such as woody-patch dynamics in drylands and the associated community changes. Field experiments are also limited by the uncontrollable environments in which they are carried out, while controlled laboratory experiments are limited by their spatial extent. Dynamic models complement empirical studies by circumventing these limitations. They help identify asymptotic behaviors, they allow the study of ecological processes on large scales, and they can be used to elucidate mechanisms of ecological processes by independently controlling biotic and abiotic factors. In addition, they can be used to study various scenarios of ecosystem dynamics, and, to a debatable extent, make predictions and forecasts (see Section 4.3.3).

The need for dynamic modeling in ecology is a direct result of ecosystem complexity. The dynamics of relatively simple systems are often describable by first principle equations, derived from fundamental theories. In that case, modeling is not required, although simple models may still be useful in gaining deeper insights into the dynamics of such systems. Simple fluids provide a good example; the dynamics of such fluids are described very well by the Navier–Stokes equations, the derivation of which is based on fluid-dynamics theory [310]. The more complex the system the less applicable the first-principle approach. Chemically reactive fluids already require some degree of modeling. Chemical reactions, such as the BZ reaction, generally consist of tens of elementary reactions. The complete breakdown of the overall reaction to elementary reactions is often not known and model reaction schemes are needed. The combination of these model schemes with the mass-action law of chemical kinetics and Fick's law of diffusion, result in reaction-diffusion type models. Ecological systems, which contain living entities, are far more complex and call for phenomenological modeling. Such modeling relates empirical observations in a way which is consistent with fundamental theory, but is not directly derived from such a theory. Yet, even complex systems of this kind often contain physical elements whose dynamics are derivable from fundamental theory, such as the flow of limiting resources. Thus, although dynamic models of ecosystems heavily rely on phenomenological modeling, opportunities to apply the first-principle approach should not be overlooked.

4.2 The modeling process

Modeling an ecosystem consists of three main steps. The first is the identification of the state variables that define the system and the parameters that characterize the system's environment. The second step is setting the

[1]LTER stands for the Long Term Ecological Research network. It is a world-wide collaborative effort to investigate ecological processes over long temporal and spatial scales.

computational algorithms or the partial differential equations that define the rules or laws that propagate the state variables in time. The third step is testing the model for consistency with the physical context it pertains to, and by confronting model predictions with empirical data. These steps are described in Sections 4.2.1, 4.2.2 and 4.2.3 below.

4.2.1 Defining the model system

We will make use of the criterion introduced in Section 3.1.1 to distinguish between an ecosystem and its environment: an ecosystem consists of biotic and abiotic elements with strong reciprocal coupling, while the environment of an ecosystem consists of elements that strongly affect the ecosystem's elements but are weakly being affected by them. Figure 4.2 shows a schematic illustration of this criterion. The ecosystem elements are quantified by a set of state variables, U_1, $U_2, \ldots$, which are mutually coupled in general. The dynamics of these variables are affected by a set of parameters, P_1, $P_2, \ldots$, that quantify the environment. The state variables are generally space and time dependent to allow for spatial instabilities and the emergence of self-organized heterogeneity, even for a uniform environment characterized by space-independent parameters. Making the distinction between the ecosystem elements and the environment elements is a major challenge. It requires an understanding of

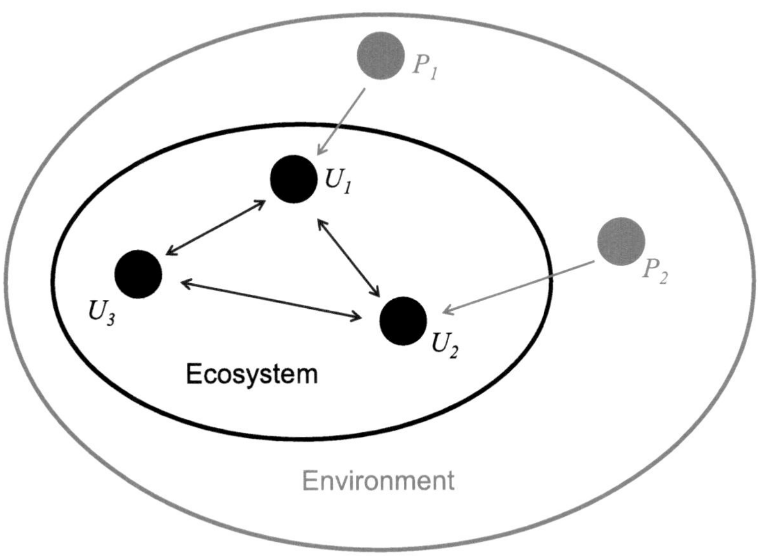

FIGURE 4.2: A schematic illustration of the distinction between an ecosystem and its environment. The inner loop represents the ecosystem whose elements are quantified by a set of mutually coupled state variables, U_i. The outer loop represents the environment whose elements and their impact on the ecosystem are quantified by a set of parameters, P_i.

the important processes at work and often involves assumptions or hypotheses that can be tested by confronting model predictions with empirical results.

A few considerations assist in defining the state variables of an ecosystem and distinguishing them from the parameters that represent the environment. The first consideration is clarifying the exact ecological and physical context of interest. The following two examples illustrate this point. In a dryland ecosystem the ground topography can be considered as an element of the environment, with strong impact on water flow and vegetation patchiness, provided the soil is stable. However, if the soil is unstable and is modified by erosion-deposition processes, the ground topography may have to be considered as a system element. Likewise, in species-diversity studies of plant communities, herbivore pressure can be considered as an element of the environment, provided other food resources are available to the herbivores. If that is not the case, the herbivores' population may significantly fluctuate with vegetation changes, and may have to be included as an additional system element.

Another consideration is related to the type of information a dynamic model of an ecosystem is expected to produce–qualitative or quantitative. Qualitative information refers to the nature of the system state, whether it is a steady, oscillatory, spatially uniform, or a patterned state; to the nature of transitions between different states, whether they are gradual or abrupt, reversible or irreversible; to the environmental changes that can induce such transitions; to species-assemblage properties, such as diversity-productivity relations, whether they are hump shaped, monotonically increasing, or monotonically decreasing, and so on. Quantitative information provides, in addition, numerical estimates of various characteristics of the system state, such as bio-productivity, patch-size distributions and species richness; estimates of transition thresholds and the degree of irreversibility, and so on.

Models aimed at providing quantitative information must take into account all elements and processes that may affect the numerical results they produce, and are generally highly elaborate. By contrast, models aimed at qualitative information can be much simpler, for despite the overwhelming complexity of ecosystems, not all system processes are equally important in any given context. In fact, substantial changes in ecosystem dynamics are often driven by very few processes [130]. Such models can contain only those state variables and processes that *generate* the qualitative information, and disregard other variables and processes that may affect numerical values but not qualitative behaviors. An example will help clarifying this point. A model aimed at revealing the possible vegetation states of dryland landscapes, and the nature of state transitions, may be based on biomass and water variables only, and on pattern-forming feedbacks that couple them. Models of this kind predict a specific sequence of vegetation states along a rainfall gradient and the existence of bistability ranges between any consecutive pair of states (see Section 10.1). These are valuable qualitative predictions that can be tested in controlled laboratory experiments. Vegetation models of this kind may not give, however, accurate estimates of transition thresholds or of bistability ranges, because additional elements and processes affecting this quantitative information, such as soil erosion and deposition, nutrient dynamics, pathogen dynamics and others, are not included in the definition of the system.

The differences between elaborate models to produce quantitative information and simple models to produce qualitative information, projects on the type of dynamic model to be used. The qualitative information referred to above is tightly related to instability phenomena. Systems states appear and disappear or exchange stability in bifurcations of various types, and the nature of state transitions is determined by the bifurcations being supercritical or subcritical (see Section 2.2.3). PDE models, which are amenable to the tools of pattern-formation theory, are superior to individual-based models in making this qualitative information transparent. Individual-based models, on the other hand, may be superior when quantitative information is needed, because of the detailed information they can accommodate in their formulation. Indeed, detailed simulation models often rely on the individual-based modeling approach. A combination of both types of models may be needed when continuous processes, such as convection or erosion, are essential.

Processes occurring on separate time scales are yet another consideration to make in defining the state variables of an ecosystem. It is often the case that state variables that appear important to include in the system can actually be eliminated because of their fast dynamics and their almost immediate equilibration on instantaneous values of slower state variables.

4.2.2 Setting up the model

Once the state variables of a dynamic model have been identified the next step is determining the laws or rules that propagate them in time. As pointed out earlier, conceptual models can be very instrumental in determining the processes that couple the state variables with one another and with parameters representing the environment. These processes are mathematically translated to terms in the equations that constitute PDE models or to computational rules in the algorithms of individual-based models. We will focus here on PDE models, because they form the basis for integrating pattern formation theory into ecological theory, and refer the reader who is interested in individual-based modeling to the vast literature that is available on this topic [143, 62, 113].

PDE models consist of equations for the state variables that describe the biotic and abiotic elements of an ecosystem. The typical structure of an equation for a state variable U is illustrated in Figure 4.3. The left side is the partial time derivative of $U(\mathbf{r},t)$ at the location $\mathbf{r}=(x,y,z)$. It describes the rate of change of U at the point $\mathbf{r}=(x,y,z)$ at time t. The right side describes processes that contribute to the temporal change of U at that point. These processes can be divided into three main groups: (i) local processes, i.e., processes that depend on the values of state variables at the point $\mathbf{r}$, (ii) short-distance nonlocal processes involving fluxes of matter through the point $\mathbf{r}$ from and to nearby points, and (iii) long-distance nonlocal processes that depend on the values of state variables at distant points $\mathbf{r}'$.

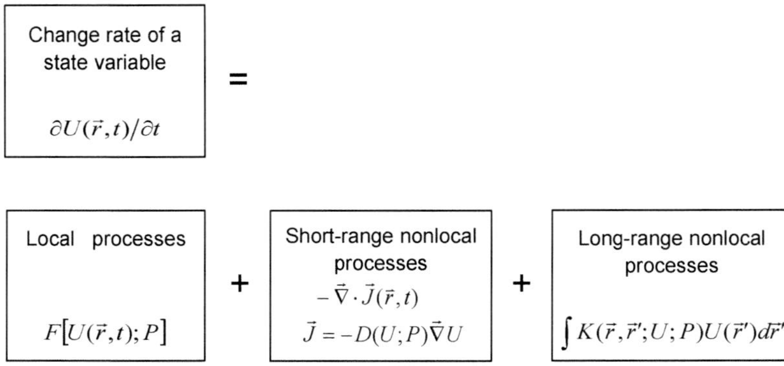

FIGURE 4.3: A schematic illustration of a PDE model. The model consists of equations for the temporal rate of change of any state variable U in the model. Shown are three types of processes that contribute to these temporal rates of change along with their mathematical forms: local processes, short-range nonlocal processes involving fluxes and long-range nonlocal processes. The mathematical forms are schematic; in general the contributions to the rate of change of a state variable U_i are nonlinear and may depend on other state variables U_j. These contributions also depend on a set of parameters P representing elements of the environment or organism-related parameters that reflect processes at lower organization levels.

The dependence of local processes on state variables is generally nonlinear. For example, biomass growth by local resource uptake is often represented by a nonlinear term involving a product of the biomass state variable and a nonlinear function of the resource state variable. Short-distance nonlocal processes involve the flux of a certain quantity. Mathematically, they are represented by divergence terms, $\nabla \cdot \mathbf{J}$, where $\mathbf{J}$ is a flux that is generally proportional to the quantity gradient. Such flux terms increase (decrease) the quantity at point $\mathbf{r}$ if the influx to an infinitesimal volume $d\mathbf{r} = dxdydz$ around that point is larger (smaller) than the outflux from that volume. An example of a flux term is the diffusion of a nutrient C. The flux is $\mathbf{J} = -D\nabla C$, where D is the diffusion constant, and the divergence term that describes the change of C in time is $-\nabla \cdot \mathbf{J} = D\nabla^2 C$. Readers unfamiliar with the mathematical concepts of gradient and divergence and their applications in biological contexts are referred to Ref. [165]. Long-distance nonlocal processes are represented by terms involving integrals over space. An example of such a process is long-range seed dispersal [224], where the integral is over a product, $K(\mathbf{r},\mathbf{r}')B(\mathbf{r}')$, of a kernel function, K, and a biomass variable, B, representing the contribution of a plant at point $\mathbf{r}' = (x',y')$ to seeds at point $\mathbf{r} = (x,y)$. The integral represents summation over the locations, $\mathbf{r}'$, of all plants that contribute seeds at point $\mathbf{r}$. Nonlocal processes are generally nonlinear too, although a linear approximation is often assumed.

The terms that appear in the equations for the state variables in a PDE model generally contain parameters that can be divided into two groups; environmental parameters, quantifying the manner by which biotic and abiotic elements of the environment affect system elements, and organismic parameters, representing processes occurring at levels of organization lower than those considered in the model. For example, in vegetation models in which the atmosphere is considered an element of the environment, the effects of all processes that contribute to the generation of rainfall are often lumped together and quantified by a precipitation rate parameter. This environmental parameter can be time dependent, or even space and time dependent, but this dependence is prescribed rather than a result of the system dynamics. Vegetation growth rate is an example of an organismic parameter in that it is related to processes at the organ and organ-system organization levels, e.g., resource allocation to roots, stem and leaves. Environmental parameters can assume stochastic forms to simulate environmental fluctuations.

As pointed out earlier, by focusing on qualitative information, PDE models simplify the ecological reality to an extent that makes them mathematically tractable. This allows unraveling the bifurcation structure of a PDE model and predicting qualitative ecosystem behaviors. But how simple PDE models can be? In order for a model to have a predictive power[2], rather than merely a descriptive value, we should avoid incorporating into the model the type of information we wish to extract from it. For example, if we wish to study conditions under which a multiplicity of stable states develops, we should avoid modeling a multiplicity of stable states. That is, we should avoid introducing "by hand" nonlinear terms that give rise to multiple stable states, as such terms will generally contain non-measurable parameters. Numerical values for these parameters can be determined by comparing model output with empirical data, but this will harm the predictive power of the model. Instead, we should model the ecological and physical processes that are deemed essential to the emergence of multiple stable states, introducing terms that contain independently measurable parameters. Studying model solutions we can then identify parameter ranges of multi-stability. The model becomes more elaborate but acquires predictive power in that assigning parameter values, in accordance with specific eco-physical contexts of interest, can provide information about multi-stability.

4.2.3 Testing the model

4.2.3.1 Consistency with physical context

The mathematical formulation of a dynamic model is not complete before the model is tested for consistency with the physical context it is intended

[2]We use the term "predictive power" to refer to the *capability* of a model to make new statements about the systems it describes, irrespective if these statements are corroborated or refuted by empirical data.

to model. For example, a model for dryland vegetation that is used to study desertification should show a decrease in the vegetation biomass as rainfall decreases but never to negative values; the appearance of negative biomass as the dynamics proceed is a consistency problem that indicates a modeling flaw. Another possible example for a consistency problem is a model that fails to capture a pattern-forming instability when the context it is intended to model includes spatial self-organization.

Quite often consistency problems of a dynamic model can be identified using a few preliminary analyses. In general, nonlinear PDE models are analyzed using combinations of analytical and numerical tools, which increasingly lean toward numerical tools as the model becomes more complex. Exact analytical solutions are generally restricted to very simple models, or to very simple states, such as stationary uniform states. Yet, useful information can already be obtained from such states using the method of linear stability analysis (see Section 5.3). This method provides information about the possible occurrence of instabilities and the types of modes that grow at the instability points (see Sections 2.3.1 and 5.3.3), which often indicate what types of time and/or space dependent states the system can evolve to. The lack of an instability or the appearance of an instability of the wrong type within the validity range[3] of a PDE model may indicate an inconsistency and a modeling flaw. Model consistency tests should therefore be conducted over the entire validity range of the model.

4.2.3.2 Confrontation with empirical data

Various assumptions are made in modeling ecosystems, both in the definition of the state variables and in modeling the processes that couple these variables among themselves and to the environment. These assumptions, and the model performance as a whole, can be tested by confronting model predictions with empirical data. To this end, numerical values should be assigned to the model parameters. Quite often such values are already available in the literature. Alternatively, experiments can be designed to measure the parameters directly, or calibration procedures can be used to infer their values from matching model outputs to empirical data. Once parameter values have been specified, comparisons of model solutions with empirical data, other than those used for calibration, can be made. How meaningful such comparisons are to model testing has been the subject of much debate and confusion [237, 263, 39]. Here we adopt the view that although good agreements between model results and empirical data cannot verify or confirm a model [237], they can increase the confidence in the model. This is especially so if a variety of substantially different tests show good agreements [39]. We note that the agreement between model results and empirical data should be assessed relative to the model complexity; while detailed models should aim at quantitative

[3]We define the validity range of a model as the range of parameters within which the model assumptions are satisfied.

agreement, simple models may aim at qualitative agreement only. The significance of qualitative model output and agreement with empirical data is discussed in Section 4.3.2 below.

In choosing the model results to be compared with empirical data care must be exercised not to rely solely on universal behaviors. Dryland vegetation is an excellent example to clarify this point. A variety of different models of dryland vegetation capture the instability of uniform vegetation to hexagonal gap pattern and produce the same sequence of basic vegetation states along the rainfall gradient: uniform vegetation, gaps, stripes, spots and bare soil (see Section 10.1). The reason that all models produce the same patterns is that they share the same instabilities, despite the fact that in each model the feedbacks that induce these instabilities are modeled differently and involve different eco-physical assumptions. The differences in modeling the feedbacks, however, will show up in the soil-water distributions in and around vegetation patches. Thus, adding soil-water information to the comparison of model output and empirical data can discern among the different models and highlight the best performing model.

4.3 Model output

Once a dynamic model has been set up and tested it can be analyzed by the methods of pattern formation theory or solved numerically. Analytical studies provide more insight but are often restricted to small parameter ranges—the validity ranges of the analytical approximation methods that are used. Exploration of the whole parameter range of interest generally requires numerical studies. Two widely used methods are briefly described in Section 4.3.1, numerical integration of PDEs in time and numerical continuation of solutions in parameter space. The output of numerical studies are numbers, but these numbers often hide important qualitative information. A few examples of such information and its significance for ecosystem dynamics are described in Section 4.3.2. A discussion of the significance of models' outputs for understanding basic mechanisms, for providing scenarios of possible behaviors, and for making short-term forecasts of ecosystem dynamics concludes this section (Section 4.3.3).

4.3.1 Analytical vs. numerical output

The information that can be extracted from a PDE model generally combines analytical and numerical results. Analytical results include in the first place the functional dependence of constant solutions on model parameters, the existence ranges of such solutions and their stability properties. This type of information is generally represented in the form of bifurcation diagrams.

The analysis of nonconstant solutions, i.e., time dependent, space dependent or both, generally calls for nonlinear approximation methods, some of which will be introduced in Chapter 6. These methods provide information about the forms, existence ranges and stability properties of periodic solutions and, to a limited extent, also about localized and disordered solutions. However, the assumptions that are made in applying these approximation methods restrict the validity of the analysis to a limited part of the parameter range of interest. In order to explore the whole range of interest complementary numerical analyzes are needed.

PDE models can always be solved numerically by propagating in time initial spatial distributions of the state variables (initial conditions) subject to constant constraints at the system's boundaries (boundary conditions). This way of solving PDEs, often called "numerical integration," provides the spatial forms of periodic and non-periodic solutions as they evolve in time. Since the long-term dynamics always proceed toward stable solutions, numerical integration does not provide information about unstable solutions. The results of numerical integration are often presented as snapshots of the dynamics at selected times, as shown in Figure 3.4. Solution dynamics in one space dimension (1d) can conveniently be presented in a continuous manner using space-time plots as Figure 7.8 shows. Numerical integration of PDE models can be used to produce bifurcation diagrams by solving the PDEs repeatedly at incrementally increasing values of a control parameter. However, the resulting diagrams are partial in that they do not include unstable solution branches. There are various methods of integrating PDEs numerically, many of which are specifically geared to particular types of PDEs. Since most of these methods are standard and extensively described in textbooks [245, 55], we do not present them here[4].

Another useful way of solving PDE models, particularly when complete bifurcation diagrams are needed, is by numerical continuation of solutions using iterative methods [163, 160, 55]. These methods require good "initial guesses" for the solutions, which are then iterated to accurate solutions. The initial guess can be an analytical approximate solution in its range of validity that is continued to the range of interest, or a solution obtained by numerically integrating the PDE model. Numerical continuation methods apply to ODEs (ordinary differential equations) but can be used to study PDEs too by reducing them to ODEs. For example, solutions describing stationary periodic patterns or constant-speed traveling waves in 1d satisfy ODEs when setting the time derivatives in the PDEs to zero or when transforming to a moving coordinate system. An example of using a numerical continuation method is shown in Figure 10.4. The figure shows a bifurcation diagram that includes solution branches of stationary periodic vegetation patterns calculated by setting

[4]One caveat is worth mentioning though; integrating PDEs numerically should be carried out with great care. Quite often numerical blowup or "interesting" dynamical behaviors are numerical artifacts produced by poor resolution in space and time or by unstable numerical codes.

the time derivatives in a vegetation PDE model to zero. This method can be used to obtain the explicit forms of the solutions and their existence ranges but not their stability properties. For that purpose numerical linear stability analysis can be carried out using the original PDEs. Numerical continuation has an advantage over numerical integration in that it provides the forms and existence ranges of both stable and unstable solutions, and can be used to obtain complete bifurcation diagrams. While unstable solutions cannot provide information about asymptotic ecosystem states, they may change the flow in phase space in ecologically significant ways. An example of such an effect will be discussed in Section 11.3.2.

4.3.2 The significance of qualitative information

Once the numerical output of a PDE model is available, various statistical methods can be used to analyze and confront it with empirical data, or make predictions with estimated uncertainties [101]. In general, however, there is much more in the numbers of the numerical data. The data often contain information about nonlinear constructs that are responsible for universal behaviors. The identification of such constructs helps associate the system with known and often well understood dynamical behaviors.

Instabilities are the most fundamental constructs responsible for universal behavior. This is grounded in the mathematical theory of normal forms which associates a set of characteristic dynamical equations with each instability type, and provides the machinery of deriving these equations from PDE models that contain this type of instability. In Section 2.3.1 we have briefly discussed instabilities of stationary uniform states that induce oscillations and spatial patterns. In the second part of this book we will present the normal form (or amplitude) equations associated with these instabilities and describe how they are derived. These equations are well studied and the knowledge that has been obtained can readily be used once the instabilities are identified. Thus, if the numerical data produced by a dynamic model or, alternatively, a linear stability analysis of that model suggest or indicate the existence of an oscillatory instability, a whole range of dynamical behaviors, including traveling waves, spiral waves, localized structures such as sources and sinks, and spatiotemporal chaos, are potentially relevant and might be found in the dynamic model beyond the instability point (see Section 7.2.2). Additional nonlinear constructs, such as bistability of states and resonant response to periodic forcing, and the associated universal behaviors will be discussed in Chapters 7 and 8.

The information about a nonlinear construct contained in numerical data can also be indirect; rather than showing the nonlinear construct itself it can show a dynamical behavior associated with it. The possible existence of this construct can guide us in exploring many of its other dynamical behaviors. As an example, consider the numerical data shown in Figure 4.4, which was obtained by simulating a dynamic model for woody and herbaceous vegetation in

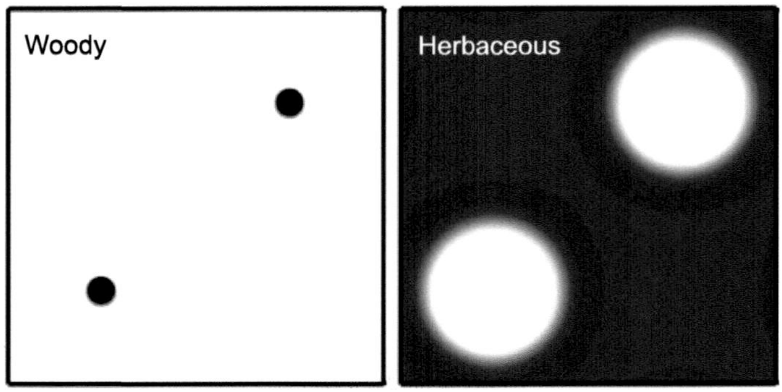

FIGURE 4.4: A numerical solution of a PDE model of a water-limited woody-herbaceous systems showing a savanna-like pattern. Shown are the biomass distributions of the woody (left) and herbaceous (right) life forms. Darker grey shades indicate higher biomass values. The solution suggests the existence of two alternative stable states, periodic woody spot pattern (shrubland) and uniform herbaceous vegetation (grassland) and many more stable hybrid states. Adopted from Ref. [102].

a water limited system. It shows a uniform distribution of herbaceous biomass, interrupted by two isolated woody patches, which exclude the herbaceous vegetation from their neighborhoods. Each isolated woody patch is a "localized structure," the appearance of which often indicates bistability of uniform and patterned states (see Section 7.2.1.3), in this case uniform herbaceous vegetation ("grassland") and a periodic woody spot pattern ("shrubland"). Applying the information that pattern-formation theory provides about this nonlinear construct, many additional vegetation states can be anticipated with appropriate initial conditions, similar to those shown in Figure 7.14.

4.3.3 What are model outputs good for?

By their very nature dynamic models can provide information about the future state of a system, given its current state. The question is how reliable this information is, and what it is good for if it does not provide reliable predictions? Natural systems are open in the sense that the elements and processes that affect their dynamics can never be fully specified. As a consequence, model predictions are always subject to uncertainties [237, 39]. In fact, the uncertainties in predicting the behaviors of complex natural systems can be so big as to make the predictions meaningless. Yet, dynamic models of complex systems can still be instrumental as inquiry tools, as a means of providing scenarios of possible system behaviors, and as a means to interpret field observations and make short-term forecasts. The following examples illustrate these values.

In Section 3.3.1 we mentioned the role that local feedbacks, operating at the sub-patch scale, play in inducing vegetation patchiness at the landscape scale. We hypothesized that in order for a local feedback to induce vegetation patchiness it should involve water transport (see also Section 9.1.2). This hypothesis can be tested by including these feedbacks in the model equations and studying the linear stability of uniform vegetation to non-uniform perturbations. Switching on each feedback separately, by appropriate parameter choices, pattern-forming feedbacks can be identified and their relations to water transport can be determined. This is an example of the utility of model analysis as an inquiry tool and of its explanatory power.

Consider now the response of spatially uniform vegetation to a gradual inter-annual rainfall drop. PDE models of dryland vegetation predict three possible scenarios [324]: (i) the vegetation remains uniform but the biomass per unit area decreases, (ii) a gap pattern develops with a narrow gap-size distribution, (iii) a scale-free pattern, characterized by a wide gap-size distribution, develops. The models also indicate what biotic and abiotic conditions favor one scenario over the others. High infiltration contrasts between bare and vegetated areas will favor, for example, the second and third scenarios over the first [105]. Fast overland flow relative to infiltration along with small root zones will further favor the third scenario over the second [324].

An example of a short-term forecast that a PDE model may make is an impending regime shift. Model studies can predict the possible occurrence of a regime shift, but will generally fail in estimating how close the regime shift is. However, the combination of such model studies with field observations of warning signals, e.g., increasing temporal variance or autocorrelations in time series (see Chapter 11), may lead to a reliable forecast.

Chapter summary

In order to apply the mathematical methods of pattern formation theory to specific ecological systems mathematical models of such systems are needed. We call such models *dynamic models* because they describe the manners by which modeled systems develop in time. Dynamic models bridge over two other modeling concepts in ecology; descriptive models, which present relationships between measured variables, either graphically or algebraically, and conceptual models, which seek to unravel basic mechanisms that underlie observed behaviors, but remain at the level of qualitative propositions. Dynamic models provide the machinery that converts the information contained in conceptual models to the information that descriptive models provide. Studies of dynamic models complement empirical studies in that they help identify asymptotic behaviors, allow the study of large-scale ecological processes, help elucidate mechanisms of ecological processes by independently controlling biotic and abiotic factors, and allow the study of various scenarios of ecosystem dynamics.

Two forms of spatially explicit dynamic models stand out, individual-based models and partial differential equation (PDE) models. The former are computational algorithms that go down to the level of individual organisms and often describe them in great detail. The latter do not address individual organisms but rather processes at small spatial scales that involve continuous organismic attributes, such as local biomass. The advantage of PDE models over individual-based models is that they are amenable to the powerful methods of pattern formation theory. For this reason we focus on PDE models.

Building up a PDE model for a particular ecological context involves three steps. The first is choosing the state variables that constitute the system and the parameters that represent the system's environment. The state variables are mutually coupled dynamic entities and their choice often reflects the feedbacks that are being modeled. The next step is formulating the PDEs for these states variables. Each PDE equates the rate of change of the state variable to the sum of mathematical terms that represent local and nonlocal processes, e.g., local biomass growth and nonlocal seed dispersal. The last step is testing the model for consistency with the ecological context that is modeled, and by confronting model results with empirical data. The consistency tests involve preliminary model analyses to identify model flaws such as nonphysical values of state variables or lack of essential instabilities.

The output of a PDE model generally consists of information about the form, existence range and stability properties of various solutions (states) obtained analytically or numerically. This information is often represented in the form of a bifurcation diagram. In addition, numerical solutions of a PDE model can show the time evolution of the state variables towards stable states. This information is often presented as a series of snapshots or space-time plots. The assumptions that are made in analytical derivations of approximate time and/or space dependent solutions generally restrict the validity of the results to a limited part of the parameter range of interest. Numerical solutions can then complement the analytical results by extending the analysis to the whole parameter range of interest. These solutions often contain information about nonlinear constructs, such as instabilities, that are responsible for universal behaviors. The identification of such constructs helps associate the system with known and often well understood dynamical behaviors.

Complex natural systems are open in the sense that the elements and processes that affect their dynamics can never be fully specified. As a consequence, model predictions are often subject to uncertainties that may be too big to make the predictions meaningful. Yet, PDE models can still be instrumental as inquiry tools, as a means of providing scenarios of possible system behaviors, and as a means to interpret field observations and make short-term forecasts.

Part II

Pattern Formation Theory

Chapter 5

Pattern formation analysis: Basic methods

5.1 Dimensional analysis

The analytical methods to be described later in this chapter and in Chapter 6 are all based on PDE models. There is one method, however, that is applicable even when the dynamic equations that describe the system are not known. The method, commonly referred to as *dimensional analysis*, is useful for deriving scaling laws that relate physical quantities of interest to other quantities that characterize the system under consideration. Although the dynamic equations that describe the system need not be known, some preliminary understanding of the problem is needed, in order to make a judicious choice of the quantities that are most relevant to the problem in question, and in some cases certain assumptions have to be made. A successful implementation of the method may require *a posteriori* justification of these assumptions by empirical studies. We begin with an overview of the method (Section 5.1.1) followed by a more detailed exposition (Sections 5.1.2, 5.1.3, 5.1.4), and conclude with another application of dimensional analysis—the simplification of model equations by expressing them in terms of dimensionless quantities consisting of rescaled state variables, time and space coordinates and a reduced set of parameters (Section 5.1.5).

5.1.1 An overview of the method

Measurable physical quantities can be divided into dimensional and non-dimensional quantities. Dimensional quantities are described by combinations of numbers and units, e.g., a distance of 10km, an oscillation period of 2s, and so on. Non-dimensional quantities do not have units and are described by numbers only, e.g., an angle of $\pi/4$, Rayleigh number of 1700 (see Section 2.1.2), etc. The attribute "dimensional" refers to the *dimension* of a physical quantity[1], a term that simply denotes the type of units used to measure that quantity. A given quantity has a dimension of length if it is measured in units of length (meter, lightyear etc.), or has a dimension of velocity if it is measured in units of length over time (meter per second, kilometer per hour etc.).

The first step in dimensional analysis is to identify a set, $\mathbf{X} = \{X_1, ..., X_n\}$, of relevant dimensional quantities that includes the variables and parameters that characterize the system under consideration, and the quantity to be described by a scaling law. Consider, as an example, the pendulum and suppose we wish to find a scaling law for the oscillation period, T_p. Our preliminary acquaintance with the system suggests that the oscillation period may depend on the gravitational acceleration, g, the pendulum length L, its mass M and the initial angle, θ, with respect to the vertical axis. Of this set of quantities, only the first four are dimensional; the angle θ is dimensionless and we may not expect dimensional analysis to unravel the way the oscillation period, T_p, depends on it. The set of relevant dimensional quantities is therefore $\mathbf{X} = \{g, L, M, T_p\}$.

The second step is to split the set $\mathbf{X}$ into a subset $\mathbf{A} = \{A_1, ..., A_k\}$ of dimensionally independent quantities and a complementary subset $\mathbf{B} = \{B_1, ..., B_m\}$, where $m = n - k$, consisting of the remaining quantities in $\mathbf{X}$ and including the quantity to be described by the scaling law. The subset $\mathbf{A}$ consists of quantities whose dimensions cannot be expressed in terms of the dimensions of other quantities in $\mathbf{A}$, and has the additional property that the dimension of any quantity in $\mathbf{B}$ is expressible in terms of the dimensions of quantities in $\mathbf{A}$. In the pendulum example, the quantities L, M and T_p are dimensionally independent because we cannot express length in terms of mass and time, and likewise for mass and time. The remaining quantity g, however, has a dimension of length over time squared that is expressible in terms of the dimensions of L (length) and T_p (time). We may therefore define the set A to consist of L, M and T_p, and the subset $\mathbf{B}$ to consist of g. Alternatively, since T_p is the quantity to be described by a scaling law, we can choose the

[1]Not to be confused with the *dimension* of a dynamical system which corresponds to the number of independent degrees of freedom.

subset **A** to consist of L, M and g, which are also dimensionally independent (see Section 5.1.2), and the subset **B** to consist of T_p.

We refer to the dimensionally independent quantities of length, mass and time as having *fundamental dimensions*, which we denote by $\mathcal{L}$, $\mathcal{M}$, and $\mathcal{T}$, respectively. Additional fundamental dimensions we often encounter are temperature, Θ, and charge, $\mathcal{Q}$. Systems in which the dimensions of all quantities are expressible in terms of a set $\{\mathcal{L}_1, ..., \mathcal{L}_k\}$ of fundamental dimensions will be called $\mathcal{L}_1 \cdots \mathcal{L}_k$ systems. The pendulum and other mechanical systems belong to the class of $\mathcal{LMT}$ systems. Fluid dynamical systems, such as the RB system, belong to the class of $\mathcal{LMT}\Theta$ systems and so on [12]. In general, none or only a few of the quantities in **X** have fundamental dimensions and the choice of the subset **A** requires some analysis. We will explain how this choice can be made in Section 5.1.2.

The final step in dimensional analysis is to use the subset **A** to define dimensionless quantities, $\Pi_1, ..., \Pi_m$, associated with the quantities in **B**, and reduce any relation among the n quantities in **X** to a relation among the m dimensionless quantities. The reduction is based on the so-called Π-theorem, and is used to obtain the scaling law for the quantity of interest. We will explain this reduction, and demonstrate it on the simple pendulum example, in Section 5.1.3. Two additional examples will be discussed in Section 5.1.4.

The significance of dimensional analysis goes beyond the derivation of scaling laws. It also eliminates redundant quantities and reveals *similar* phenomena, i.e., phenomena that share the same numerical values of the dimensionless quantities, $\Pi_1, ..., \Pi_m$, even though the values of the dimensional quantities, $X_1, ..., X_n$, are different [12]. This aspect of dimensional analysis is useful in transforming the dynamic equations of a system to a dimensionless form. We will explain this procedure in Section 5.1.5 using a simple example.

5.1.2 Dimensionally independent quantities

The subset $\mathbf{A} = \{A_1, ..., A_k\} = \{X_1, ..., X_k\}$ of the set **X** consists of dimensionally independent quantities that span all other quantities in **X**. More specifically, it is characterized by the following two properties:

1. Dimensional independence: None of the quantities in **A** has a dimension that is expressible in terms of the dimensions of other quantities in **A**, or mathematically

$$[A_1]^{c_1} \cdots [A_k]^{c_k} = 1 \quad \text{implies} \quad \mathrm{c_1} = ... = \mathrm{c_k} = 0\,, \tag{5.1}$$

   where the square brackets denote the dimension of the quantity enclosed in the brackets.

2. Spanning: All other quantities in **X**, i.e the quantities in the complement subset, $\mathbf{B} = \{B_1, ..., B_m\} = \{X_{k+1}, ..., X_{k+m}\}$, where $k + m = n$, have dimensions that are expressible in terms of the dimensions of the

quantities in **A**. This dependence can be expressed mathematically as

$$[B_j] = [A_1]^{\alpha_{1j}} \cdots [A_k]^{\alpha_{kj}} \quad j = 1, ..., m\,, \tag{5.2}$$

where the exponents α_{ij} are rational numbers.

As pointed out in the previous section, the set **X** often contains quantities that do not have fundamental dimensions and the choice of the subset **A** requires some analysis. Consider as an example an $\mathcal{LMT}$ system with a set **X** that includes the quantities velocity, v, density, ρ, and energy, E. The question that we ask is: does the set $\mathbf{A} = \{v, \rho, E\}$ satisfy the property of dimensional independence associated with (5.1) and the spanning property associated with (5.2)?

To test the first property we look for rational numbers c_1, c_2 and c_3 that satisfy

$$[v]^{c_1}[\rho]^{c_2}[E]^{c_3} = 1\,. \tag{5.3}$$

Inserting in this equation the expressions $[v] = \mathcal{LT}^{-1}$, $[\rho] = \mathcal{ML}^{-3}$ and $[E] = \mathcal{ML}^2\mathcal{T}^{-2}$, we find

$$(\mathcal{LT}^{-1})^{c_1}(\mathcal{ML}^{-3})^{c_2}(\mathcal{ML}^2\mathcal{T}^{-2})^{c_3} = \mathcal{L}^0\mathcal{M}^0\mathcal{T}^0\,. \tag{5.4}$$

Comparing the powers of $\mathcal{L}$, $\mathcal{M}$, and $\mathcal{T}$ on both sides of (5.4) we obtain a homogeneous system of linear equations for c_1, c_2 and c_3 whose matrix form is

$$\mathbf{Kc} = \mathbf{0}\,, \tag{5.5}$$

where

$$\mathbf{K} = \begin{pmatrix} 1 & -3 & 2 \\ 0 & 1 & 1 \\ -1 & 0 & -2 \end{pmatrix} \quad \text{and} \quad \mathbf{c} = \begin{pmatrix} c_1 \\ c_2 \\ c_3 \end{pmatrix}. \tag{5.6}$$

Since the determinant of the matrix **K** is non-zero the only solution of this system is the trivial solution $\mathbf{c} = \mathbf{0}$ or $c_1 = c_2 = c_3 = 0$, which proves the dimensional independence of v, ρ and E and the validity of the first property.

To test the second property it is sufficient to show that the fundamental dimensions, $\mathcal{L}$, $\mathcal{M}$, and $\mathcal{T}$ are expressible in terms of $[v]$, $[\rho]$ and $[E]$. Since, by assumption, the dimension of any quantity in the complement subset **B** is expressible in terms of $\mathcal{L}$, $\mathcal{M}$, $\mathcal{T}$, it will also be expressible in terms of $[v]$, $[\rho]$ and $[E]$, which is what we want to prove. The dimension $\mathcal{L}$ is expressible in terms of $[v]$, $[\rho]$ and $[E]$ if there exist rational numbers d_1, d_2, d_3 such that $[v]^{d_1}[\rho]^{d_2}[E]^{d_3} = \mathcal{L}$, or

$$(\mathcal{LT}^{-1})^{d_1}(\mathcal{ML}^{-3})^{d_2}(\mathcal{ML}^2\mathcal{T}^{-2})^{d_3} = \mathcal{L}^1\mathcal{M}^0\mathcal{T}^0\,. \tag{5.7}$$

This equation leads to the inhomogeneous system of linear equations

$$\mathbf{Kd} = \mathbf{l}\,, \tag{5.8}$$

where, $\mathbf{K}$ is given by (5.6), $\mathbf{d} = (d_1, d_2, d_3)^T$, $\mathbf{l} = (1, 0, 0)^T$, and the superscript T denotes the transpose. Since the matrix $\mathbf{K}$ has a non-zero determinant the inhomogeneous system (5.8) has a unique solution and therefore $\mathcal{L}$ is expressible in terms of $[v]$, $[\rho]$ and $[E]$. Similar arguments hold for the dimensions $\mathcal{M}$ and $\mathcal{T}$, with the right side of (5.8) being replaced by the vectors $\mathbf{m} = (0, 1, 0)^T$ and $\mathbf{t} = (0, 0, 1)^T$, respectively. Solving (5.8) for $\mathbf{d}$ and the corresponding equations with $\mathbf{l}$ replaced by $\mathbf{m}$ and $\mathbf{t}$ give the explicit expressions: $\mathcal{L} = [v]^{-2/3}[\rho]^{-1/3}[E]^{1/3}$, $\mathcal{M} = [v]^{-2}[E]$ and $\mathcal{T} = [v]^{-5/3}[\rho]^{-1/3}[E]^{1/3}$.

While it never harms to perform the test of the second property for any candidate for a subset $\mathbf{A}$, we do not need this test if the candidate subset contains a number k of dimensionally independent quantities that is equal to the number of fundamental dimensions, that is, $k = 3$ for $\mathcal{LMT}$ systems, $k = 4$ for $\mathcal{LMT}\Theta$ systems, etc. When this condition is satisfied the subset $\mathbf{A}$ "spans" the class of systems considered. This statement can be made more rigorous by defining a vector space associated with that class of systems, and relating dimensionally independent quantities to linearly independent vectors, as we now show.

Consider for concreteness the $\mathcal{LMT}$ class. We associate with the dimension $[X] = \mathcal{L}^a\mathcal{M}^b\mathcal{T}^c$ of any quantity X a vector, $\mathbf{x}$, whose elements are the exponents a, b, c. The independent dimensions $\mathcal{L}$, $\mathcal{M}$, and $\mathcal{T}$, are represented by the unit vectors $\mathbf{l} = (1, 0, 0)$, $\mathbf{m} = (0, 1, 0)$ and $\mathbf{t} = (0, 0, 1)$, that have already been introduced. Vector addition and substraction correspond to the multiplication and division of quantities; given two quantities X_1 and X_2 whose dimensions are represented by the vectors $\mathbf{x}_1 = (a_1, b_1, c_1)$ and $\mathbf{x}_2 = (a_2, b_2, c_2)$, the dimension of the product $[X_1X_2] = [X_1][X_2]$, is represented by the vector sum

$$\mathbf{x}_1 + \mathbf{x}_2 = (a_1 + a_2, b_1 + b_2, c_1 + c_2)\,, \tag{5.9}$$

and the dimension of the ratio $[X_1/X_2] = [X_1]/[X_2]$ is represented by the vector difference

$$\mathbf{x}_1 - \mathbf{x}_2 = (a_1 - a_2, b_1 - b_2, c_1 - c_2)\,. \tag{5.10}$$

The dimension of a quantity X raised to the power d is represented by the product of the scalar d and the vector $\mathbf{x}$ that represents the dimension of X. It is straightforward to see now that the condition (5.1) is equivalent to the condition of linearly independent vectors,

$$c_1\mathbf{x}_1 + c_2\mathbf{x}_2 + c_3\mathbf{x}_3 = \mathbf{0} \quad \text{implies} \quad c_1 = c_2 = c_3 = 0\,, \tag{5.11}$$

and that a subset $\mathbf{A}$ that satisfies (5.1) and (5.2) corresponds to a basis that spans the whole vector space.

5.1.3 The Π-theorem

Define now the dimensionless quantities

$$\Pi_j = \frac{B_j}{A_1^{\alpha_{1j}} \cdots A_k^{\alpha_{kj}}}\,, \tag{5.12}$$

where the exponents α_{ij} are chosen such that

$$[B_j] = [A_1]^{\alpha_{1j}} \cdots [A_k]^{\alpha_{kj}} .$$

Any relation among the set, $\mathbf{X}$, of relevant dimensional quantities that can be written in the form

$$f(\mathbf{X}) = f(A_1, ..., A_k, B_1, ..., B_m) = 0 , \tag{5.13}$$

can be rewritten as

$$\tilde{f}(A_1, ..., A_k, \Pi_1, ..., \Pi_m) = 0 . \tag{5.14}$$

Suppose now that B_m is the quantity of interest and solve (5.14) for Π_m:

$$\Pi_m = h(A_1, ..., A_k, \Pi_1, ..., \Pi_{m-1}) . \tag{5.15}$$

Since the quantities in the subset $\mathbf{A}$ are dimensional, changing their units will change the value of Π_m. But Π_m is non-dimensional and cannot be affected by unit changes. It therefore follows that the function h in (5.15) cannot depend on the quantities A_j, $j = 1, ..., k$ [12]. This leads to the result known as the Π-*theorem* [32], namely that the relation (5.13) among the n relevant dimensional quantities can be reduced to a relation among a smaller number, m, of dimensionless quantities,

$$\phi(\Pi_1, ..., \Pi_m) = 0 , \tag{5.16}$$

or

$$\Pi_m = \psi(\Pi_1, ..., \Pi_{m-1}) , \tag{5.17}$$

where m is equal to the number of relevant dimensional quantities, n, minus the number k of independent dimensions. For $m = 1$, ψ is a constant number independent of any quantity in the set $\mathbf{X}$.

The final result of dimensional analysis is obtained by using (5.12) to write (5.17) as

$$B_m = A_1^{\alpha_{1m}} \cdots A_k^{\alpha_{km}} \psi(\Pi_1, ..., \Pi_{m-1}) , \tag{5.18}$$

thus obtaining a scaling law for the quantity B_m in terms of the quantities that belong to the subset $\mathbf{A}$. Note that the function ψ remains unknown and in cases where $m \geq 2$ some assumptions are often needed, as we will see in the next section.

Let us return now to the pendulum, as an example of an $\mathcal{LMT}$ system, and derive a scaling law for the pendulum's period T_p. It is straightforward to show (see Section 5.1.2) that the set $\mathbf{X} = \{L, M, g, T_p\}$ can be split into the subset $\mathbf{A} = \{L, M, g\}$ of dimensionally independent quantities and the dependent quantity $B_1 = T_p$. Thus $n = 4$, $k = 3$ and $m = 1$. The dimensionless quantity associated with T_p is $\Pi_1 = T_p\sqrt{g/L}$, and according to (5.17) $\Pi_1 = \psi = const.$ Thus $T_p = c\sqrt{L/g}$, where c is an undetermined constant, which may depend on dimensionless relevant quantities, the initial angle θ in this case.

It is interesting to compare this result with a calculation of the oscillation periodic, based on the equation of motion of the pendulum, for small angles θ [99]:

$$T_p = 2\pi\sqrt{\frac{L}{g}}\left(1 + \frac{1}{4}\sin^2\frac{\theta}{2} + ...\right) , \tag{5.19}$$

where the ellipses denote higher order contributions in $\sin^2(\theta/2)$. The dimensional analysis result, readily obtained without knowing the pendulum equation, correctly captures the independence of the oscillation period on the pendulum mass and the ways it scales with the pendulum length and the gravitational acceleration.

5.1.4 Examples of dimensional analysis

A classical example of dimensional analysis is the evaluation of the radius of a spherical shockwave that develops as a result of a very intense explosion [12]. The analysis was done by G. I. Taylor during the Second World War [300, 301]. The release of a large amount of energy in a small area creates an immense pressure behind the shockwave, several orders of magnitude larger than the pressure ahead of shockwave. This pressure gap holds during the early stages of the shockwave propagation and keep intact the shockwave structure. The relevant dimensional quantities that form the set $\mathbf{X}$ include the energy released, E, the initial air density, ρ, the shockwave velocity, v, and the shockwave radius, r. Since the dimensions of these quantities can all be expressed in terms of $\mathcal{L}$, $\mathcal{M}$ and $\mathcal{T}$, we regard the system as belonging to the $\mathcal{LMT}$ class. We have already seen in Section 5.1.1 that the the quantities $A_1 = v$, $A_2 = \rho$ and $A_3 = E$ form a subset $\mathbf{A}$ that satisfies the two properties (5.1) and (5.2) for $\mathcal{LMT}$ systems. We therefore identify $B_1 = r$ with the complement subset and conclude that $n = 4$, $k = 3$ and $m = 1$. The dimensionless quantity associated with r is

$$\Pi_1 = \frac{r v^{2/3} \rho^{1/3}}{E^{1/3}} , \tag{5.20}$$

and according to the Π-theorem $\Pi_1 = c$, where c is constant. This gives the following relation for r:

$$r = c\left(\frac{E}{\rho v^2}\right)^{1/3} . \tag{5.21}$$

In the original analysis the set of independent dimensional quantities, $\mathbf{A}$, was chosen to include the time t that has elapsed since the explosion, instead of the shockwave velocity v. Expressing v as r/t in (5.21) and solving for r yields the original result

$$r = c\left(\frac{t^2 E}{\rho}\right)^{1/5} . \tag{5.22}$$

This scaling relation has been confirmed by data obtained from a series of high-speed photographs [300, 301, 12].

Both in the pendulum and in the intense explosion examples the number of dimensionally dependent quantities was $m = 1$. As a consequence, the derived scaling laws contained an undetermined constant, c, rather than an undetermined function ψ of $m-1$ dimensionless quantities (see (5.18)). The last example we consider is the RB system of thermal convection for which $m > 1$. This system has been discussed in some length in Section 2.1.2. Suppose we wish to derive a scaling relation for the pressure difference across the vertical dimension of the cell. The relevant dimensional quantities include the fluid density, ρ, e.g., at the bottom of the cell, the temperature difference across the cell, ΔT, the coefficient of thermal expansion, α, the gravitational acceleration, g, the cell height, d, the thermal diffusivity, κ, and the shear viscosity[2], η. To this set of quantities we add the quantity for which we wish to find a scaling relation, the pressure difference, Δp. Thus, the set of relevant dimensional quantities we consider is $\mathbf{X} = \{\rho, \alpha, g, d, \Delta T, \kappa, \eta, \Delta p\}$.

The quantities in $\mathbf{X}$ have the following dimensions:

$$[\rho] = \mathcal{L}^{-3}\mathcal{M}, \quad [\alpha] = \Theta^{-1}, \quad [g] = \mathcal{L}\mathcal{T}^{-2}, \quad [d] = \mathcal{L}, \quad [\Delta T] = \Theta,$$
$$[\kappa] = \mathcal{L}^2\mathcal{T}^{-1}, \quad [\eta] = \mathcal{L}^{-1}\mathcal{M}\mathcal{T}^{-1}, \quad [\Delta p] = \mathcal{L}^{-1}\mathcal{M}\mathcal{T}^{-2}.$$

These dimensions suggests that the RB system belongs to the $\mathcal{LMT}\Theta$ class of systems, and therefore we should look for a subset $\mathbf{A}$ of four dimensionally independent quantities. The first four quantities, ρ, α, g and d involve all four independent dimensions $\mathcal{L}, \mathcal{M}, \mathcal{T}$, and Θ and are therefore a good candidate for the subset $\mathbf{A}$. Looking for rational numbers c_1, c_2, c_3, c_4 that satisfy

$$[\rho]^{c_1}[\alpha]^{c_2}[g]^{c_3}[d]^{c_4} = 1, \tag{5.23}$$

we readily find that $c_1 = c_2 = c_3 = c_4 = 0$, establishing the dimensional independence of ρ, α, g and d. We thus identify the subset $\mathbf{A}$ with the $k = 4$ quantities $\{\rho, \alpha, g, d\}$ and the complement subset $\mathbf{B}$ with the $m = 4$ quantities $\{\Delta T, \kappa, \eta, \Delta p\}$. Defining the dimensionless quantities,

$$\Pi_1 = \alpha\Delta T, \quad \Pi_2 = \frac{\kappa}{\sqrt{gd^3}}, \quad \Pi_3 = \frac{\eta}{\rho\sqrt{gd^3}}, \quad \Pi_4 = \frac{\Delta p}{\rho g d}, \tag{5.24}$$

we obtain the relation

$$\Pi_4 = \psi(\Pi_1, \Pi_2, \Pi_3). \tag{5.25}$$

It is interesting to note that the dimensionless Rayleigh number (2.1) introduced in Section 2.1.2 is given by $R = \Pi_1\Pi_2^{-1}\Pi_3^{-1}$ and that another dimensionless parameter characterizing RB convection, the Prandtl number, is given

[2]The shear viscosity is related to the kinematic viscosity, ν, introduced in Section 2.1.2 through the relation $\eta = \rho\nu$.

by $P_r = \Pi_3\Pi_2^{-1}$. Equation (5.25) gives the scaling relation for the pressure difference, which can be written as

$$\Delta p = \rho g d \varphi(R, P_r, \Pi_1)\,, \tag{5.26}$$

where φ is an undetermined function. This result may be regarded as useless in the sense that the function φ contains the quantities ρ, g and d, and, therefore, the scaling relation (5.26) is not conclusive. Quite often, however, undetermined functions, such as φ, approach constant values, particularly when the dimensionless quantities they depend on assume large or small values [12]. In this particular case, a calculation of the pressure difference, based on the solution of the thermal convection equations which describes the conduction state (fluid at rest), gives

$$\Delta p = \rho g d \left(1 + \frac{1}{2}\Pi_1\right)\,, \tag{5.27}$$

which is consistent with the dimensional analysis result (5.26), and implies that for $\alpha\Delta T \ll 1$, φ indeed approaches a constant value.

5.1.5 Non-dimensional forms of dynamic equations

We have been concerned so far with the derivation of scaling laws in cases where the dynamic equations that describe the system are not known. Dimensional analysis, however, remains useful even when such equations are available, for several reasons: (i) it provides a means of reducing the number of parameters, (ii) it reveals similarities in the roles different parameters play, and (iii) it helps identify small parameters which may be useful for applying perturbation techniques in analytical studies. Applying dimensional analysis to dynamic equations results in non-dimensional forms of these equations. In implementing this analysis we have to distinguish between variables and parameters and include in the subset **A** parameters only. The obvious reason is that we cannot eliminate the dependent and independent variables in terms of which the solutions to the dynamic equations are expressed. Consider, as an example, a periodically forced, damped spring, whose dynamic equation (derived from the second law of Newton) is

$$m\frac{d^2x}{dt^2} + b\frac{dx}{dt} + kx = f\cos(\omega t)\,. \tag{5.28}$$

The equation contains an independent time variable, t, a dependent displacement variable, x, and five parameters: the spring mass, m, the spring force constant, k, the damping constant, b, the force amplitude, f, and the force frequency, ω.

Since the system belongs to the $\mathcal{LMT}$ class, we need to choose for the subset **A** three parameters with independent dimensions. A possible choice is $\mathbf{A} = \{m, k, f\}$. It is straightforward to check that $[m]^{c_1}[k]^{c_2}[f]^{c_3} = 1$ implies

$c_1 = c_2 = c_3 = 0$ and, thus, verify the dimensional independence of these parameters. The complement subset is $\mathbf{B} = \{x, t, b, \omega\}$, and the dimensionless quantities associated with the elements of $\mathbf{B}$ are

$$\begin{aligned} \chi &= \frac{x}{k^{-1}f}, \\ \tau &= \frac{t}{k^{-1/2}m^{1/2}}, \\ \beta &= \frac{b}{k^{1/2}m^{1/2}}, \\ \nu &= \frac{\omega}{k^{1/2}m^{-1/2}}. \end{aligned} \tag{5.29}$$

Expressing the dimensional quantities $\{x, t, b, \omega\}$ in terms of their dimensionless counterparts $\{\chi, \tau, \beta, \nu\}$, and using these expressions in (5.28) we obtain the non-dimensional counterpart of (5.28),

$$\frac{d^2\chi}{d\tau^2} + \beta\frac{d\chi}{d\tau} + \chi = \cos(\nu\tau). \tag{5.30}$$

Note that (5.30) contains only two parameters, β and ν, instead of the five parameters that (5.28) contains, m, b, k, f and ω. This reduction in the number of parameters is responsible for the similarity of different spring systems that share the same numerical values of β and ν, as they are all described by exactly the same equation.

The example described above is simple in that it involves a single equation and a relatively small number of parameters. Models describing more complex systems generally involve several state variables and more parameters that share the same dimensions. The increased scaling freedom that this redundancy provides can be used to further simplify the non-dimensional equations. For example, in Section 9.2.2 we will use different scaling for the biomass and water variables in order to eliminate an additional parameter.

5.2 Two canonical models

The model equations that describe various ecosystems are often too complicated for applying advanced methods of pattern-formation theory. In such cases much insight can still be obtained by complementing these models with simpler models that share the same universal pattern formation phenomena but are more amenable to mathematical analysis. The latter models are also most instrumental for introducing the methods of pattern formation theory and for studying basic mechanisms of pattern formation, as we will see in Chapters 6 and 7.1.

We introduce here two models of this kind, the Swift–Hohenberg (SH) equation and the FitzHugh–Nagumo (FHN) model, which we choose to present in non-dimensional forms from the start. The SH equation is simpler in that it contains a single state variable and has a variational form. This property excludes asymptotic dynamic behavior, but allows for instabilities that induce stationary spatial patterns. The FHN model consists of two state variables and captures asymptotic dynamics, such as oscillations, traveling waves and spatio-temporal chaos, in addition to stationary patterns. Despite their simplicity, the two models, and various variants thereof, are rich enough to capture a variety of pattern-formation behaviors found in more elaborate and realistic models. In the following two sections we briefly describe a few behaviors of this kind, postponing more detailed accounts of these behaviors to Section 5.3.2 and Chapters 6 and 7.

5.2.1 The Swift–Hohenberg model

The Swift–Hohenberg (SH) equation is the simplest PDE model that captures an instability of a stationary uniform state to stationary periodic patterns. It can be cast in the non-dimensional form,

$$\partial_t u = \lambda u - u^3 - \left(\nabla^2 + k_c^2\right)^2 u\,, \tag{5.31}$$

where $u = u(x, y, t)$ is a real-valued state variable, λ is a control parameter, ∇^2 is the Laplacian in two space dimensions, and k_c is a critical wavenumber to be discussed below. Since we are mostly interested in large systems we will analyze the SH model on the infinite domain[3] $-\infty < x < \infty$, $-\infty < y < \infty$. The SH equation has originally been used to study the extent to which nonequilibrium transitions are similar to equilibrium phase transition [298, 55], but has become a canonical model for pattern formation studies in general. Note that the SH model does not have explicit space or time dependence; the only spatio-temporal dependence comes through the solution $u(x, t)$. This is a mathematical reflection of the system being uniform and stationary, i.e., symmetric to translations in space and in time. The state of the system, however, can break this symmetry as we shall soon see.

The variational nature of the SH equation is manifested by the existence of a real, space-independent quantity that never increases in time. This quantity extends the concept of a Lyapunov function, introduced in Section 2.2.5, to spatially extended systems where it becomes a functional of the spatially dependent state variables $\mathbf{u}(\mathbf{x}, t)$[4] and involves spatial integration over the

[3] In order to solve it numerically, however, we need to consider a finite domain and specify boundary conditions. Because of the fourth-order spatial derivatives in the equation two boundary conditions should be specified for any boundary point. A common choice is that u and its normal derivative, $\nabla u \cdot \hat{n}$, are both zero on the boundaries.

[4] A functional is a map that associates a number with a function, such as the value of a definite integral of some function.

entire system. The Lyapunov functional of the SH model is given by

$$L[u] = \int\int_{\Omega} dxdy \left[-\frac{\lambda}{2}u^2 + \frac{1}{4}u^4 + \frac{1}{2}\left(\nabla^2 u + k_c^2 u\right)^2 \right], \tag{5.32}$$

where Ω is the system's domain. To verify that $L[u]$ is indeed a Lyapunov functional of the SH equation we calculate its time derivative:

$$\dot{L} = \int\int_{\Omega} dxdy \Big\{ \left(-\lambda u + u^3\right)\partial_t u + \left[\left(\nabla^2 + k_c^2\right)u\right]\left[\left(\nabla^2 + k_c^2\right)\partial_t u\right] \Big\}. \tag{5.33}$$

Integrating twice by parts the term $[(\nabla^2 + k_c^2)u]\nabla^2\partial_t u$ to shift the operation of ∇^2 from $\partial_t u$ to $(\nabla^2 + k_c^2)u$ we find [55]:

$$\dot{L} = -\int\int_{\Omega} dxdy \left(\partial_t u\right)^2 \leq 0\,, \tag{5.34}$$

which proves that $L[u]$ cannot increase in time and is therefore a Lyapunov functional. Note that since

$$\dot{L} = \int\int_{\Omega} dxdy \frac{\delta L}{\delta u}\partial_t u\,, \tag{5.35}$$

where $\delta L/\delta u$ is the functional derivative of L[5], we can write the SH equation in the variational form (compare (5.34) with (5.35)):

$$\partial_t u = -\frac{\delta L}{\delta u}\,. \tag{5.36}$$

The existence of a Lyapunov functional rules out asymptotic dynamics, but not spatial patterns. Equation (5.31) has a uniform zero solution $u = 0$ for any constant value of λ. For $\lambda > 1$ it has in addition a symmetric pair of uniform non-zero solutions, $u = \pm\sqrt{\lambda - 1}$, which bifurcate from the zero solution in a stationary uniform pitchfork bifurcation at $\lambda = 1$. The zero solution, however, loses stability already at $\lambda = 0$ to the growth of periodic perturbations with wavenumber k_c in a non-uniform stationary instability, as we show in Section 5.3.1. This instability results in periodic stationary stripe patterns that minimize L. Because of the system's isotropy random initial perturbations about the unstable zero solution will generally lead to disordered stripe patterns as the left-most panel in Figure 5.1 shows. The subsequent dynamics show a relaxation process that culminates in an ordered stripe pattern (right most panel in Figure 5.1). The relaxation involves gradual elimination of grain boundaries and defects that "cost energy," i.e., increase L.

In the following chapters we will use several modifications of the SH equation to study pattern formation problems that are motivated by ecological

[5] Equation (5.35) is simply the counterpart of (2.18) when viewing the spatial coordinates x, y as continuous indices.

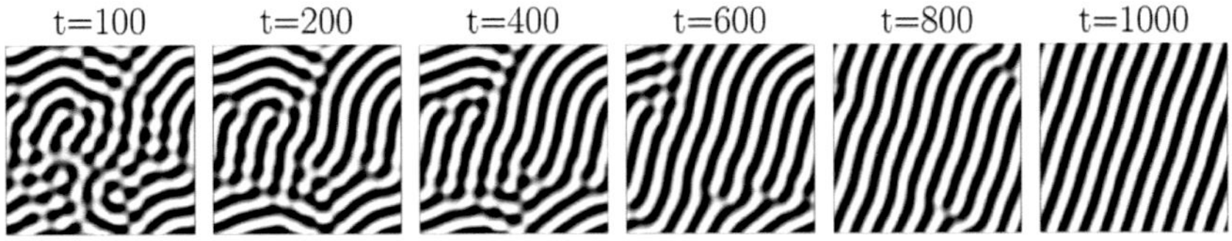

FIGURE 5.1: A simulation of the SH equation showing the relaxation of a disordered stripe pattern, consisting of stripe domains oriented in different directions and defects, to an ordered stationary stripe pattern. Parameters: $\lambda = 0.1$, $k_0 = 1$.

questions. One modification consists of adding a quadratic term, αu^2, to the right side of (5.31). This term breaks the inversion symmetry, $u \to -u$, of the equation and allows for hexagonal spot and gap patterns (see Section 7.1.1). It also induces a bistability range of the zero and spot-pattern solutions, within which a variety of hybrid solutions exist (see Sections 3.3.2 and 7.2.1.3). We will use the modified SH equation to study transitions from patterned to uniform states, the roles hybrid states play in such transitions, and the possible implications for desertification (see Section 11.2.2). Another modification of the SH equation to be studied involves periodic modulation of the parameter λ in space (see Section 8.1.2). These studies are relevant for understanding rehabilitation of degraded vegetation by means of water harvesting methods (see Section 3.3.3).

5.2.2 The FitzHugh–Nagumo model

The FitzHugh–Nagumo (FHN) model represents one step up in model complexity with respect to the SH equation. In addition to a uniform stationary instability leading to bistability of uniform states, and to a non-uniform stationary instability leading to stationary patterns, it also captures a uniform oscillatory instability that leads to long-wavelength traveling waves (see Section 2.3.1). The non-variational nature of the FHN model shows up not only in periodic oscillations and traveling waves, but also in the possible onset of spatio-temporal chaos. Originally, the model has been proposed as a simplified description of spike generation in axons of nerve cells [93, 221], but its relevance to a variety of other biological, chemical and physical contexts has made it another canonical model of pattern formation.

The basic form of the FHN model to be considered here is

$$\partial_t u = u - u^3 - v + \nabla^2 u\,, \tag{5.37a}$$

$$\partial_t v = \epsilon\,(u - a_1 v - a_0) + \delta\nabla^2 v\,. \tag{5.37b}$$

This non-dimensional form contains four parameters, a_0, a_1, ϵ and δ, of which the last three are assumed to be non-negative. The two real-valued state

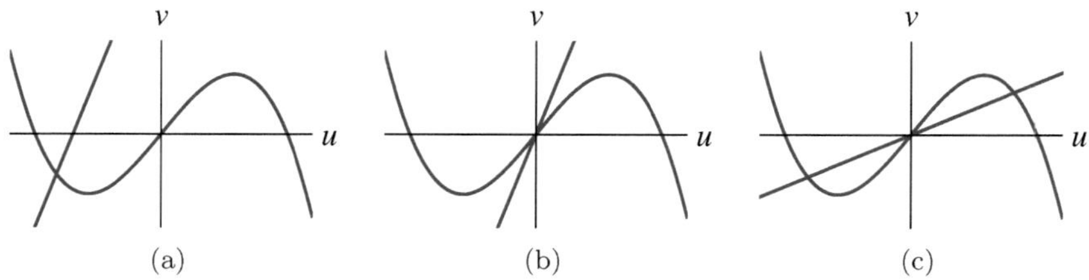

FIGURE 5.2: Three cases of nullcline intersection in the FHN model. The intersection points represent stationary uniform states.

variables u and v are often referred to as the *activator* and *inhibitor*, respectively. The variable u activates the growth of itself and of v, while the variable v inhibits the growth of itself and, most importantly, of u. Focusing on large systems, we will analyze the FHN model on the infinite domain $-\infty < x < \infty$, $-\infty < y < \infty$. Numerical studies on finite domains will generally use either periodic boundary conditions or Neumann boundary conditions[6]. Like the SH model, the FHN model too does not have explicit space or time dependence and therefore represents systems that are symmetric to translations in space and in time.

The stationary uniform states of (5.37) are determined by the intersection points of the two curves $v = u - u^3$ and $v = (u - a_0)/a_1$ in the (u, v) plane. Such curves are often called *nullclines*. Three basic cases, illustrated in Figure 5.2, can be distinguished according to the values of a_0 and a_1: (a) the nullclines intersect at a single point lying on one of the outer branches of the cubic nullcline $v = u - u^3$, (b) the nullclines intersect at a single point lying on the middle branch of the cubic nullcline, (c) the nullclines intersect at three points, each lying on a different branch of the cubic nullcline. Note that the interscetion points and the states they represent are independent of the two other parameters, ϵ and δ. These parameters, however, affect the stability of these states. With appropriate choices of their values case (a) can describe an excitable system [211], case (b) can describe a system undergoing either a uniform oscillatory instability or a non-uniform stationary instability, and case (c) can describe a bistable system.

We will be concerned in this book with cases (b) and (c) and refer the reader to Refs. [211, 55] for studies of case (a). In both cases the parameter a_0 can be set to zero or chosen to be small in order to facilitate the analysis, for then the FHN model is symmetric or nearly symmetric to the inversion transformation, $u \to -u$, $v \to -v$. In case (b) the single stationary uniform state is linearly stable when $\epsilon > \epsilon_H$ and $\epsilon > \epsilon_T$, where ϵ_H denotes a Hopf

[6]Neumann boundary conditions refer to the values of the normal derivatives of the state variables at the boundaries. Quite often the normal derivatives are set to zero to account for the absence of fluxes through the boundaries.

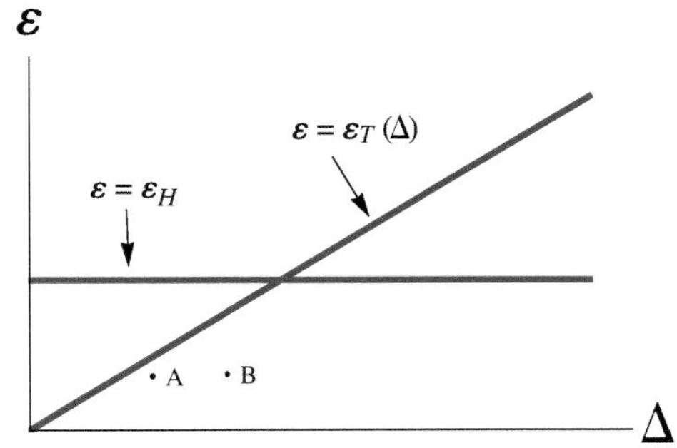

FIGURE 5.3: Codimension-2 Hopf–Turing bifurcation in the FHN model. Shown are the threshold lines of the Hopf bifurcation, $\epsilon = \epsilon_H$, and of the Turing bifurcation, $\epsilon = \epsilon_T(\delta)$, and their codimension-2 intersection point at which both bifurcations coincide. The zero state is stable when $\epsilon > \epsilon_H$ and $\epsilon > \epsilon_T$.

bifurcation to uniform oscillations, and ϵ_T denotes a Turing bifurcation[7] to stationary periodic patterns. For $a_0 = 0$, the thresholds are given by $\epsilon_H = 1/a_1$ and $\epsilon_T = \delta/(2 - a_1 + 2\sqrt{1 - a_1})$. Figure 5.3 shows these thresholds in the (δ, ϵ) plane. Beyond the onset of the Hopf bifurcation, i.e., at $\epsilon < \epsilon_H$, spiral-waves often appear. The spiral wave is a rotating wave pattern consisting of a core around which contours of equal oscillation phase rotate, and a far field in the form of nearly periodic traveling waves propagating in the radial direction. Figure 5.4(a) shows a snapshot of a spiral wave (see also Section 7.2.2).

The intersection point of the two thresholds, $\epsilon = \epsilon_H$ and $\epsilon = \epsilon_T$, i.e., the point where the two instabilities coincide, is called a codimension-2 *Hopf–Turing bifurcation*[8]. The spatio-temporal dynamics beyond this point can be quite rich as we shall see in Section 7.2.3. Panels b and c of Figure 5.4 show an example of an intricate spatial structure beyond the Hopf–Turing bifurcation, where a rotating spiral wave of the Hopf mode hosts at its core a spot of the Turing mode.

Case (c) describes bistability of two stationary uniform states, to be denoted (u_-, v_-) and (u_+, v_+). They correspond to the two intersection points on the left and right branches of the cubic nullcline in Figure 5.2(c), respectively. Patterns in this case can appear as alternate spatial domains of the two states separated by transition zones, commonly referred to as *fronts* or *domain walls.*

[7]The term "Turing bifurcation" refers to a non-uniform stationary instability of a uniform state, induced by diffusive processes [313].

[8]The term "codimension-n" refers to the number of parameters, n, that need to be tuned in order to reach a bifurcation threshold. In a k-dimensional parameter space a codimension-n threshold forms a surface of dimension $k - n$ or codimension n. Referring to the codimension of the surface is convenient because it does not require specifying the dimension of the parameter space.

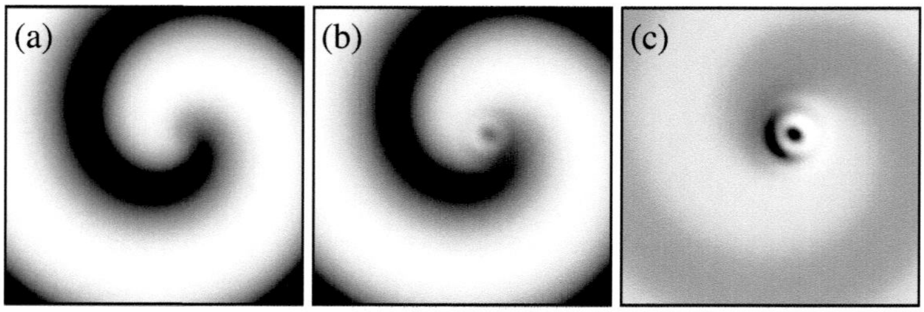

FIGURE 5.4: Hosting of a Turing mode at the core of a Hopf spiral in the FHN model. (a) A Hopf spiral above the Hopf–Turing bifurcation but close to the Turing threshold $\epsilon = \epsilon_T$ (point A in Figure 5.3). (b) A Hopf spiral with a Turing spot at its core farther away from the Turing threshold (point B in Figure 5.3). (c) The difference between (a) and (b). From [203].

For the symmetric model ($a_0 = 0$) the two stable states are related to one another by the inversion transformation and satisfy $(u_-, v_-) = (-u_+, -v_+)$. In that case, fronts that separate the two states can be stationary, but are not necessarily so; stationary fronts can destabilize in a pitchfork bifurcation to a pair of counter-propagating fronts upon decreasing ϵ or δ. This instability can induce rich spatio-temporal dynamics, such as the break up of a spiral wave and the development of an irregular spatio-temporal state in the vicinity of the instability, as Figure 5.5 shows. The irregular state involves repeated events of spiral-wave creation and annihilation and is a form of spatio-temporal chaos (see Section 7.3.1).

Unlike the SH equation, the FHN model cannot be variational anywhere in the (ϵ, δ) plane because it includes a Hopf bifurcation to temporal oscillations and a pitchfork front bifurcation to a pair of counter-propagating fronts[9]. This does not rule out the possibility that the FHN is variational in a limited parameter range. Indeed, the functional

$$L[u,v] = \int\int_{\Omega} dxdy \left[V[u,v] + \frac{a_1\epsilon}{2}|\nabla u|^2 - \frac{a_1\delta}{2}|\nabla v|^2 \right], \tag{5.38}$$

where,

$$V[u,v] = \frac{1}{2}(\partial_t u)^2 + \frac{1}{2\epsilon}(\partial_t v)^2 - \epsilon a_1 F(u,v), \tag{5.39}$$

and

$$F(u,v) = \frac{1}{2}u^2 - \frac{1}{4}u^4 - uv + \frac{1}{2}a_1\left(v + \frac{a_0}{a_1}\right)^2,$$

acts as a Lyapunov functional for the FHN model in a limited ϵ range [95].

[9]Coexistence of stable counter-propagating fronts rules out a Lyapunov functional, because if the functional decreases in time for one front solution it must increase in time for the other front solution.

FIGURE 5.5: Spiral break up and the development of spatio-temporal chaos in the vicinity of the pitchfork front bifurcation in the FHN model (case (c)). From [117].

This can be shown by taking the time derivative of L. Using the FHN model equations and integration by parts, assuming Neumann boundary conditions, we obtain

$$\dot{L} = -\int\int_{\Omega} dxdy \left[|\nabla(\partial_t u)|^2 + \frac{\delta}{\epsilon}|\nabla(\partial_t v)|^2 + (\epsilon a_1 + 3u^2 - 1)(\partial_t u)^2 \right]. \quad (5.40)$$

Clearly L cannot increase in time in the parameter range $\epsilon > a_1^{-1}$, implying that the FHN model is variational in this range. As pointed out in Section 2.2.5, the failure of L to be a Lyapunov functional is often associated with the appearance of non-variational behaviors. Indeed, the condition $\epsilon > a_1^{-1}$ guarantees that Hopf bifurcation to uniform oscillations, discussed earlier in this section, does not occur.

We will use the FHN model and variants thereof to study various pattern formation problems that bear on questions of population and landscape ecology. We note in this regard that models of dryland vegetation can be viewed as activator-inhibitor systems, like the FHN model, with the vegetation biomass playing the role of the activator and the lack of soil-water the role of the inhibitor. This analogy will be used to gain deeper insight into the dynamics of dryland vegetation.

5.3 Linear stability analysis of uniform states

A powerful method that provides information about the possible instabilities a uniform state of a spatially extended system can go through is *linear stability analysis.* We have already encountered this method when we studied instabilities in low-dimensional systems. Here, we extend this analysis to spatially extended systems in which a uniform state can lose stability to the growth of non-uniform perturbations. We use the two canonical models presented in the previous section as simple case studies for illustrating the method. The SH model is used to study a non-uniform stationary instability that involves the monotonic growth of spatially periodic perturbations and generally results in stationary periodic patterns (Section 5.3.1). The FHN model is used to study a uniform oscillatory instability that involves the oscillatory growth of uniform and long-wavelength periodic perturbations and leads to uniform oscillations or long-wavelength traveling waves (Section 5.3.2). We conclude in Section 5.3.3 with a general discussion of instabilities in spatially extended systems, revisiting the classification of instabilities made in Section 2.3.1 and commenting on the existence of a marginal mode associated with translation symmetry for spatially dependent states.

5.3.1 SH model: A non-uniform stationary instability

Consider the Swift–Hohenberg (SH) model on the one-dimensional infinite domain, $-\infty < x < \infty$,

$$\partial_t u = \lambda u - u^3 - \left(\partial_x^2 + k_0^2\right)^2 u\,, \tag{5.41}$$

and its stationary uniform solution $u_s = 0$. This solution is *linearly stable* if any infinitesimally small perturbation decays in the course of time, and it is *linearly unstable* if there exists a perturbation that grows with time. To determine the linear stability of this stationary solution we write $u(x,t) = u_s + \delta u(x,t)$, where $\delta u(x,t)$ is an infinitesimally small perturbation and linearize the equation in $\delta u(x,t)$. This gives

$$\partial_t \delta u = \lambda \delta u - \left(\partial_x^2 + k_0^2\right)^2 \delta u\,. \tag{5.42}$$

Since any perturbation can be represented as a Fourier integral and since the equation for δu is linear, we need only consider the stability of $u_s = 0$ to the growth of any Fourier mode,

$$\delta u(x,t) = \alpha_k(t) e^{ikx} + c.c.,$$

where $\alpha_k(t)$ is a complex valued amplitude of the mode k and *c.c.* denotes the complex conjugate of $\alpha_k e^{ikx}$ since u is assumed to be real valued. In other words we study the growth or decay of sinusoidal perturbations with wavenumber k or wavelength $2\pi/k$. Inserting this form into (5.42) we find that the amplitude satisfies the linear ODE

$$\dot{\alpha_k} = \sigma\alpha_k \,,$$

whose solution is

$$\alpha_k(t) = \alpha e^{\sigma(k;\lambda)t} \,,$$

where $\alpha = \alpha_k(0)$ is an arbitrary small constant representing the initial amplitude of a perturbation along the mode k, and

$$\sigma(k;\lambda) = \lambda - (k_0^2 - k^2)^2 \tag{5.43}$$

is the growth rate of the mode k. For $\sigma(k;\lambda) < 0$ a perturbation along the mode k decays in time, while for $\sigma(k;\lambda) > 0$ it grows. In general σ might be a complex valued function of k, in which case, the growth rate is defined to be its real part, $\sigma_R = Re\{\sigma\}$.

Figure 5.6(a) shows plots of σ_R vs. k, for negative, zero and positive values of λ. Such plots are called *growth-rate curves*. When $\lambda < 0$ we find $\sigma(k) < 0$ for all k, implying the decay of all modes and therefore of any perturbation. We conclude that the zero state $u_s = 0$ is linearly stable for $\lambda < 0$. This is no longer the case when $\lambda > 0$, for in that case there is a band of growing modes. The instability threshold and the critical mode that grows past this threshold value can be calculated using the condition that the growth rate is zero,

$$\sigma(k;\lambda) = 0 \,, \tag{5.44}$$

and that it has a maximum,

$$\frac{d\sigma}{dk}(k,\lambda) = 0 \,, \qquad \frac{d^2\sigma}{dk^2}(k,\lambda) < 0 \,. \tag{5.45}$$

Applying these conditions to (5.43) we find $\lambda = 0$ and $k = k_0$. Thus, the first mode to grow, as λ crosses the instability point, $\lambda = 0$, has the form

$$\delta u = \alpha e^{ik_0 x} + c.c. \,. \tag{5.46}$$

This is an example of a non-uniform stationary instability; the first mode to grow is non-uniform and the growth is monotonic in time (see Section 2.3.1). The growth of this mode breaks the space-translation symmetry of the system and dictates the wavenumber of the pattern that emerges close to the instability point, as we shall see in Chapter 6.

5.3.2 FHN model: A uniform oscillatory instability

As another example of linear stability analysis in spatially extended systems, consider the FHN model (5.37) in one space dimension. For simplicity

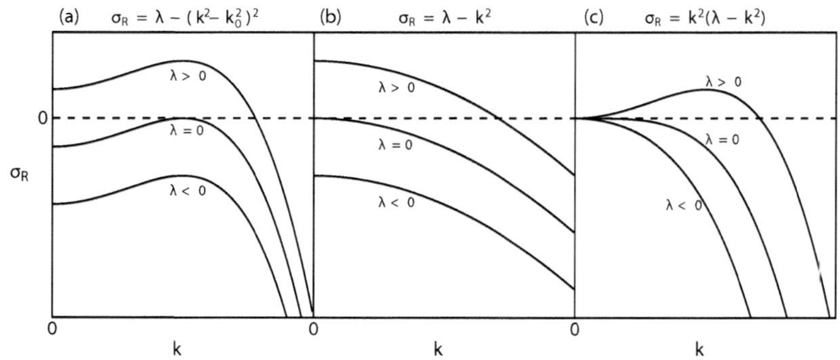

FIGURE 5.6: Three types of growth-rate curves for different values of λ, highlighting the growth or decay of the different Fourier modes: (a) the first mode to grow at $\lambda = 0$ has a finite wavenumber k_0, (b) the first mode to grow at $\lambda = 0$ is uniform ($k = 0$), (c) the first mode to grow at $\lambda = 0$ is uniform, but unlike case (b) the fastest growing mode at positive λ has a finite wavenumber. Case (c) is characterized by having a uniform marginal mode at all λ values and is typical to systems with a conservation law.

we take $a_0 = 0$ and $\delta = 0$, and choose $a_1 < 1$ (case (b) in Section 5.2.2). We study the linear stability of the zero solution $u_s = 0$, $v_s = 0$, regarding ϵ as the control parameter. As in the previous example we consider the stability of the zero solution to the growth of a Fourier mode k. That is, we write $u(x,t) = u_s + \delta u(x,t)$ and $v(x,t) = v_s + \delta v(x,t)$, where

$$\left[\begin{array}{c} \delta u(x,t) \\ \delta v(x,t) \end{array}\right] = \left[\begin{array}{c} \alpha_k(t) \\ \beta_k(t) \end{array}\right] e^{ikx} + c.c.\,, \tag{5.47}$$

insert these forms for u and v in (5.37), and linearize in α_k and β_k. This yields the linear ODE system

$$\dot{\alpha_k} = (1-k^2)\alpha_k - \beta_k\,, \tag{5.48a}$$
$$\dot{\beta_k} = \epsilon\,(\alpha_k - a_1\beta_k)\,. \tag{5.48b}$$

Looking for solutions of the form

$$\left[\begin{array}{c} \alpha_k(t) \\ \beta_k(t) \end{array}\right] = \left(\begin{array}{c} \alpha \\ \beta \end{array}\right) e^{\sigma(k;\epsilon)t}\,, \tag{5.49}$$

we obtain the eigenvalue problem

$$\mathbf{J}\left(\begin{array}{c} \alpha_k \\ \beta_k \end{array}\right) = \sigma\left(\begin{array}{c} \alpha_k \\ \beta_k \end{array}\right), \qquad \mathbf{J} = \left(\begin{array}{cc} 1-k^2 & -1 \\ \epsilon & -\epsilon a_1 \end{array}\right), \tag{5.50}$$

where $\mathbf{J}$ is the Jacobian matrix (see Section 2.2.2) associated with (5.48). Note that we could also obtain this eigenvalue problem by combining (5.47) with

(5.49) and use the combined form directly in (5.37). The eigenvalues, $\sigma(k;\epsilon)$, satisfy the characteristic equation $det|\mathbf{J}-\sigma\mathbf{I}|=0$ or

$$\sigma^2+(k^2-1+\epsilon a_1)\sigma+\epsilon a_1(k^2-1)+\epsilon=0\,, \tag{5.51}$$

and are given by

$$\sigma=\frac{1}{2}\left(1-k^2-\epsilon a_1\right)\pm\frac{1}{2}\sqrt{(1-k^2+\epsilon a_1)^2-4\epsilon}\,. \tag{5.52}$$

The square root term in (5.52) suggests that σ can be complex valued. Assuming that this is indeed the case, i.e., that $\omega_0^2\equiv\epsilon-(1-k^2+\epsilon a_1)^2/4>0$, we search for a possible instability of the zero solution, $u_s=0$, $v_s=0$, by applying the conditions (5.44) and (5.45) to the real part of σ,

$$\sigma_R=\frac{1}{2}\left(1-k^2-\epsilon a_1\right)\,. \tag{5.53}$$

The condition $\sigma_R(k;\epsilon)=0$ gives $\epsilon(k)=(1-k^2)/a_1$, and the condition $d\sigma_R/dk(k;\epsilon)=0$ gives $k=0$. This provides the instability threshold $\epsilon_c=a_1^{-1}$, denoted in Section 5.2.2 by ϵ_H, and the wavenumber $k_c=0$ of the critical mode that begins to grow at the instability point. We can test now the assumption that σ is complex valued in the vicinity of the instability point by evaluating the sign of ω_0^2. Inserting $\epsilon=\epsilon_c=a_1^{-1}$ and $k=k_c=0$ into the expression for ω_0^2, and recalling that $a_1<1$, we find indeed that $\omega_0^2=a_1^{-1}-1>0$ as assumed[10].

Thus, the first mode to grow as ϵ crosses the instability point, $\epsilon_c=a_1^{-1}$, has the form

$$\begin{pmatrix}\delta u\\ \delta v\end{pmatrix}=\begin{pmatrix}\alpha\\ \beta\end{pmatrix}e^{i\omega_0 t}+c.c.\,, \tag{5.54}$$

where

$$\omega_0=\sqrt{\epsilon_c-1}\,. \tag{5.55}$$

This is an example of a uniform Hopf bifurcation, which is a uniform oscillatory instability; the first mode to grow is uniform and the growth is oscillatory in time. The growth of this mode breaks the time-translation symmetry of the system and dictates the frequency, ω_0, of the oscillatory state that emerges close to the instability point, as we shall see in Chapter 6.

It is useful to define a small parameter λ that describes the deviation from the instability point. In the present case the definition is $\lambda=(\epsilon_c-\epsilon)/\epsilon_c$. In terms of λ the growth rate (5.53) is given by

$$\sigma_R=\frac{1}{2}\left(\lambda-k^2\right)\,. \tag{5.56}$$

[10]When this in not the case and σ is real valued, the conditions (5.44) and (5.45) should be applied to σ as a whole. Such a situation arises when δ is not zero and leads to the Turing instability discussed in Section 5.2.2.

The wavenumber dependence of σ_R is of the form depicted in Figure 5.6(b). At any finite distance below ϵ_c or above $\lambda = 0$, a band of long-wavelength modes (small k) grow along with the uniform mode. We shall see in Chapter 6 that this band of modes gives rise to long-wavelength traveling-wave phenomena.

5.3.3 Instability types revisited

We can generalize the linear stability analysis described above for the zero solutions of the SH and FHN models as follows. Let $\mathbf{u}(\mathbf{r},t) = (u_1(\mathbf{r},t), u_2(\mathbf{r},t), \ldots, u_n(\mathbf{r},t))$ be a vector function of $\mathbf{r} = (x,y)$ and t, representing the state variables of a system described by the nonlinear evolution equation

$$\partial_t \mathbf{u} = \mathcal{N}(\mathbf{u};\mathbf{p}) \tag{5.57}$$

where $\mathcal{N}(\mathbf{u};\mathbf{p})$ is a nonlinear operator acting on $\mathbf{u}$ and which depends on the parameters $\mathbf{p} = (p_1, p_2, ..., p_m)$. Assume that $\mathbf{u}_s$ is a steady-state solution of this system, i.e., $\mathcal{N}(\mathbf{u}_s;\mathbf{p}) = 0$, not necessarily constant and satisfying any boundary conditions, and consider an infinitesimal perturbation, $\delta\mathbf{u}$, of this steady-state, so that the perturbed state is $\mathbf{u}(\mathbf{r},t) = \mathbf{u}_s + \delta\mathbf{u}(\mathbf{r},t)$. Expanding $\mathcal{N}(\mathbf{u_s} + \delta\mathbf{u};\mathbf{p})$ as a Taylor series in powers of $\delta\mathbf{u}$ we obtain

$$\mathcal{N}(\mathbf{u_s} + \delta\mathbf{u};\mathbf{p}) = \mathcal{N}(\mathbf{u}_s;\mathbf{p}) + \mathcal{L}(\mathbf{u}_s;\mathbf{p})\delta\mathbf{u} + \ldots,$$

where $\mathcal{L}(\mathbf{u}_s;\mathbf{p})$ is a linear operator acting on $\delta\mathbf{u}$. The linearized equation for $\delta\mathbf{u}$ follows from neglecting terms of second and higher orders, and is given by

$$\partial_t \delta\mathbf{u} = \mathcal{L}(\mathbf{u}_s;\mathbf{p})\delta\mathbf{u}\,. \tag{5.58}$$

We next look for separable solutions to this equation in the form $\delta\mathbf{u}(\mathbf{r},t) = e^{\sigma t}\mathbf{v}(\mathbf{r})$, which after substitution into (5.58) leads to the eigenvalue problem

$$\mathcal{L}(\mathbf{u}_s;\mathbf{p})\mathbf{v} = \sigma\mathbf{v}\,. \tag{5.59}$$

Typically, problems of this type will admit an infinite set of eigenvalues σ. If the spatial domain is finite we expect a countably infinite set of eigenvalues, while if the domain is infinite we expect a continuous spectrum of σ values (e.g., (5.43) in the SH example). The general solution to (5.58) will be a superposition, either as a sum or an integral, of the functions $e^{\sigma t}\mathbf{v}$. If all eigenvalues have real part strictly less than zero the solution decays to zero exponentially and the solution $\mathbf{u}_s$ is linearly stable. Naturally, the stability properties of the system will depend on the values of the parameters $\mathbf{p}$. If, for a given set of parameters, there are eigenmodes $\mathbf{v}$ whose eigenvalues have positive real parts, the steady-state solution $\mathbf{u}_s$ is linearly unstable. These eigenmodes, or possibly a subset of them, will represent the dominant modes in the new state that the system evolves toward.

We can reformulate now the four instability types of uniform states introduced in Section 2.3.1. The distinction between the different types is based on two properties of the eigenvalues σ with the largest real parts in the vicinity of the instability point. The first property is whether these eigenvalues are real or complex valued, and the second is whether they correspond to a zero or to a finite wavenumber k. Writing $\sigma(k) = \lambda(k) \pm i\omega(k)$, the four types are:

- *Uniform stationary instability*: In this case $\omega = 0$ and $\max(\lambda)$ occurs at $k = 0$. The state of the system loses stability to uniform perturbations that grow monotonically, and generally evolves to another stationary uniform state.

- *Non-uniform stationary instability*: Here $\omega = 0$ as in the previous case, but $\max(\lambda)$ occurs at $k > 0$. The state of the system loses stability to spatially periodic perturbations that grow monotonically, and generally evolves to a stationary patterned state.

- *Uniform oscillatory instability*: In this case $\omega \neq 0$ and $\max(\lambda)$ occurs at $k = 0$. The state of the system loses stability to uniform perturbations that grow in an oscillatory manner, and generally evolves to a state of uniform oscillations.

- *Non-uniform oscillatory instability*: Here $\omega \neq 0$ as in the previous case, but $\max(\lambda)$ occurs at $k > 0$. The state of the system loses stability to spatially periodic perturbations that grow in an oscillatory manner, and generally evolves to a traveling-wave state.

The growth rate curves associated with uniform and non-uniform instabilities have maxima at zero wavenumber and at finite wavenumber, respectively, for all values of the control parameter near the instability point (Figure 5.6(a,b)). There is also an intermediate case, shown in Figure 5.6(c), in which the maximum moves from zero to a finite wavenumber as the instability point is traversed. Associated with this behavior is a uniform mode with a zero eigenvalue for *all* control-parameter values. This is the case, for example, when a conservation law exists [11].

5.3.4 A marginal translation mode

Linear stability analysis is not restricted to uniform states. In the next chapter we will apply the method to periodic solutions that emerge beyond pattern forming instabilities. Linear stability analysis of non-periodic non-uniform solutions is harder and is generally done numerically. One aspect

---

[11] Consider as an example the Cahn–Hilliard model for phase separation in binary fluids [231],

$$\partial_t u = \nabla^2 \left(\lambda u - u^3 + \nabla^2 u\right) , \tag{5.60}$$

which has the form of a conservation law, $\partial_t u + \nabla \cdot J = 0$, where $J = -\nabla\left(u^3 - \lambda u - \nabla^2 u\right)$. A linear stability analysis of the zero solution of (5.60) yields the growth rate $\sigma = k^2(\lambda - k^2)$ whose form is shown in Figure 5.6(c). Positive growth rates appear for as soon as λ becomes positive, but as λ increases the fastest growing mode changes from a uniform mode to a mode with a finite wavenumber, $k = \sqrt{\lambda/2}$.

of such solutions, however, should be noted: in uniform systems they are marginally stable to spatial translations. To see this, take the space derivatives (or gradient) of $\mathcal{N}(\mathbf{u}_s;\mathbf{p}) = 0$. For a uniform system, the spatial dependence of $\mathcal{N}$ comes only through the spatial dependence of the stationary solution $\mathbf{u}_s$. We therefore get

$$\mathcal{L}(\mathbf{u}_s;\mathbf{p})\nabla\mathbf{u}_s = 0\,. \tag{5.61}$$

Comparing this form with the eigenvalue problem (5.59) we see that the spatial gradient of the stationary solution, $\mathbf{u}_s$, is an eigenmode of $\mathcal{L}$ with a zero eigenvalue that always exists, for any set of parameters $\mathbf{p}$. This eigenmode describes space translations, as can be seen by perturbing $\mathbf{u}_s$ along this mode:

$$\mathbf{u}(\mathbf{r}) = \mathbf{u}_s(\mathbf{r}) + \delta\mathbf{r}\cdot\nabla\mathbf{u}_s \approx \mathbf{u}_s(\mathbf{r}+\delta\mathbf{r})\,, \tag{5.62}$$

where $\delta\mathbf{r}$ is a constant small displacement vector. The marginal nature of the *translation mode*, $\nabla\mathbf{u}_s$, results from the existence of a continuous family of solutions—$\mathbf{u}_s$ and any translated form thereof. Any perturbation along this family results in another asymptotic solution and therefore does not decay or grow.

As an example consider the FHN model (5.37) in one space dimension, and assume that $\mathbf{u}_s(x) = (u_s(x), v_s(x))$ is a stationary non-uniform solution of (5.37). Such a solution can represent a stationary pulse-like solution (see Section 7.2.1.2) or, for $a_0 = 0$, a stationary front solution (see Section 7.3.1). An infinitesimal perturbation $\delta\mathbf{u}(x,t) = (\delta u(x,t), \delta v(x,t))$ about this solution satisfies (5.58) where

$$\mathcal{L}(\mathbf{u}_s;\mathbf{p}) = \begin{pmatrix} 1-3u_s^2+\partial_x^2 & -1 \\ \epsilon & -\epsilon a_1+\delta\partial_x^2 \end{pmatrix}, \tag{5.63}$$

and $\mathbf{p} = (a_1, \epsilon, \delta)$. Inserting now the solution $\mathbf{u}_s(x) = (u_s(x), v_s(x))$ into (5.37) and taking the derivative of the resulting equations with respect to the spatial coordinate x gives

$$\mathcal{L}(\mathbf{u}_s;\mathbf{p})\begin{pmatrix} \partial_x u_s \\ \partial_x v_s \end{pmatrix} = \mathbf{0}\,, \tag{5.64}$$

which is equation (5.61) for the particular case that is considered—the FHN model in one space dimension. Note that the linear operator $\mathcal{L}$ in (5.63) coincides with the Jacobian matrix in (5.50) when $\mathbf{u}_s(x)$ is the (uniform) zero state and the operator $\mathcal{L}$ is acting on an eigenmode of wavenumber k.

The translation mode plays an important role in the nonlinear analysis of localized structures. It is a soft mode that can responde to small perturbations that break translational symmetry, either external or developing from the proximity and the interaction with other localized structures. It is worth noting that other continuous symmetries, if they exist, such as Galilean invariance, introduce additional soft modes that can interact with one another [84, 153].

Chapter summary

The methods of pattern formation theory are generally applied to the dynamic equations that describe the particular system of interest. There is a simple method, however, that is applicable even when these dynamic equations are not known. The method, *dimensional analysis*, provides a way of obtaining scaling laws that is based on the identification of a set of physical quantities with dimensions that are (a) independent of one another, (b) can be combined to form the dimension of any other dependent quantity of interest. The method is demonstrated using a few examples, including the Rayleigh–Bénard system described in Chapter 2. When the dynamic equations are known, dimensional analysis can also be used to eliminate redundant parameters and transform the equations into simpler dimensionless equations.

Even when the dynamic equations are known, they are often too complicated for applying advanced methods of pattern-formation theory. In such cases much insight can still be obtained by complementing the dynamic equations with simple models that share the same universal pattern formation phenomena and are more amenable to mathematical analysis. Two simple models of this kind are introduced, the Swift–Hohenberg (SH) equation that captures a non-uniform stationary instability of a uniform state to spatially periodic patterns, and the FitzHugh–Nagumo (FHN) model that captures both a non-uniform stationary instability and a uniform oscillatory instability. While the SH equation is variational and therefore can only produce stationary asymptotic patterns, the FHN model is not variational and shows oscillatory asymptotic patterns, such as long-wavelength traveling waves, and more complex dynamical behaviors, including dual-mode spiral waves and spatio-temporal chaos. These models will be used in Chapters 6 and 7.1 to explain several methods of pattern-formation theory and study basic mechanisms of pattern formation.

The most basic method of analyzing pattern forming systems is *linear stability analysis*. The method provides information about the possible instabilities a uniform state can go through, the instability thresholds and the types of modes that grow beyond the instabilities. We encountered this method in Section 2.2.2 in the context of low dimensional dynamical systems and generalize it here to spatially extended systems, using the SH and the FHN models as examples. Acquainted with this method we revisit the four instability types of uniform states introduced in Section 2.3.1 and reformulate them more rigorously. We conclude this chapter pointing out the existence of a marginal translation mode for non-uniform solutions that break the translational symmetry of the system, and discuss its significance.

Chapter 6

Pattern formation analysis: Advanced methods

6.1 Amplitude equations

The linear stability analysis discussed in the previous chapter provides information about the onset of an instability and about its nature. It tells us which parameter can drive the instability, the threshold value at which the instability occurs, and the type of mode that grows at the instability point. Since the analysis addresses infinitesimally small perturbations about the state whose stability is studied, it does not provide information about the new state the system evolves toward. Indeed, the general solution of the system (5.58) of linear ODEs describes exponentially growing perturbations which soon break the assumption of small perturbations.

To obtain information about the asymptotic dynamics we must go beyond the linear approximation. Including in the analysis terms that are nonlinear in the perturbations can saturate the initial exponential growth and lead to convergent asymptotic behavior. In the following sections we will describe a nonlinear analysis of this kind, which culminates in an equation for the amplitude of the mode that grows at the instability point. We first discuss general aspects of this analysis (Sections 6.1.1 and 6.1.2) and then present the derivation of amplitude equations for a non-uniform stationary instability and for a uniform oscillatory instability, using the SH and FHN models as examples (Section 6.1.3). Under certain conditions, amplitude equations are reducible to phase equations (Section 6.1.4), which describe the dynamics of the phase of the periodic pattern or the oscillations that appear beyond the instability point. While amplitude and phase equations are simpler to analyze and provide much insight about pattern formation and dynamics, they are limited in the scope of problems they can be used to study, as discussed in Section 6.1.5.

6.1.1 The general concept

At the core of the nonlinear analysis to be described lies the derivation of a dynamic equation, the so called *amplitude equation*, for the amplitude of the mode that grows at an instability point. Denoting the amplitude by A, a typical state variable, $u(\mathbf{r},t)$, of the system is approximated as

$$u(\mathbf{r},t) \approx A(\mathbf{r},t)e^{ik_0 x} + c.c.\,, \tag{6.1}$$

in the case of a non-uniform stationary instability, or as

$$u(\mathbf{r},t) \approx A(\mathbf{r},t)e^{i\omega_0 t} + c.c.\,, \tag{6.2}$$

in the case of a uniform oscillatory instability. These are the leading-order forms in asymptotic expansions, where higher-order contributions generally contain harmonics of the leading order forms. We assumed here, without loss of generality, that the uniform state that loses stability is the zero state. Although the derivation of the amplitude equation takes into account nonlinear deviations from the unstable state, it still assumes proximity to the instability point. This assumption has three important consequences for the derivation: (i) the amplitude is small in absolute value, (ii) the eigenvalue of the growing mode has small real part, implying slow temporal changes of the amplitude, and (iii) the band of the growing modes is narrow, implying weak spatial changes. The derivation makes use of the smallness of A and its time and space derivatives in a perturbation theory analysis that culminates in a nonlinear PDE for the amplitude A to a given order in the distance from the instability point. The derivation of amplitude equations is often called "weak nonlinear analysis," because of the assumed proximity to the instability point and the small amplitudes involved.

The amplitude A is considered to be a complex-valued variable. The reason for that becomes evident when we write it in polar form, $A = \rho \exp(i\phi)$, where $\rho = |A|$ is the magnitude (or modulus) of A and $\phi = \arg(A)$ is its argument or phase. Inserting this form in (6.1) we obtain

$$u \approx \rho e^{ik_0(x+x_0)} + c.c.\,, \tag{6.3}$$

where $x_0 = \phi/k_0$. Thus, the phase ϕ describes space translations. Recall now that uniform systems possess a translation mode with a zero eigenvalue (see Section 5.3.3). This marginal mode plays a significant role in the asymptotic dynamics of spatial patterns, and the phase ϕ provides a means of taking this role into account. Moreover, as the distance from the instability point increases, the time-scale associated with the magnitude, ρ, becomes short relative to the time scale of the phase, which is independent of this distance. As a result, the magnitude becomes a fast variable, capable of relaxing to instantaneous phase values and following the phase adiabatically, $\rho = \rho[\phi]$ (see Section 6.1.4). Had we not considered A to be complex valued it would have been impossible to study these asymptotic phase dynamics.

Quite often two amplitudes are needed to describe the asymptotic dynamics near an instability point. One example of such a case is a non-uniform oscillatory instability, in which a traveling-wave mode begins to grow. In a system with a parity symmetry, $x \to -x$, in which along with a wave that propagates to the right there also exists a symmetric wave that propagates to the left, a typical variable $u(\mathbf{r}, t)$ is expressed as

$$u \approx A_R e^{ik_0x - i\omega_0t} + A_L e^{-ik_0x - i\omega_0t} + c.c.\,, \tag{6.4}$$

where A_R and A_L are the amplitudes of the right and left propagating waves, respectively. Other examples are codimension-2 instabilities where two modes simultaneously grow from an unstable uniform state. In a Hopf–Turing bifurcation, for example (see Section 5.2.2), a typical state variable, $u(\mathbf{r}, t)$, is approximated as

$$u \approx A_H e^{i\omega_0t} + A_T e^{ik_0x} + c.c.\,. \tag{6.5}$$

Note that the two forms (6.4) and (6.5) describe different patterns and wave phenomena despite the similar elements they share—periodicity in space and time. In particular, note the difference between standing waves, obtained for $A_R = A_L$ in (6.4), and non-propagating oscillating patterns, obtained for constant, nonzero values of A_H and A_T in (6.5). Standing waves are characterized by having nodes that separate domains that oscillate out of phase, whereas no nodes exist for a non-propagating oscillating pattern—the pattern oscillates in phase as a whole.

The amplitude equation approach provides the mathematical basis for the universal nature of pattern formation phenomena near instability points. Systems that share the same type of instability are described by the same amplitude equation, and therefore behave similarly in the vicinity of the instability point. For example, we will soon see that all systems that go through a non-uniform stationary instability are described by the Newell–Whithead–Segel (NWS) equation (6.36), all systems that go through a uniform oscillatory instability are described by the Complex Ginzburg–Landau (CGL) equation (6.53), and so on. The specifics of a given system come through the coefficients of the various terms in the amplitude equation, and their particular dependence on system parameters. The practical significance of this universality is that amplitude equations, such as the NWS and CGL equations, have thoroughly been studied, and the information that has been acquired (see Chapters 7 and 8) is readily available once an amplitude equation of this kind has been derived for a particular system. This information includes existence ranges for the newly formed states beyond an instability point, secondary instabilities of these states, multiplicity of stable states, conditions for the onset of spatio-temporal chaos, and so on. Amplitude equations derived for the first time, or amplitude equations that have been studied only to a limited extent, are still very instrumental in that they are more amenable to analysis.

6.1.2 Derivation of amplitude equations—general considerations

6.1.2.1 Symmetry considerations

Significant information about the structural form of amplitude equations can be deduced without lengthy calculations, using symmetry arguments. Consider a uniform system, i.e., a system whose dynamic equations (5.57) do not have explicit space dependence and is therefore symmetric to space translations. The state of the system can break this symmetry by forming a spatially periodic pattern of the form (6.1). Because of the translational symmetry of the system, if $u(x) = A\exp[ik_0x] + c.c.$ is an approximate solution, $u(x+x_0) = A\exp[ik_0(x+x_0)] + c.c.$ is an approximate solution too, where x_0 is an arbitrary constant displacement. But this solution can be rewritten as $u(x+x_0) = A\exp(i\phi)\exp(ik_0x) + c.c.$, where $\phi = k_0x_0$, that is, in the form (6.1) with the amplitude A replaced by $A(\phi) = A\exp(i\phi)$. As a consequence, if A satisfies some amplitude equation,

$$\partial_t A = \mathcal{F}[A, A^*]\,, \tag{6.6}$$

$A(\phi)$ should satisfy the same equation, and therefore the amplitude equation should be invariant under the transformation $A \to A\exp(i\phi)$ for any constant phase ϕ. This invariance allows for terms in $\mathcal{F}$ of the form $A^n A^{*m} = |A|^m A^{n-m}$, where n and m are integers satisfying $n = m+1$, but excludes terms that violate this condition, such as the linear term A^*, all quadratic terms, A^2, ${A^*}^2$, $|A|^2$, the cubic terms, A^3, ${A^*}^3$, $|A|^2A^*$, etc.. It does, however, allow for the linear term A and the cubic term $|A|^2A$. The invariance under $A \to A\exp(i\phi)$ also allows for spatial derivative terms, such as $\partial_x^2 A$

and $|A|^2\partial_x^2 A$. A similar argument with exactly the same conclusion holds for systems that are uniform in time, i.e., symmetric to time translations and have oscillatory states of the form (6.2) that break this symmetry.

As another example for a symmetry consideration consider a system that remains invariant under the parity transformation $x \to -x$, near a non-uniform stationary instability. Then, if $u(x) = A\exp[ik_0x] + A^*\exp[-ik_0x]$ is an approximate solution, $u(-x) = A^*\exp[ik_0x] + A\exp[-ik_0x]$ is an approximate solution too. This implies that any amplitude equation that A satisfies should be invariant under the transformation $A \to A^*$. This in turn implies that the coefficients of all terms in the amplitude equation should be real valued. Additional examples of symmetry arguments will be encountered in Chaps. 7 and 8.

Symmetry considerations, however, cannot tell us which of the allowed terms will actually appear, nor can it tell us anything about the coefficients of the terms. This system-specific information can only be obtained by a systematic derivation of the amplitude equations. Before embarking on the derivation of amplitude equations it will be helpful to discuss two additional concepts that will be used in the derivation, multiple scales and solvability conditions.

6.1.2.2 Multiple scales

We have already pointed out that the spatial and temporal variations of the amplitude A are weak. In the case of a non-uniform stationary instability there exists a length scale, k_0^{-1}, associated with the critical mode that begins to grow at the instability point. In addition, there exists a much longer length scale, the length scale over which the amplitude varies, which is related to the narrow band of modes that grow along with the critical mode, k_0. This is basically the spatial counterpart of the beats phenomenon in temporal oscillations [9]. Likewise, in a uniform oscillatory instability there exists a fast time scale, ω_0^{-1}, associated with the oscillatory instability, and a much longer time scale associated with the slow dynamics in the vicinity of the instability point. In both cases the two distinct scales show up as a slow amplitude modulation of fast oscillations (in time or space). The perturbation theory to be described in the next two sections makes use of this scale separation by defining slow space and time variables,

$$X = \lambda^p x\,, \qquad T = \lambda^q t\,, \tag{6.7}$$

where $\lambda \ll 1$ is a small parameter, often taken to be the distance from an instability point, p and q are positive rational numbers, and x and t are the fast space and time variables. The slow space and time dependence of the amplitude is then accounted for by assuming it is a function of the slow variables only, $A = A(X,T)$, and that $|\partial_X A|$ and $|\partial_T A|$ are of the same order as $|A|$. This way

$$|\partial_x A| = \lambda^p|\partial_X A| \ll |A| \ll 1\,, \qquad |\partial_t A| = \lambda^q|\partial_T A| \ll |A| \ll 1\,, \tag{6.8}$$

where we used the chain rule.

The multiplicity of time scales, however, is not limited to oscillatory systems, nor to two time scales. Even in the absence of a fast time scale there still exist processes occurring on different time scales. For example, a system that goes through a non-uniform stationary instability to periodic stripe patterns generally develops, on a relatively short time scale, stripe patterns with defects. The system, however, continues evolving in time, slowly eliminating the defects, as Figure 5.1 demonstrates. The existence of multiple time scales can also be understood mathematically. The right side of the amplitude equation (6.6) contains terms that differ in the powers of A and A^* and in the order of space derivatives, and are therefore of different magnitudes. It makes sense then to write the right side of (6.6) as a power series in the small parameter λ^q, which determines the weak temporal dependence of the amplitude:

$$\partial_t A = \lambda^q \mathcal{F}_1[A, A^*] + \lambda^{2q} \mathcal{F}_2[A, A^*] + \lambda^{3q} \mathcal{F}_3[A, A^*] + \dots\,, \tag{6.9}$$

where $F_1, F_2, \dots$ are all of the same order of magnitude. As the system proceeds in time, terms of a given order of magnitude act to balance one another up to small differences of higher order of magnitude. As a result the dynamics slow down. The balancing of the next higher order terms slows down the dynamics even further, revealing the role of yet higher order terms in the slow asymptotic dynamics. An equation of the form (6.9) can formally be obtained by resolving the slow time $T = \lambda^q t$ into multiple slow time variables, $T_1 = \lambda^q t$, $T_2 = \lambda^{2q} t$, $T_3 = \lambda^{3q} t, \dots$, and regarding the amplitude $A(X, T)$ as dependent on all these slow variables, $A = A(X, T_1, T_2, T_3, \dots)$ [21]. Using the chain rule we can write the time derivative of A as

$$\partial_t A = \lambda^q \partial_{T_1} A + \lambda^{2q} \partial_{T_2} A + \lambda^{3q} \partial_{T_3} A + \dots\,, \tag{6.10}$$

which has the form of (6.9) once we identify $\mathcal{F}_i[A, A^*]$ with $\partial_{T_i} A$. This establishes the formal relation between the general amplitude-equation form (6.9) and multiple slow-time variables, and allows deriving the amplitude equation by calculating the partial slow-time derivatives $\partial_{T_i} A$ up to a given order.

6.1.2.3 Solvability conditions

Along with the derivation of the slow time derivatives, $\partial_{T_i} A$, order by order we will also derive higher order corrections to the approximate solution forms (6.1), (6.2) etc. That is, we will expand

$$u = \lambda^s u_1 + \lambda^{2s} u_2 + \lambda^{3s} u_3 + \dots\,, \tag{6.11}$$

where $u_1 \propto e^{ik_0 x} + c.c.$ in the case of a non-uniform stationary instability, $u_1 \propto e^{i\omega_0 t} + c.c.$ in the case of a uniform oscillatory instability, and so on, and s is a positive rational number. At each order i we will obtain a linear equation for u_i of the form

$$\mathcal{L} u_i = R_i\,, \tag{6.12}$$

where $\mathcal{L}$ is a linear operator[1], and R_i depends on lower order contributions to the solution, that is, on $u_1, ..., u_j$ with $j < i$, and are therefore known at order i. Equations (6.12) for $i = 1, 2, ...,$ will not provide periodic solutions of the forms discussed in Section 6.1.1 and small corrections in the form of higher harmonics, if they fail to satisfy *solvability conditions*. To get a grip on the origin of these conditions let us consider a simple example—a periodically forced harmonic oscillator,

$$\ddot{u} + \omega^2 u = F(t)\,, \tag{6.13}$$

where $u(t)$ represents the displacement from the equilibrium position, ω is the oscillator's frequency and $F(t)$ is the periodic force (per unit mass). Note that (6.13) has the form of (6.12) with $\mathcal{L} = d^2/dt^2 + \omega^2$. The crucial point to observe is that if $F(t)$ has a component that oscillates with the oscillator's frequency, ω, the solution grows with no bound. As an example consider

$$F(t) = C\cos\Omega t\,, \tag{6.14}$$

where C is an arbitrary constant. If $\Omega = \omega$, equation (6.13) has the particular solution, $u_p = (C/2\omega)t\sin\omega t$, and the general solution of (6.13) is

$$u(t) = A\cos(\omega t + \phi) + \frac{C}{2\omega}t\sin\omega t\,, \tag{6.15}$$

where A and ϕ are constants. This solution contains a term that diverges as $t \to \infty$. On the other hand, if $\Omega \neq \omega$, the general solution is

$$u(t) = A\cos(\omega t + \phi) + \frac{C}{\omega^2 - \Omega^2}\cos\Omega t\,, \tag{6.16}$$

which does not diverge. A term in $F(t)$ that oscillates with the oscillator's frequency is called a *resonant* or *secular* term. Such terms often appear in the right side, R_i, of (6.12), and lead to divergent contributions to the approximate solution (6.11). Keeping such contributions limits the validity of the approximate solution to short times[2] (or to short distances). In order to extend the solution's validity we need to avoid resonant terms in R_i. The solvability conditions provide a means of projecting these terms out of the approximate solution without discarding their effects. This is achieved through constraints on the dynamics of the amplitude A, which are the terms that appear in the amplitude equation.

The solvability conditions are choices that we make to guarantee that the contributions to the approximate solution fall within a certain class of

[1]We are refering to situations in which the operator $\mathcal{L}$ has a non-zero null space and therefore is not invertible.

[2]This can be seen by considering, as an illustrative example, a contribution $u_2 \propto t\exp(i\omega_0 t)$ to the approximate solution u. As the expansion (6.11) implies, this contribution is no longer small compared to the contribution $u_1 \propto \exp(i\omega_0 t)$ at times $t \sim \lambda^{-s}$ and longer.

functions, e.g., higher harmonics of the leading-order form. These choices should be made consistently in a sense that will become clear in Section 6.1.3.1. Note that although the approximate solution appears to be periodic (a periodic leading-order form and its harmonics) it may not be so since the amplitude has its own spatio-temporal dynamics. The concept of solvability conditions is useful not only in the derivation of amplitude equations, but also in analyzing front dynamics as we will see in Section 7.1.3.

Returning to (6.12), the solvability condition associated with this equation amounts to the removal of terms in R_i that resonate with solutions of the homogeneous problem, $\mathcal{L}u_i = 0$, by setting their coefficients to zero. These conditions provide the expressions for the slow-time derivatives, $\partial_{T_i}A$, that are needed to construct the amplitude equation (see equation (6.10)). In general, the identification of resonant terms may not be as straightforward as in the forced oscillator example. We briefly describe now a formal way of applying solvability conditions, that follows from the *Fredholm alternative theorem.* Let us assume that the linear operator $\mathcal{L}$ acts on some vector space on which an inner product is defined. Denoting the inner product of two vectors f and g by $\langle f, g\rangle$, the theorem says that, within the same vector space, (6.12) has a solution if and only if[3]

$$\langle R_i, h\rangle = 0 \tag{6.17}$$

for any vector h that satisfies

$$\mathcal{L}^\dagger h = 0\,. \tag{6.18}$$

The dagger sign in (6.18) denotes the adjoint operator, defined by the relation $\langle \mathcal{L}f, g\rangle = \langle f, \mathcal{L}^\dagger g\rangle$. The orthogonality relation (6.17) provides the solvability conditions associated with equation (6.12); in order for this equation to be solvable for functions within the vector space considered, its right side, R_i, should be orthogonal to any vector in the null space of the adjoint operator $\mathcal{L}^\dagger$.

The projection of the null space out of the right side of (6.12) is equivalent to removing resonant terms. This can readily be seen using the forced oscillator example (6.13). Consider the vector space of complex valued periodic functions of time with period $T = 2\pi/\omega$, and define the inner product between any pair of functions f and g in that space by

$$\langle f, g\rangle = \int_0^T f^* g dt\,. \tag{6.19}$$

Using integration by parts it is easy to see that the operator $\mathcal{L} = d^2/dt^2 + \omega^2$ is self-adjoint, i.e., that $\langle \mathcal{L}f, g\rangle = \langle f, \mathcal{L}g\rangle$. It is also easy to see that the null space of $\mathcal{L}$, and therefore of $\mathcal{L}^\dagger$, is spanned by $h_\pm = \exp(\pm i\omega t)$. According to the Fredholm alternative theorem the functions $h_\pm$ should be orthogonal

[3]It is easy to see that (6.17) is a necessary condition for solvability, for if u_i is a solution of (6.12) then $\langle R_i, h\rangle = \langle \mathcal{L}u_i, h\rangle = \langle u_i, \mathcal{L}^\dagger h\rangle = 0$.

to the periodic forcing $F(t)$. If this forcing contains resonant terms, such as $C\cos(\omega t)$, this condition amounts to removing them by setting $C = 0$. In the following subsections we will see two concrete examples of applying the Fredholm alternative theorem.

6.1.3 Two examples

6.1.3.1 Amplitude equation for stripe patterns

We describe the derivation of an amplitude equation for stripe patterns using the SH equation. We could also use the FHN model to derive this equation but the SH equation is simpler to analyze and better serves to illustrate the method. The linear stability analysis described in Section 5.3.1 for the SH equation in a one-space dimension can easily be extended to its two-dimensional form (5.31). The main difference is that now the perturbation δu about the zero state is taken to be along a Fourier mode with any wave vector, $\mathbf{k} = k_x\hat{\mathbf{x}} + k_y\hat{\mathbf{y}}$, in the plane, where $\hat{\mathbf{x}}$ and $\hat{\mathbf{y}}$ are unit vectors in the x and y directions. The growth rate, σ, of such modes is still given by (5.43), except that now $k^2 = \mathbf{k}\cdot\mathbf{k}$. It is instructive to consider the *neutral stability surface* $\sigma(\lambda;\mathbf{k}) = 0$, or

$$\lambda = ({k_0}^2 - k^2)^2 = ({k_0}^2 - k_x^2 - k_y^2)^2\,, \tag{6.20}$$

in the space spanned by k_x, k_y and λ. Because of the system's isotropy the surface has an azimuthal symmetry and is independent of the polar angle $\theta = \arctan(k_y/k_x)$. Figure 6.1(a) shows a cut of this surface with a constant angle plane. For a given λ, all modes below this neutral-stability curve decay and all modes above it grow. At $\lambda = 0$ all modes decay except for the family of symmetric critical modes with $k = k_0$. Slightly above this threshold a narrow band of modes grows along with the critical modes.

To study the growth of these modes and the subsequent system dynamics we assume that the system is close to the instability point, so that $0 < \lambda \ll 1$, and pursue a perturbation theory analysis in the small parameter λ. We first approximate a solution of the SH equation by a power series of the form (6.11), where u_1 is given by

$$u_1 = \tilde{A}(X, Y, T_1, T_2, \ldots)e^{ik_0x} + c.c.\,. \tag{6.21}$$

In the following we use the tilde notation to denote a rescaled amplitude, $\tilde{A} = \lambda^{-s}A$, whose magnitude is of order unity. For a constant amplitude $\tilde{A}$ the form (6.21) describes a periodic stripe pattern with a constant wavevector, $\mathbf{k} = k_0\hat{\mathbf{x}}$. The slow space and time dependence of the amplitude describes weak modulations of this periodic pattern. The analysis to be described culminates in an equation for the amplitude A and in expressions for the first few terms in the expansion (6.11).

To pursue this analysis we first need to determine the exponents, s, q, p_x and p_y, in the scaling factors λ^s for u, λ^{nq} for T_n, λ^{p_x} for X and λ^{p_y} for Y. Note that because the amplitude modulates a pattern oriented in the x

direction we should expect X and Y to scale differently in λ. The exponent s can be determined by the condition that the terms λu and u^3 should be of the same order of magnitude since u^3 acts to saturate the exponential growth that λu induces. This gives $u \sim \sqrt{\lambda}$ or $s = 1/2$. The exponent q is determined by observing that the maximal growth rate $\sigma(k = k_0)$ scales like λ (see (5.43)). This implies that the fastest time dependence of u is through the combination σt, or that $T_1 = \lambda t$ and $q = 1$. The remaining exponents p_x and p_y are determined by the band-width of growing modes along the x and y directions, around the first mode to grow, $\mathbf{k} = k_0\hat{\mathbf{x}}$, as Figure 6.1(b) illustrates. The band width along the x direction is determined by setting $k_y = 0$ in (6.20). This gives $\lambda = (k_0 + k_x)^2(k_0 - k_x)^2$ or $|k_x - k_0| \sim \lambda^{1/2}$. Similarly, the band width along the y direction is determined by setting $k_x = k_0$, which gives $|k_y| \sim \lambda^{1/4}$. Thus, a wave-vector within the band of growing modes, $\mathbf{k} = k_0\hat{\mathbf{x}} + \delta\mathbf{k}$, will contribute a spatial dependence to the amplitude of the form $\delta\mathbf{k}\cdot\mathbf{x}$ which can be expressed in terms of the slow variables $X = \lambda^{1/2}x$ and $Y = \lambda^{1/4}y$, that is, with $p_x = 1/2$ and $p_y = 1/4$.

We begin the derivation of the amplitude equation by inserting the expansion

$$u = \lambda^{1/2}u_1 + \lambda u_2 + \lambda^{3/2}u_3 + \dots\,, \tag{6.22}$$

into (5.31), where u_1 is given by (6.21) and

$$X = \lambda^{1/2}x\,, \qquad Y = \lambda^{1/4}y\,, \qquad T_n = \lambda^n t\,, \quad n = 1, 2, \dots\,. \tag{6.23}$$

Since the time and space derivatives act now on functions of multiple time and space variables, we use the chain rule to convert them as follows:

$$\partial_x \to \partial_x + \lambda^{1/2}\partial_X\,, \qquad \partial_y \to \lambda^{1/4}\partial_Y\,, \tag{6.24}$$

$$\partial_t \to \lambda\partial_{T_1} + \lambda^2\partial_{T_2} + \dots\,. \tag{6.25}$$

Note that since the amplitude, $\tilde{A}$, describes weak modulations of a stationary stripe pattern oriented in the x direction, we do not have fast y dependence nor fast time dependence in the functions that the derivatives are expected to act on. As a result the derivatives with respect to y and t do not have contributions of order unity as the derivative with respect to x has. Also note that the derivative with respect to x does not act on the amplitude $\tilde{A}$ and the derivative with respect to X does not act on $\exp(ik_0x)$.

We will consider contributions up to order $\lambda^{3/2}$. This is the lowest order that captures essential aspects of the dynamics such as nonlinear saturation of the exponential growth beyond the instability point and the band of modes that grow along with the critical mode. Since the first order in u contributes a factor $\lambda^{1/2}$, time derivatives appear at order $\lambda^{3/2}$ and higher. We therefore do not need to consider multiple time scales beyond T_1 and consider only a single slow time variable, $T \equiv T_1 = \lambda t$.

Collecting all terms that contribute to order $\lambda^{1/2}$ we obtain

$$\mathcal{L}u_1 = 0\,, \tag{6.26}$$

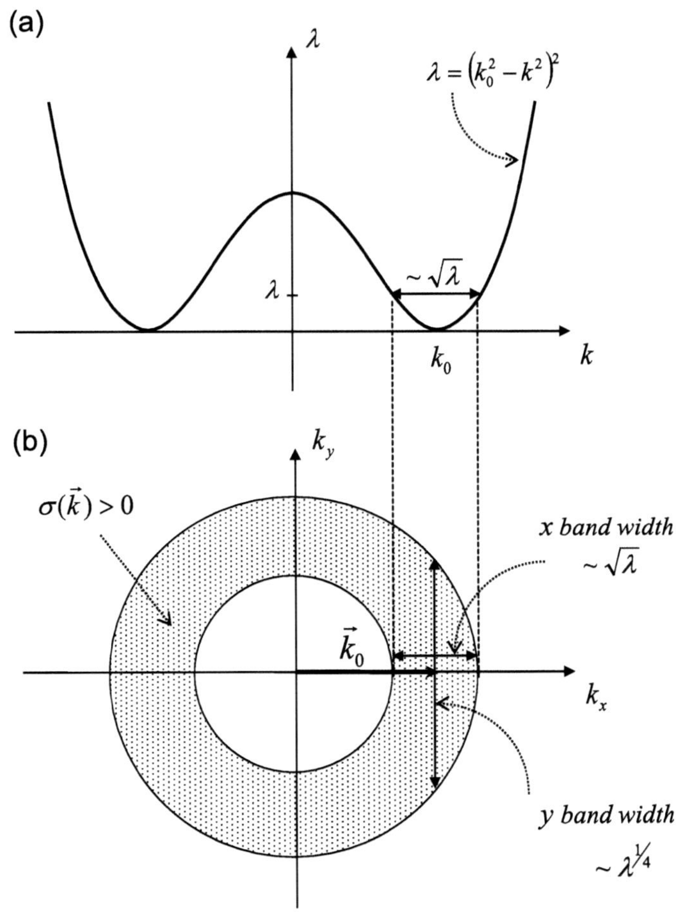

FIGURE 6.1: The neutral stability surface for stripe patterns, $\sigma(\lambda; \mathbf{k}) = 0$, $\lambda > 0$. (a) A cut at constant angle. (b) A cut at constant λ. For a given λ, the band of growing modes (hatched annulus) around the critical mode $k_0\hat{\mathbf{x}}$ has a width of order $\lambda^{1/2}$ in the x direction and of order $\lambda^{1/4}$ in the y direction.

where $\mathcal{L}$ is the linear operator

$$\mathcal{L} = \left(\partial_x^2 + {k_0}^2\right)^2 . \tag{6.27}$$

Indeed, the form (6.21) that we proposed for u_1 is a solution of (6.26). Collecting all terms that contribute to order λ we obtain

$$\mathcal{L}u_2 = 0 . \tag{6.28}$$

Since u_2 satisfies the same equation as u_1, it has the same solution form $u_2 = \tilde{B} \exp(ik_0 x) + c.c.$, where $\tilde{B}$ is an undetermined amplitude. It is easy to verify that keeping this contribution in the analysis will not change the amplitude equation up to the order $\lambda^{3/2}$ considered here. Since it anyway represents a small correction of order $\lambda^{1/2}$ to the amplitude $\tilde{A}$ we can discard it by choosing the solution $u_2 = 0$[4]. Finally, at order $\lambda^{3/2}$ we obtain

$$\mathcal{L}u_3 = -\left(2\partial_x\partial_X + \partial_Y^2\right)^2 u_1 - \partial_T u_1 + u_1 - u_1^3 . \tag{6.29}$$

Inserting into (6.29) the form (6.21) for u_1, using

$$u_1^3 = \left(\tilde{A}e^{ik_0x} + \tilde{A}^*e^{-ik_0x}\right)^3 = \tilde{A}^3e^{3ik_0x} + 3\tilde{A}^2\tilde{A}^*e^{ik_0x} + c.c. , \tag{6.30}$$

we find

$$\mathcal{L}u_3 = \left(-\left(2ik_0\partial_X + \partial_Y^2\right)^2 \tilde{A} - \partial_T\tilde{A} + \tilde{A} - 3|\tilde{A}|^2\tilde{A}\right)e^{ik_0x} - \tilde{A}^3e^{3ik_0x} + c.c. . \tag{6.31}$$

We now project out resonant terms from the right side of (6.31) by demanding that it is orthogonal to the null space of $\mathcal{L}^\dagger$ (see Section 6.1.2.3). To implement this condition we consider the vector space of complex-valued periodic functions with period $2\pi/k_0 = L$ with the inner product

$$\langle f, g\rangle = \int_0^L f^*(x)g(x)dx . \tag{6.32}$$

Using integration by parts it is straightforward to show that the operator $\mathcal{L}$ is self-adjoint, that is $\mathcal{L}^\dagger = \mathcal{L}$. Thus, the right side of (6.31) should be orthogonal to the null space of $\mathcal{L}$, or to the functions $h_\pm = \exp(\pm ik_0x)$. This implies that the coefficients of $h_\pm$ in the right side of (6.31) should be set to zero which yields the amplitude equation

$$\partial_T\tilde{A} = \tilde{A} - 3|\tilde{A}|^2\tilde{A} - \left(2ik_0\partial_X + \partial_Y^2\right)^2 \tilde{A} . \tag{6.33}$$

Expressing this equation in terms of the amplitude $A = \sqrt{\lambda}\tilde{A}$ and going back to the original space and time variables, x, y, t we find

$$\partial_t A = \lambda A - 3|A|^2A + \left(2k_0\partial_x - i\partial_y^2\right)^2 A . \tag{6.34}$$

In terms of A the leading order approximation for u is

$$u \approx Ae^{ik_0x} + c.c. . \tag{6.35}$$

[4]Choosing this solution amounts to setting $\tilde{B} = 0$, and thus solving the problem of an undetermined amplitude. In general, however, we should be careful in making such choices in order to avoid inconsistencies between different solvability conditions at higher orders [159, 145, 146]. For example, going to order $\lambda^{5/2}$ necessitates two slow time scales, T_1 and T_2, and a consistency condition that the time derivatives commute, i.e., $\partial_{T_2}(\partial_{T_1}A) = \partial_{T_1}(\partial_{T_2}A)$. Satisfying this condition may require a particular choice of $\tilde{B}$.

The next order contribution to u can be obtained by solving (6.31) (with the resonant terms removed). This yields a a third-harmonic contribution proportional to $A^3 \exp(3ik_0x)$. Equation (6.34) is an example of the *Newell–Whitehead–Segel* (NWS) equation, named after Newell and Whitehead [228] and Segel [276] who derived an equation of the same structural form to describe weak modulations of roll patterns in thermal convection. The NWS equation is often written in the form

$$\partial_t A = \lambda A - |A|^2 A + \left(\partial_x - i\partial_y^2\right)^2 A\,, \tag{6.36}$$

which is obtainable from (6.34) by rescaling the spatial coordinate x and the amplitude A ($x \to 2k_0 x,\ A \to A/\sqrt{3}$).

6.1.3.2 Amplitude equation for uniform oscillations

To describe the derivation of an amplitude equation for uniform oscillations we resort to the symmetric FHN model with non-diffusing inhibitor, that is, equation (5.37) with $a_0 = 0$ and $\delta = 0$. The zero solution of this equation goes through a uniform oscillatory instability at $\epsilon_c = a_1^{-1}$ (see Section 5.3.2). The result (5.53) or (5.56) for the growth rate applies also to the case of two space dimensions with $k^2 = \mathbf{k}\cdot\mathbf{k} = k_x^2 + k_y^2$. The neutral stability curve $\sigma_R = 0$ or $\lambda = k^2$ is shown in Figure 6.2(a). We determine the scaling factors for the slow space and time variables and for the state variables u and v using arguments similar to those we used in analyzing the SH equation. Beyond the instability point, that is, for $\lambda = (\epsilon_c - \epsilon)/\epsilon_c > 0$, there is a band of growing modes around the critical uniform mode, $k_0 = 0$, whose width scales like $\sqrt{\lambda}$ both in the x and y directions, as Figure 6.2 illustrates. This dictates the scaling exponents for the space variables, $p_x = p_y = 1/2$. Since the maximal growth rate scales like λ, the scaling exponent for the slow time variable is $q = 1$. Finally, equating the order of magnitude of u^3 with that of $\nabla^2 u$, and the order of magnitude of u with that of v give the scaling exponent $s = 1/2$ for both u and v.

We thus expand u and v as

$$\begin{pmatrix} u \\ v \end{pmatrix} = \lambda^{1/2}\begin{pmatrix} u_1 \\ v_1 \end{pmatrix} + \lambda\begin{pmatrix} u_2 \\ v_2 \end{pmatrix} + \lambda^{3/2}\begin{pmatrix} u_3 \\ v_3 \end{pmatrix} + \dots, \tag{6.37}$$

where

$$\begin{pmatrix} u_1 \\ v_1 \end{pmatrix} = \begin{pmatrix} 1 \\ c \end{pmatrix}\tilde{A}(X,Y,T)e^{i\omega_0 t} + c.c.\,, \tag{6.38}$$

and $\tilde{A}$ is the amplitude of the oscillatory uniform mode, which depends on the slow space and time variables

$$X = \lambda^{1/2}x\,, \qquad Y = \lambda^{1/2}y\,, \qquad T = \lambda t\,. \tag{6.39}$$

As in the derivation of the NWS equation, we carry out the nonlinear analysis to order $\lambda^{3/2}$ at which time variables slower than λt do not show up. We

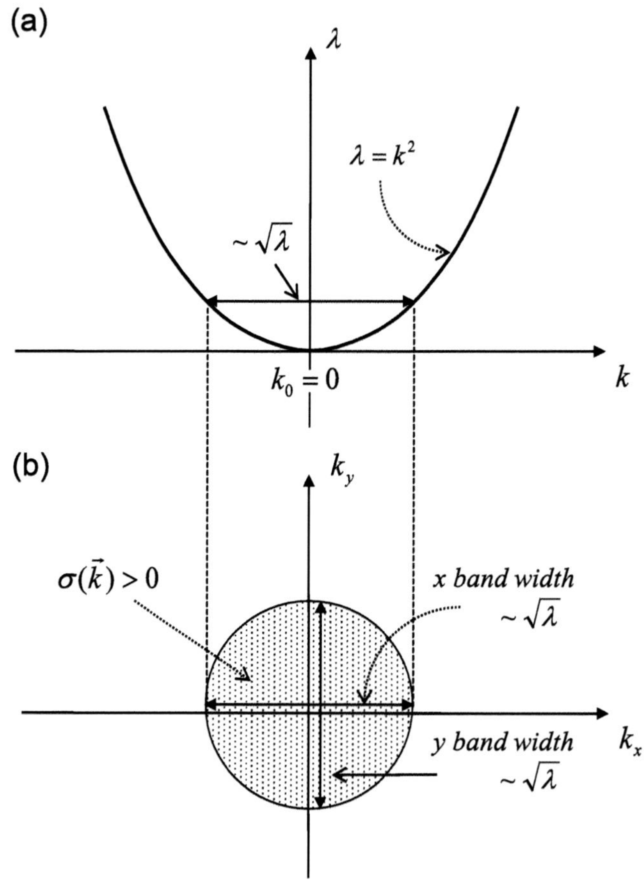

FIGURE 6.2: The neutral stability surface for uniform oscillations, $\sigma_R(\lambda; \mathbf{k}) = 0$, $\lambda > 0$. (a) A cut at constant angle. (b) A cut at constant λ. For a given λ, the band of growing modes (hatched circle) around the critical mode $k_0 = 0$ has a width of order $\lambda^{1/2}$ both in the x and in the y directions.

therefore consider a single slow time variable. Expressions for the constant c and for the Hopf frequency, ω_0, will be determined by the first order analysis. The result for ω_0 will coincide, as we will see, with the value (5.55) obtained using a linear stability analysis.

We now insert the expansion (6.37) into the symmetric FHN model and collect all contributions to a given order up to $\lambda^{3/2}$. In doing so we express ϵ in terms of λ, and use the following conversions of the space and time derivatives

$$\partial_x \to \lambda^{1/2}\partial_X\,, \qquad \partial_y \to \lambda^{1/2}\partial_Y\,, \qquad \partial_t \to \partial_t + \lambda\partial_T\,. \tag{6.40}$$

At order $\lambda^{1/2}$ we obtain

$$\mathcal{L}\begin{pmatrix} u_1 \\ v_1 \end{pmatrix} = 0\,, \qquad \mathcal{L} = \begin{pmatrix} \partial_t - 1 & 1 \\ -\epsilon_c & \partial_t + 1 \end{pmatrix}. \tag{6.41}$$

Inserting the form (6.38) in (6.41) we find

$$c = 1 - i\omega_0\,, \qquad \omega_0 = \sqrt{\epsilon_c - 1}\,. \tag{6.42}$$

At order λ we obtain

$$\mathcal{L}\begin{pmatrix} u_2 \\ v_2 \end{pmatrix} = 0\,. \tag{6.43}$$

Since u_2, v_2 satisfy the same equation as u_1, v_1 they merely contribute a small correction to the amplitude $\tilde{A}$ in (6.38). Moreover, as in the case of the SH equation (see Section 6.1.3.1), u_2, v_2 do not contribute to the equation for u_3, v_3 and therefore do not affect the amplitude equation up to order $\lambda^{3/2}$. We thus discard this second order contribution, setting $u_2 = v_2 = 0$, as we did for the SH equation.

Finally, at order $\lambda^{3/2}$ we obtain

$$\mathcal{L}\begin{pmatrix} u_3 \\ v_3 \end{pmatrix} = \tag{6.44}$$
$$\begin{pmatrix} -\partial_T\tilde{A} + \partial_X^2\tilde{A} + \partial_Y^2\tilde{A} - 3|\tilde{A}|^2\tilde{A} \\ -c\partial_T\tilde{A} + (c - \epsilon_c)\tilde{A} \end{pmatrix} e^{i\omega_0 t} + \begin{pmatrix} -\tilde{A}^3 \\ 0 \end{pmatrix} e^{3i\omega_0 t} + c.c.\,.$$

In order to project out resonant terms from the right side of (6.44) we consider the vector space of complex-valued periodic vector functions with period $T = 2\pi/\omega_0$, and define an inner product between any pair of vector functions in this space, $\mathbf{f} = (f_1, f_2)^T$ and $\mathbf{g} = (g_1, g_2)^T$, as

$$\langle \mathbf{f}(t), \mathbf{g}(t)\rangle = \langle f_1, g_1\rangle + \langle f_2, g_2\rangle\,, \tag{6.45}$$

where

$$\langle f_i, g_i\rangle = \int_0^T f_i^* g_i dt\,. \tag{6.46}$$

The adjoint operator is then

$$\mathcal{L}^\dagger = \begin{pmatrix} -\partial_t - 1 & -\epsilon_c \\ 1 & -\partial_t + 1 \end{pmatrix}, \tag{6.47}$$

and its null space is spanned by

$$\mathbf{h} = \begin{pmatrix} -c \\ 1 \end{pmatrix} e^{i\omega_0 t}\,, \tag{6.48}$$

and its complex conjugate $\mathbf{h}^*$. The requirement that the right side of (6.44) is orthogonal to $\mathbf{h}$ and $\mathbf{h}^*$ gives the amplitude equation

$$(c^* - c)\partial_T\tilde{A} = (|c|^2 - c)\tilde{A} - 3c^*|\tilde{A}|^2\tilde{A} + c^*(\partial_X^2\tilde{A} + \partial_Y^2\tilde{A})\,, \tag{6.49}$$

where we used the identity $\epsilon_c = (1+\omega_0^2) = |c|^2$. In terms of the amplitude $A = \sqrt{\lambda}\tilde{A}$ and the original space and time variables, x, y, t the amplitude equation reads

$$\partial_t A = \frac{1}{2}\lambda c A - \frac{3c^*}{c^* - c}|A|^2 A + \frac{c^*}{c^* - c}\nabla^2 A \,. \tag{6.50}$$

We can simplify this equation (6.50) by using the following transformations:

$$B = \sqrt{3} e^{\frac{1}{2} i\omega_0 \lambda t} A \,, \qquad \tau = t/2 \,. \tag{6.51}$$

This gives

$$\partial_\tau B = \lambda B + (1 - i\omega_0^{-1})\nabla^2 B - (1 - i\omega_0^{-1})|B|^2 B \,. \tag{6.52}$$

Equation (6.52) has the form

$$\partial_t A = \lambda A + (1 + i\alpha)\nabla^2 A - (1 + i\beta)|A|^2 A \,, \tag{6.53}$$

which is known as the *Complex Ginzburg–Landau* (CGL) equation[5] [6]. Note that in one space dimension, and for $\alpha = \beta = 0$, the CGL equation reduces to the NWS equation (6.36).

We illustrated the derivation of amplitude equations using two simple examples, the NWS equation for an amplitude which modulates stationary parallel stripes, and the CGL equation for an amplitude which modulates uniform oscillations. However, the method can be used to derive amplitude equations for more complicated situations, such as hexagonal patterns or patterns arising near codimension-2 bifurcations. The utility of this analytical approach for understanding pattern formation phenomena will be demonstrated in Chapters 7.1 and 8.

6.1.4 Phase equation

The significance of the phase, $\arg A$, in the dynamics of patterns beyond instability points has already been pointed out in Section 6.1. While the time scale associated with the magnitude $|A|$ scales like λ^{-1} and therefore gets shorter as the distance from the instability point increases, the time scale associated with phase changes remains long due to the existence of a marginal translation mode. In fact, this time scale is infinitely long for uniform phase changes, because they amount to spatial translations of the whole pattern which do not decay. As a consequence, at any finite distance, λ, from the instability point we can conceive weakly modulated phase fields whose time evolution is much slower than that of the magnitude $|A|$. Under this condition

[5]The control parameter λ is often eliminated from (6.53) by rescaling time. This parameter, however, represents the distance from the oscillatory instability and can change its sign. We therefore prefer the explicit form (6.53).

the magnitude $|A|$ rapidly relaxes to the value that corresponds to the instantaneous phase field, and the amplitude equation is reducible to an equation for the phase only, as we will now show.

We describe the derivation of a phase equation using the CGL equation in one space dimension. The simplest solutions appearing beyond the instability of the zero state describe traveling waves of the form

$$A_0 = \rho_0 e^{iqx - i\nu t + i\Phi_0}\,, \tag{6.54}$$

where Φ_0 is an arbitrary constant phase. Inserting this form into (6.53) we find

$$\rho_0 = \sqrt{\lambda - q^2}\,, \tag{6.55}$$

and the dispersion relation

$$\nu = \lambda\beta + (\alpha - \beta)q^2\,. \tag{6.56}$$

Consider now a more general traveling-wave solution, with a weakly modulated phase Φ, which we write as

$$A = \rho[\Phi]e^{iqx - i\nu t + i\Phi}\,. \tag{6.57}$$

Here, $\rho[\Phi]$ is the phase dependent magnitude of the amplitude, and the square brackets denote that ρ is a functional of the phase that includes spatial phase derivatives. Obviously, for constant phase $\Phi = \Phi_0$, ρ should reduce to ρ_0.

The weak space and time dependence of Φ is captured by introducing slow space and time variables. In deriving amplitude equations we defined the slow variables using λ as the small parameter (see (6.7)). In the present context λ is too big as it represents the rate of change associated with the fast variable ρ. We therefore introduce an auxiliary small parameter, ϵ, that satisfies $0 < \epsilon \ll \lambda$ and use the method of multiple-time scales (see Section 6.1.2.2) to write the phase as

$$\Phi = \Phi(X, T_1, T_2, ...)\,, \quad \text{with} \quad X = \epsilon x\,, \quad T_1 = \epsilon t \;\; T_2 = \epsilon^2 t, \dots. \tag{6.58}$$

The phase equation can then be written in the formal form

$$\partial_t \Phi = \epsilon \partial_{T_1}\Phi + \epsilon^2 \partial_{T_2}\Phi + \dots. \tag{6.59}$$

Note that since ϵ is not an intrinsic parameter we should be able to scale it out from the final phase equation to be derived. We used here the same scaling in ϵ for X and T_1, despite the fact that the amplitude equation contains a first derivative with respect to time and a second derivative with respect to space. The reason is that the phase modulates a traveling wave solution with a relatively high wavenumber $q \sim \sqrt{\lambda}$. Thus, the second space derivative of the amplitude contributes a term $q\partial_x\Phi$, that is much larger than $\partial_x^2\Phi$. Balancing $\partial_t\Phi$ with $q\partial_x\Phi$ then gives the chosen scaling.

To calculate the partial time derivatives $\partial_{T_n}\Phi$ we expand ρ as

$$\rho[\Phi] = \rho_0 + \epsilon\rho_1[\Phi] + \epsilon^2\rho_2[\Phi] + \ldots, \tag{6.60}$$

and insert (6.57) into the CGL equation (6.53), converting space and time derivatives as

$$\partial_x \to \partial_x + \epsilon\partial_X , \qquad \partial_t \to \partial_t + \epsilon\partial_{T_1} + \epsilon^2\partial_{T_2} + \ldots . \tag{6.61}$$

Collecting all terms that contribute to order unity and equating the real and imaginary parts we obtain the expressions for ρ_0 and ν given above. Repeating this procedure for all terms that contribute to order ϵ we obtain

$$\rho_1 = -\rho_0^{-1} q\partial_X\Phi . \tag{6.62}$$

and

$$\partial_{T_1}\Phi = 2(\beta - \alpha)q\partial_X\Phi , \tag{6.63}$$

At order ϵ^2 we obtain

$$\rho_2 = -\frac{1}{2\rho_0^3}\left[\lambda(\partial_X\Phi)^2 + (\alpha\rho_0^2 - 2\beta q^2)\partial_X^2\Phi\right] , \tag{6.64}$$

and

$$\partial_{T_2}\Phi = (\beta - \alpha)(\partial_X\Phi)^2 + D\partial_X^2\Phi , \tag{6.65}$$

where

$$D = 1 + \alpha\beta - \frac{2q^2(1+\beta^2)}{\rho_0^2} . \tag{6.66}$$

Inserting (6.63) and (6.65) into (6.59) and going back to the fast spatial coordinate x we obtain the phase equation up to 2nd order in the small parameter:

$$\partial_t\Phi + V_g\partial_x\Phi = (\beta - \alpha)(\partial_x\Phi)^2 + D\partial_x^2\Phi , \tag{6.67}$$

where

$$V_g = \frac{d\nu}{dq} = 2q(\alpha - \beta) , \tag{6.68}$$

is the group velocity of the traveling-wave solutions. The amplitude's magnitude is given in terms of the phase by

$$\rho[\Phi] = \rho_0 - c_1\partial_x\Phi - c_2(\partial_x\Phi)^2 - c_3\partial_x^2\Phi , \tag{6.69}$$

where

$$c_1 = \frac{q}{\rho_0} , \qquad c_2 = \frac{\lambda}{2\rho_0^3} , \qquad c_3 = \frac{1}{2\rho_0^3}(\alpha\rho_0^2 - 2\beta q^2) . \tag{6.70}$$

This dependence reflects the immediate response of ρ to slow phase variations with no time delay. Equation (6.57), with the phase Φ and the magnitude ρ satisfying (6.67) and (6.69), constitutes the phase approximation to traveling-wave solutions of the CGL equation. We will use this approximation in Section 6.2.2 to study the stability of traveling-wave solutions beyond a uniform oscillatory instability. Similar analyses can be applied to other types of amplitude equations.

6.1.5 Limitations of amplitude and phase equations

The amplitude equation approach has a few significant limitations. First and foremost is the assumed proximity to an instability point, $\lambda \ll 1$, which is needed to justify the perturbation expansion and the truncation at a given power of λ. The farther we go from an instability point the less accurate the amplitude equation, especially in capturing the long term dynamics, which are controlled by resonant terms that are beyond the truncation order and therefore are not considered. The study of pattern formation far from onset requires different analytical methods, such as the singular perturbation theory to be discussed in the next section. Another limitation is the scope of phenomena that amplitude equations can describe. The NWS equation provides a good example. It describes weak modulations of a parallel stripe pattern oriented along a given direction. However, because of the system's isotropy and the absence of any preferred direction, the actual patterns beyond a non-uniform stationary instability often consist of many stripe-domains oriented in different directions that cannot be described as weak modulations of a stripe pattern along a single direction. Lastly, there are situations in which the amplitude equation approach fails to capture important qualitative behaviors, such as hybrid states (see Sections 3.3.2 and 7.2.1.3).

The validity of phase equations such as (6.67) is not only subject to the limitations of the amplitude equations from which they are derived, but is also restricted by the condition $\epsilon \ll \lambda$ where ϵ quantifies the weak variation of the phase in space and time. This condition requires the absence of mechanisms that create steep phase gradients or fast temporal changes. An example of such a mechanism is negative phase diffusion, $D(k) < 0$, to be discussed in Section 6.2.2. Since the translation mode in uniform systems remains marginal also far from instability points, the concept of phase dynamics is still applicable when the condition $\lambda \ll 1$ is not met. Indeed, phase equations can still be derived directly from the original equations, rather than from the relevant amplitude equations which are valid only close to instability points [226, 227]. Derivations of such phase equations for the SH and FHN models appear in Refs. [55, 120], respectively.

Despite these limitations the derivations of amplitude and phase equations near instability points provide a powerful approach to understanding pattern formation phenomena and accounting for experimental observations, as we will see in the next section and in Chapters 7 and 8.

6.2 Linear stability analysis of periodic states

The amplitude equation is a powerful tool to study the existence and stability of periodic solutions. We have already described the method of linear stability analysis in the context of uniform stationary states in Section 5.3. Direct extension of this analysis to periodic solutions, in time or in space, is hard, first because the periodic solutions are generally not known, and secondly because the linear equations for the infinitesimally small deviations from the solutions are not autonomous. The amplitude equation approach circumvents both difficulties; it provides a simple approximate solution for the periodic state and allows the derivation of autonomous linear equations for appropriately defined small deviations. We will demonstrate the method in Section 6.2.1 for stationary spatially periodic states using the NWS amplitude equation (6.34).

Stability analyses that are based on amplitude equations can identify phase instabilities, in which the growing perturbations involve phase modulations, and amplitude instabilities in which the growing perturbations modulate the absolute value of the amplitude. Phase instabilities can also be studied by analyzing the phase equations that are derivable from the amplitude equations. We will describe such an analysis in Section 6.2.2 for traveling-wave solutions of the CGL equation (6.53) using the phase equation (6.67).

6.2.1 Amplitude equation analysis

Consider the SH equation (5.31). Close to the instability point of the zero solution, the amplitude A of the periodic solution, $u \approx A\exp(ik_0x) + c.c.$, satisfies the NWS equation (6.34). Anticipating a narrow band of periodic solutions with wavenumbers k close to k_0, we look for stationary periodic solutions of the NWS equation of the form

$$A_0 = \rho_0 e^{iqx} \,. \tag{6.71}$$

Inserting this form in (6.34) we find

$$\rho_0 = \sqrt{\lambda - \lambda_P}/\sqrt{3}\,, \qquad \lambda_P = 4k_0^2 q^2 \,. \tag{6.72}$$

The parabola $\lambda = \lambda_P(q)$ (see Figure 6.4) designates the value of λ at which the periodic solution $A = A_0$, for a particular q, appears. For any λ the relation $\lambda = \lambda_P(q)$ defines the band width,

$$-\sqrt{\lambda}/(2k_0) < q < \sqrt{\lambda}/(2k_0), \tag{6.73}$$

of periodic solutions, A_0, that exist for that particular λ value. This band coincides with the band of the growing modes found in the linear stability analysis of the zero state. Indeed, demanding non-negative growth rate, $\sigma \geq 0$, in (5.43) leads to $-\sqrt{\lambda} \leq k^2 - k_0^2 \leq \sqrt{\lambda}$. Using $k^2 - k_0^2 = (k + k_0)(k - k_0) \approx 2k_0(k - k_0)$ and identifying q with $k - k_0$ we obtain (6.73). We now show how the NWS equation, and in general any other amplitude equation, can be used to study the linear stability of periodic solutions within the band of existing solutions.

To study the linear stability of the stationary periodic solutions (6.71) we consider a perturbed solution of the form [56]

$$A(\mathbf{r},t) = A_0(x) + \delta A(\mathbf{r},t)\,,$$

where

$$\delta A(\mathbf{r},t) = e^{iqx}\left[\delta a_+(t)e^{i\mathbf{Q}\cdot\mathbf{r}} + \delta a_-^*(t)e^{-i\mathbf{Q}\cdot\mathbf{r}}\right]\,, \tag{6.74}$$

is a complex valued perturbation representing a pair of Fourier modes with wave-vectors $q\hat{\mathbf{x}} \pm \mathbf{Q}$ ($\mathbf{Q} = Q_x\hat{\mathbf{x}} + Q_y\hat{\mathbf{y}}$), δa_+ and δa_-^* are complex-valued functions of time, and $\mathbf{r} = (x,y)$. The choice to represent δa_-^* as the complex conjugate of another function, δa_-, is for convenience only. The form of δA can be motivated by writing the amplitude in a polar form, $A = \rho\exp(i\phi)$, and defining a local wave vector, $\mathbf{K} = \nabla\phi$, and a local frequency, $\Omega = -\partial_t\phi$. It then follows (assuming the mixed second-order partial derivatives of ϕ are continuous) that the local wave vector satisfies the conservation equation:

$$\partial_t\mathbf{K} + \nabla\Omega = 0\,. \tag{6.75}$$

This implies that the line integral, $\int_C \mathbf{K}\cdot d\mathbf{r}$, of $\mathbf{K}$ along any trajectory C in the x,y plane must be constant in time (provided Ω assumes the same value at both edges of C). Thus, if $\mathbf{K}$ increases in time somewhere along the trajectory it must decrease somewhere else, and therefore if the spectrum of A contains a Fourier component $q\hat{\mathbf{x}} + \mathbf{Q}$ we should expect it to contain also a Fourier component $q\hat{\mathbf{x}} - \mathbf{Q}$.

The perturbation δA, assumed to be infinitesimal, satisfies the linear equation

$$\partial_t\delta A = \lambda\delta A + \left(2k_0\partial_x - i\partial_y^2\right)^2\delta A - 6|A_0|^2\,\delta A - 3A_0^2\,\delta A^*\,. \tag{6.76}$$

Using the form (6.74) in (6.76) we find the following linear ODEs for the Fourier amplitudes:

$$\begin{aligned}\dot{\delta a}_+ &= -(p^2 + U_+)\delta a_+ - p^2\delta a_- \\ \dot{\delta a}_- &= -p^2\delta a_+ - (p^2 + U_-)\delta a_-\,,\end{aligned} \tag{6.77}$$

where $p^2 = 3\rho_0^2 = \lambda - \lambda_P$ and

$$U_\pm = (2k_0(q \pm Q_x) + Q_y^2)^2 - \lambda_P\,. \tag{6.78}$$

Equations (6.77) have solutions of the form

$$\begin{pmatrix}\delta a_+ \\ \delta a_-\end{pmatrix} = \begin{pmatrix}\delta a_{0+} \\ \delta a_{0-}\end{pmatrix} e^{\sigma t} \tag{6.79}$$

with real eigenvalues given by

$$\sigma_\pm(\mathbf{Q}) = -p^2 - \frac{1}{2}(U_+ + U_-) \pm \sqrt{p^4 + \frac{1}{4}(U_+ - U_-)^2}\,. \tag{6.80}$$

Consider first the stability of periodic solutions to perturbations for which $\mathbf{Q} = \mathbf{0}$, that is, perturbations that do not change the solutions' wave-vectors $q\hat{\mathbf{x}}$. In that case

$$\sigma_+ = 0\,, \qquad \sigma_- = -2p^2\,. \tag{6.81}$$

The zero eigenvalue, σ_+, represents the marginal translation mode which affects the phase of the periodic pattern (see Section 5.3.4). This can be seen by calculating the eigenvector associated with σ_+ and realizing that the perturbation δA of A_0 can be written as $\delta x(dA_0/dx)$ where $\delta x \propto \delta a_{0+}$ (see (5.62)). The negative eigenvalue, σ_-, represents the decay of magnitude perturbations which becomes faster as the magnitude $|A_0| = \rho_0$ increases. Thus, for finite ρ_0, the first perturbations to grow are those associated with long wavelength phase modulations.

Let us consider now two limiting cases, corresponding to perturbations that vary in the x-direction only, i.e., $\mathbf{Q} = Q\hat{\mathbf{x}}$, and perturbations that vary in the y-direction only, i.e., $\mathbf{Q} = Q\hat{\mathbf{y}}$. In the former case the condition $\sigma_+(\mathbf{Q}) > 0$ leads to

$$\lambda > \frac{3p^2}{2} + k_0^2 Q^2\,, \tag{6.82}$$

and for this to hold for some real Q we must have $\lambda > 3p^2/2$ or

$$\lambda < \lambda_E\,, \qquad \lambda_E = 3\lambda_P = 12k_0^2 q^2\,. \tag{6.83}$$

Periodic solutions outside the parabola $\lambda = \lambda_E$ in the λ, q plane (see point B in Figure 6.4) are unstable to longitudinal perturbations, that is, to perturbations involving spatial variations in the direction of the pattern's wave vector $q\hat{\mathbf{x}}$. The instability that occurs as λ is decreased below λ_E is called the *Eckhaus instability* [75]. Figure 6.3(a) shows an illustration of a stripe pattern undergoing an Eckhaus instability.

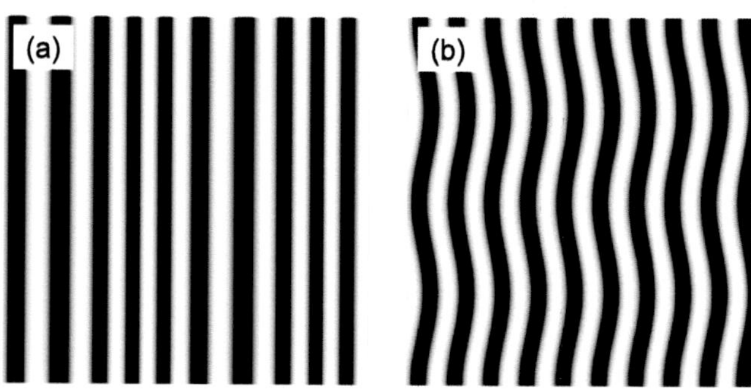

FIGURE 6.3: Illustrations of the deformation a stripe pattern undergoes in the course of (a) longitudinal Eckhaus instability, and (b) transverse zigzag instability. Adopted from [55].

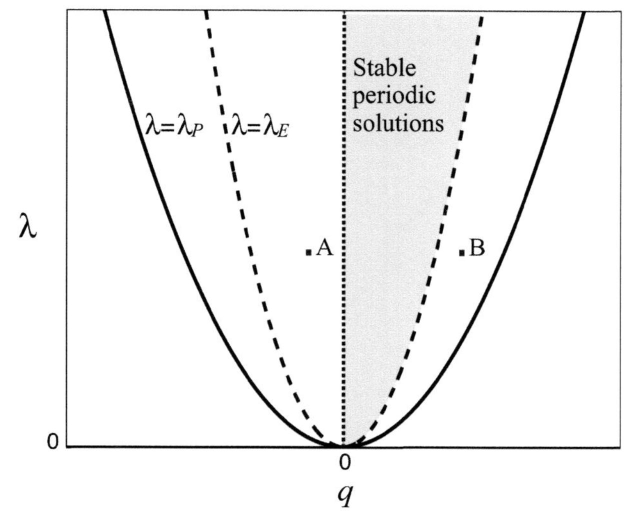

FIGURE 6.4: Existence and stability ranges of periodic stripe solutions of the NWS equation. Shown are the existence boundary $\lambda - \lambda_P$ (solid line), the Eckhaus instability boundary $\lambda = \lambda_E$ (dashed line), and the zigzag instability boundary $q = 0$ (dotted line). Stable periodic stripe solutions exist within the grey area (the Busse balloon).

The Eckhaus instability is characterized by the growth of two modes, $q \pm Q_{max}$, on both sides of the central mode at $q\hat{\mathbf{x}}$. These side modes grow faster than all other modes and can be calculated by maximizing σ_+ with respect to Q [228, 56]. Such a calculation leads to

$$Q_{max}^2 = q^2 \left[1 - \left(\frac{\lambda - \lambda_P}{\lambda_E - \lambda_P} \right)^2 \right] .$$

Note that Q_{max} vanishes at the onset, $\lambda = \lambda_E$, of the Eckhaus instability, implying it is a long-wavelength instability. The picture depicted above for the NWS equation carries over to the SH equation by shifting the wave-vectors $q\hat{\mathbf{x}} \pm \mathbf{Q}$ to $(k_0 + q)\hat{\mathbf{x}} \pm \mathbf{Q}$.

Consider now perturbations that vary in the y-direction, i.e., $\mathbf{Q} = Q\hat{\mathbf{y}}$. In this case the condition for an instability is

$$\sigma_+(\mathbf{Q}) = -(Q^2 + 4k_0 q)Q^2 > 0 , \tag{6.84}$$

implying that such perturbations can grow only when $q < 0$. Thus, for any $\lambda > \lambda_P$ (existence condition of A_0), periodic solutions with $q < 0$ (see point A in Figure 6.4) are unstable to transverse perturbations, i.e., to perturbations involving spatial variations in a direction perpendicular to the pattern's

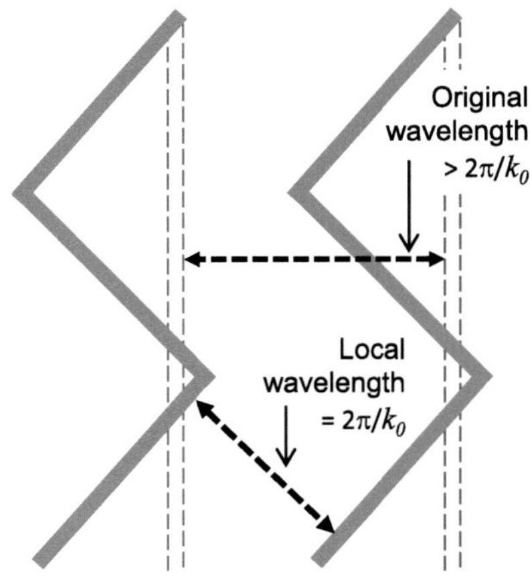

FIGURE 6.5: A schematic illustration of the zigzag-instability mechanism. A periodic pattern with an original wavenumber $k < k_0$ (or $q < 0$) can lower its energy (Lyapunov-functional) value by a zigzag-like deformation, which creates domains with the energetically preferred wavenumber k_0. Such a deformation can only increase the wavenumber and therefore cannot lower the energy of a pattern with $k > k_0$ ($q > 0$).

wave vector $q\hat{\mathbf{x}}$. This is the so-called *zigzag instability*, a name that reflects the zigzag-like patterns that develop beyond the instability point. Periodic modulations in the y direction is a mechanism by which the system increases the local wavenumber to the stable regime as Figure 6.5 illustrates. Figure 6.3(b) shows an illustration of a stripe pattern undergoing a zigzag instability.

The existence and stability ranges of periodic solutions of the form (6.71) are shown in Figure 6.4. The solutions exist within the parabola $\lambda = \lambda_P = 4k_0^2q^2$ but are stable to longitudinal perturbations only within the inner parabola $\lambda = \lambda_E = 12k_0^2q^2$. Within this parabola only solutions with $q > 0$ (or $k > k_0$ in the SH equation) are stable to both longitudinal and transverse perturbations. The stability domain of periodic solutions in the (q, λ) plane is often called the *Busse balloon* [33]. In principle, we should have studied also the general case of an arbitrary perturbation wave-vector $\mathbf{Q} = Q_x\hat{\mathbf{x}} + Q_y\hat{\mathbf{y}}$. It follows from (6.80), however, that increasing $|q|$ from zero the first instability to occur is either the Eckhaus or the zigzag instability [55].

It is important to stress that although here we considered a specific model, the SH equation, the analysis applies to any system that goes through a non-uniform stationary instability to stripe or roll patterns, because it is based on the universal NWS amplitude equation. A similar analysis can be carried out for traveling-wave patterns too, using the CGL equation. Since the Eckhaus and zigzag instabilities are *phase instabilities* in that they involve the growth of phase modulations, we should expect the phase equation (6.67) to capture them too. In the following section we show how this equation can be used to study longitudinal instabilities of traveling-wave solutions.

6.2.2 Phase equation analysis

The nonlinear phase equation (6.67) can be transformed to a linear diffusion equation by means of a Cole–Hopf transformation [134, 48]. Writing the phase as

$$\Phi(x,t) = \frac{D}{\beta - \alpha} \ln \Psi(z,t)\,, \qquad z = x - V_g t\,, \tag{6.85}$$

and inserting this form into (6.67) we find

$$\partial_t \Psi = D\partial_z^2 \Psi\,. \tag{6.86}$$

The traveling-waves solutions (6.54) of the CGL equation correspond to constant phase solutions $\Phi = \Phi_0$ of (6.67) or $\Psi = \Psi_0$ of (6.86). To study their stability we perturb these solutions, $\Psi = \Psi_0 + \delta\Psi$, and follow the perturbation $\delta\Psi$ in time. Obviously, any component in the Fourier expansion of $\delta\Psi$ decays to zero for $D > 0$ and diverges for $D < 0$, indicating a phase instability at $D = 0$.

Note that the validity of the phase equation (6.67) breaks down for negative D values; higher orders in the expansion (6.59) must be taken into account to saturate the perturbation growth. Also note that for negative D the perturbation $\delta\Psi(z) = \delta\Psi(x - V_g t)$ is growing in a coordinate frame that moves with the group velocity V_g. When the group velocity is sufficiently high relative to the perturbation growth rate, a localized perturbation may still decay at any fixed point in space because of its fast propagation away from this point. We say in this case that the state is *convectively unstable*; perturbations grow as they move, but not at fixed points in space. When moving perturbations also grow at fixed points we say that the system's state is *absolutely unstable*.

Inspecting the form (6.66) of $D(q)$ we can immediately infer that for $1 + \alpha\beta < 0$, $D(q) < 0$ for any q. That is, when this inequality is satisfied, the periodic traveling-wave solutions (6.54) are *all* unstable, and indeed the dynamics of the CGL equation become chaotic. This inequality is known as the *Newell criterion* and the chaotic state that sets in when this criterion is satisfied is termed *Benjamin–Feir turbulence* [6]. When the Newell criterion is not satisfied a band of stable modes exists for $\lambda > \lambda_E$ where λ_E is determined by the value of λ at which $D(q) = 0$, and is given by

$$\lambda_E = \left[1 + \frac{2(1+\beta^2)}{1+\alpha\beta}\right] q^2\,. \tag{6.87}$$

This is the Eckhaus instability threshold. Indeed, setting the conditions under which the CGL equation reduces to the NWS equation (6.36), that is $\alpha = 0$, $\beta = 0$, we recover the Eckhaus instability threshold $\lambda_E = 3q^2$ (or $\lambda_E = 12k_0^2 q^2$ for (6.34)).

6.3 Singular perturbation theory

The equations that describe a dynamical system often contain a small parameter which can be used to approximate solutions as perturbation expansions. The leading order form in such an expansion is a solution of the "unperturbed system," obtained by setting the small parameter to zero, and the higher order terms provide small corrections to the leading order approximation. There are situations, however, in which the solution of the unperturbed system cannot provide a good approximation uniformly in time or in space, no matter how small is the small parameter. This is generally the case when the small parameter multiplies a high derivative term in the equations. Singular perturbation theory provides methods for handling such situations. In this section we provide a glimpse of this theory. We begin with a description of a few basic ideas (Section 6.3.1), and then apply the method to approximating stationary periodic solutions far from their onset (Section 6.3.2).

6.3.1 The general concept

Consider, as an example, the symmetric FHN model for uniform fields,

$$\frac{du}{dt} = u - u^3 - v\,, \tag{6.88a}$$

$$\frac{dv}{dt} = \epsilon\,(u - a_1 v)\,, \tag{6.88b}$$

with $a_1 < 1$ (Figure 5.2(b)). Decreasing ϵ below $\epsilon_H = a_1^{-1}$ leads to a Hopf bifurcation of the zero state and to sinusoidal oscillations (see Section 5.3.2). Decreasing ϵ far below the Hopf threshold, i.e., to very small positive ϵ values, leads to relaxation oscillations similar to those observed in the BZ reaction (see Figure 2.4). However, setting ϵ strictly to zero results in a one-dimensional dynamical system (since v is constant) that cannot have oscillatory solutions. Thus, the limit $\epsilon \to 0$ is singular; the behavior of the system at $\epsilon = 0$ is qualitatively different from the behavior as ϵ approaches zero.

To see how this singular nature of (6.89) arises let us rescale the time coordinate by introducing a slow time, $\tau = \epsilon t$, in terms of which (6.89) reads

$$\epsilon\frac{du}{d\tau} = u - u^3 - v\,, \tag{6.89a}$$

$$\frac{dv}{d\tau} = u - a_1 v\,. \tag{6.89b}$$

In this form of the equations, the small parameter multiplies the time derivative of u. As long as the time derivative $du/d\tau$ is of order unity the term $\epsilon du/d\tau$ is small and can be neglected. However, there might exist time intervals in which $du/d\tau \sim \epsilon^{-1}$. During these intervals the term $\epsilon du/d\tau$ cannot be neglected. Singular perturbation theory has been developed to handle such situations where small parameters multiply derivatives of state variables. It does so by dividing the time or space axis into *outer regions* where all state variables vary gradually and *inner regions* where some of the state variables vary abruptly. The equations are then solved in each region separately, employing appropriate approximations, and the solutions are matched in the overlap domains of outer and inner regions [225], or at the contact point of two outer regions enclosing a diminishingly small inner region [315]. Inner regions that involve abrupt spatial changes in some state variables are also called "transition layers." Inner regions that involve abrupt temporal changes are often called "critical transitions" or "catastrophic shifts."

Unlike amplitude equations, singular perturbation theory can be used to study patterns far from their onset. Figure 6.6 shows stationary and traveling-wave solutions of the FHN equations (5.37) near and far from the Turing and Hopf bifurcations that induce them. Far from these bifurcations two types of regions develop within a single wavelength interval, an inner region where the u field changes abruptly and an outer region where it changes gradually (like the v field). In the following section we will use the example of stationary patterns far from the Turing bifurcation (6.6(b)) to explain how singular perturbation theory can be used to find approximate solutions for such patterns.

6.3.2 Singular perturbation analysis of stationary periodic patterns

Stationary, one-dimensional solutions of the FHN model (5.37) satisfy the equations

$$\mu\frac{d^2u}{d\chi^2} + u - u^3 - v = 0\,, \tag{6.90a}$$

$$\frac{d^2v}{d\chi^2} + u - a_1 v - a_0 = 0\,, \tag{6.90b}$$

where u and v are functions of the rescaled spatial coordinate $\chi = \sqrt{\mu}x$ and $\mu = \epsilon/\delta$. Taking μ to be a small parameter, makes (6.90) a singular perturbation problem since the small parameter multiplies a spatial derivative term.

Consider first outer regions where both u and v vary gradually in space (see Figure 6.6b). In these regions the term $\mu d^2u/d\chi^2$ is negligible for $\mu \ll 1$ and we are left with a cubic equation, $u - u^3 - v = 0$, which we can solve for u in terms of v. The solutions of this equation give the three branches of the cubic nullcline shown in Figure 5.2. Since the middle branch is unstable we only consider the two other branches which we denote $u = u_\pm(v)$. These branches dictate two types of outer regions characterized by high ($u = u_+(v)$) and low ($u = u_-(v)$) values of u, as Figure 6.7(a) illustrates. Inserting the solutions $u = u_\pm(v)$ in (6.90b) we obtain closed equations for v in the two

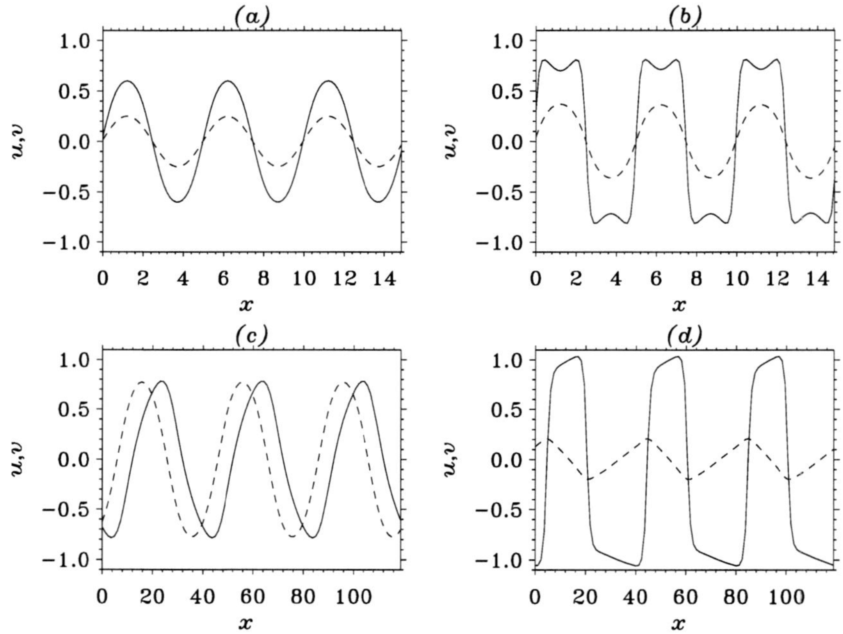

FIGURE 6.6: The development of two spatial scales far from the onset of Turing and Hopf bifurcations in the FHN model: stationary periodic pattern near the Turing bifurcation (a) and far from it (b). Traveling pattern near the Hopf bifurcation (c) and far from it (d). In all panels the solid line is the u field and the dashed line is the v field. From [118].

types of outer regions. In these regions v and u both change on the same length scale, $dv/d\chi \sim du/d\chi \sim \mathcal{O}(1)$.

Consider now the inner region that must exist between the two types of outer regions, in which u changes from u_+ to u_-, or vice versa, over a short distance. In this region the term $\mu d^2u/d\chi^2$ is no longer negligible. Balancing this term with the other terms in the equation for u gives $du/d\chi \sim \mathcal{O}(\mu^{-1/2})$. That is, changes of order unity in u occur over short distances of order $\sqrt{\mu} \ll 1$. Over such diminishingly small distances v hardly changes and can be approximated by a constant value independent of χ. Mathematically, we analyze the inner region by stretching the spatial coordinate back to $x = \mu^{-1/2}\chi$:

$$\frac{d^2u}{dx^2} + u - u^3 - v = 0\,, \tag{6.91a}$$

$$\frac{d^2v}{dx^2} + \mu(u - a_1 v - a_0) = 0\,. \tag{6.91b}$$

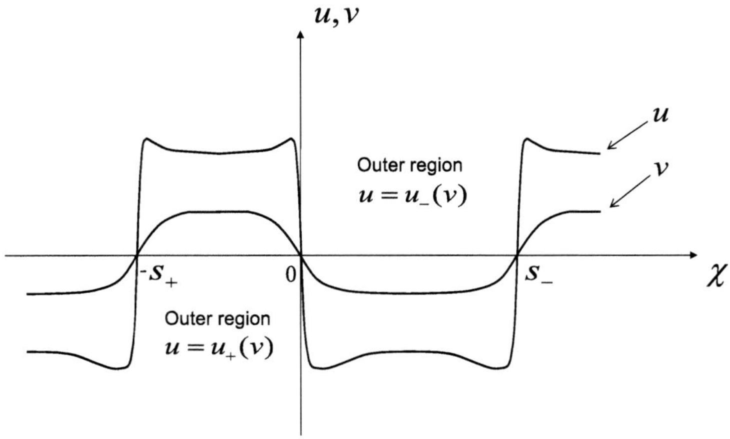

(a) Two outer regions in a single-wavelength interval: A high-u region of width s_+ where $u = u_+(v)$, and a low-u region of width s_- where $u = u_-(v)$. In between the outer regions is a narrow inner region whose stretched form is shown below.

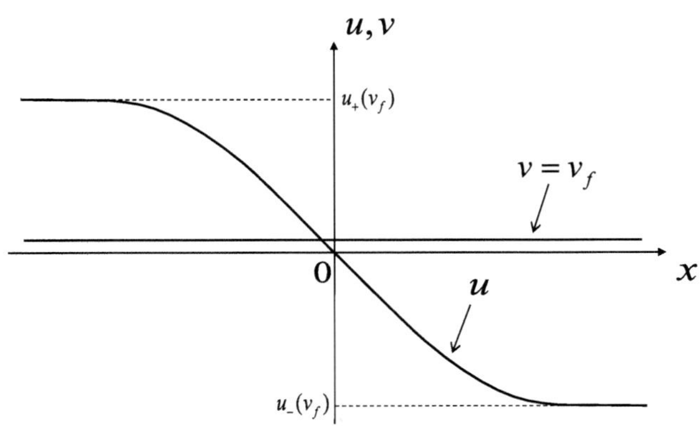

(b) An inner region in a stretch coordinate system where $v = v_f$ is constant. For a stationary solution $v_f = 0$.

FIGURE 6.7: A schematic illustration of singular perturbation analysis of stationary periodic solutions in the FHN model.

Neglecting the term proportional to μ in (6.91b) we find $d^2v/dx^2 = 0$ and the solution $v = v_f = constant$ as Figure 6.7(b) illustrates. Inserting this solution in (6.91a) we obtain a closed equation for u.

We can proceed now to solving the equations of the inner and outer regions in the limit $\mu \to 0$. Since we are looking for periodic solutions we consider the interval of one wavelength as depicted in Figure 6.7(a). In the inner region u

satisfies the equation

$$\frac{d^2u}{dx^2} + u - u^3 - v_f = 0\,, \qquad -\infty < x < \infty\,, \tag{6.92}$$

where v_f is yet an undetermined constant. We look for solutions that asymptote to $u_+(v_f)$ as $x \to -\infty$ and to $u_-(v_f)$ as $x \to \infty$, or vice versa. Such solutions exist only for $v_f = 0$. An intuitive way to see this is to write (6.92) as

$$\frac{d^2u}{dx^2} = -\frac{dV}{du}\,, \tag{6.93}$$

where

$$V = \frac{1}{2}u^2 - \frac{1}{4}u^4 - v_f u\,, \tag{6.94}$$

and to draw an analogy between (6.93) and the equation of motion of a unit-mass particle whose position, u, is changing in time, x, as a result of a conservative force, $-dV/du$, where V is a double-hump potential energy. The only way for the particle to start arbitrarily close to the top of one hump at time $x = -\infty$ with a zero initial velocity du/dx, and asymptotically approach the top of the second hump at time $x = \infty$ with a zero final velocity, is that the two humps are equally high, or that $v_f = 0$. Mathematically, this result can be derived by relating v_f to the velocity of a propagating wavefront, as is done in Section 7.2.1.1. Equation (6.92) with $v_f = 0$ has two symmetric solutions

$$u = \mp \tanh\left(\frac{x}{\sqrt{2}}\right)\,, \tag{6.95}$$

satisfying, respectively, the asymptotic conditions $u \to u_+(0)$ as $x \to -\infty$ and $u \to u_-(0)$ as $x \to \infty$ and vice versa, where $u_\pm(0) = \pm 1$. These solutions account for the sharp decline of u at $x = 0$ and the sharp surge of u at $x = -s_-/\sqrt{\mu}$ and $x = s_+/\sqrt{\mu}$ (see Figure 6.7(a)).

In the outer regions $u = u_\pm(v)$, and v satisfies the equations

$$\frac{d^2v}{d\chi^2} + u_+(v) - a_1 v - a_0 = 0\,, \qquad -s_+ < \chi < 0\,, \tag{6.96a}$$

$$\frac{d^2v}{d\chi^2} + u_-(v) - a_1 v - a_0 = 0\,, \qquad 0 < \chi < s_-\,, \tag{6.96b}$$

where $s_\pm$ are undetermined yet. The solutions of these equations should be matched to the constant value, $v_f = 0$, of v in the diminishingly narrow inner regions centered at $\chi = 0$, $\chi = -s_+$ and $\chi = s_-$. These continuity conditions dictate the boundary conditions for (6.96), $v(-s_+) = v(0) = 0$ for the outer region $-s_+ < \chi < 0$, and $v(0) = v(s_-) = 0$ for the outer region $0 < \chi < s_-$. We further require that the solutions are continuously differentiable at $\chi = 0$, $\chi = -s_+$ and $\chi = s_-$. This will lead to a relation between s_- and s_+, as we will shortly see. Note that we cannot require the continuity of the second and higher derivatives at these points as the second derivative is discontinuous;

the jump $[d^2u/d\chi^2]_{\chi=0}$ in the second derivative at $\chi = 0$ (or at $\chi = -s_+$ or at $\chi = s_-$) is $|u_+(0) - u_-(0)| = 2$, as equations (6.96) imply.

Approximate solutions to (6.96) can be obtained by linearizing the cubic nullcline around $v = 0$, assuming v is sufficiently small: $u_\pm(v) \approx \pm 1 - (1/2)v$. This leads to the following linear boundary value problems:

$$\frac{d^2v}{d\chi^2} - q^2(v - v_+) = 0\,, \quad -s_+ < \chi < 0 \qquad (6.97a)$$
$$v(0) = v(-s_+) = 0\,,$$
$$\frac{d^2v}{d\chi^2} - q^2(v - v_-) = 0\,, \quad 0 < \chi < s_- \qquad (6.97b)$$
$$v(0) = v(s_-) = 0\,,$$

where $q^2 = a_1 + 1/2$ and $v_\pm = q^{-2}(\pm 1 - a_0)$. The solutions to these boundary value problems are

$$v(\chi) = v_+\left(1 + \frac{\sinh q\chi - \sinh q(\chi + s_+)}{\sinh qs_+}\right), \quad -s_+ < \chi < 0\,, \qquad (6.98a)$$
$$v(\chi) = v_-\left(1 + \frac{\sinh q(\chi - s_-) - \sinh q\chi}{\sinh qs_-}\right), \quad 0 < \chi < s_-\,. \qquad (6.98b)$$

The requirement that v is continuously differentiable at $\chi = 0$ is met by deriving the solutions (6.98) with respect to χ, taking the limit $\chi \to 0$ and equating the two limits. This leads to the following relation between s_- and s_+:

$$v_+ \tanh\left(\frac{qs_+}{2}\right) = -v_- \tanh\left(\frac{qs_-}{2}\right). \qquad (6.99)$$

The requirements that v is continuously differentiable at $\chi = -s_+$ and at $\chi = s_-$ lead to the same relation.

The singular perturbation analysis described above provides approximate stationary periodic solutions for the FHN model when $\mu = \epsilon/\delta \ll 1$. The inhibitor v is given by (6.98). The activator u is given by $u_+(v)$ in the range $-s_+ < \chi < 0$ and by $u_-(v)$ in the range $0 < \chi < s_-$. At $\chi = 0$, $\chi = -s_+$ and $\chi = s_-$ u goes through sharp changes described by (6.95) with $x = \chi/\sqrt{\mu}$. Equation (6.99) dictates a relation between the width of the up-state domain, s_+ and the width of the down-state domain, s_-, leaving the wavelength $s_+ + s_-$ undetermined. This result implies the existence of a family of stationary periodic solutions with variable wavenumber.

The singular perturbation method allows studying not only periodic patterns, but also localized structures. A stationary pulse solution, consisting e.g., of a single up-state domain in a system otherwise occupied by the down state, is obtainable from (6.99) in the limit $s_- \to \infty$. Imposing this limit leads to a pulse solution whose width is given by

$$s_+ = -q^{-1}\ln(-a_0)\,, \qquad -1 < a_0 < 0\,. \qquad (6.100)$$

A necessary condition for such a solution to exist is the existence of a uniform down-state solution, i.e., parameters a_0 and a_1 that correspond to cases (a) or (c) in Figure 5.2. In Section 7.3.1 we will extend the singular perturbation analysis described above to the study of propagating fronts and traveling waves.

Chapter summary

The analytical methods described in this chapter provide tools for studying the asymptotic dynamics and the patterns that appear beyond an instability point. Unlike linear stability analysis, these methods take into account nonlinear contributions in the deviations from the unstable state, which are needed to counter-act the initial exponential growth of the deviations. Two main methods are described; weak nonlinear analysis that results in dynamic equations for the amplitudes of the growing modes near an instability point, and singular perturbation analysis, which is not limited to the vicinity of an instability point, and handles situations where the asymptotic patterns consist of transition layers, where some state variables change abruptly.

The general concept behind the amplitude equation approach is that at any small but finite distance from the instability point the amplitude of the first mode to grow is small and has weak temporal and spatial dependence. These properties allow the derivation of the amplitude equation using a multiple-scale analysis. In this analysis the solution is approximated as an asymptotic expansion up to a given order in the small deviation from the instability point, where at each order a solvability condition is imposed. The solvability conditions are then translated into a constraint on the amplitude dynamics, which is the amplitude equation.

Amplitude equations assume the same structural form for any system that goes through the same instability type. Thus, the NWS equation (6.36), which describes weak spatio-temporal modulations of stripe patterns, is derivable for any system that goes through a non-uniform stationary instability. Likewise, the CGL equation (6.53), which describes weak spatio-temporal modulations of uniform oscillations, is derivable for any system beyond a uniform oscillatory instability. The differences between distinct systems are reflected through the coefficients of the various terms that appear in the amplitude equation, which are expressible in terms of parameters that are specific to the system in question. This universality is a powerful result; identifying an instability of a certain type in a specific system provides at once much information about potential dynamical behaviors.

The amplitude equation is most useful for studying the existence and stability of periodic solutions. Quite often the instabilities of stationary periodic patterns or periodic traveling waves are phase instabilities, that is, they modulate the amplitude's argument. Examples of such instabilities are the Eckhaus and zigzag instabilities, which modulate the pattern's phase along and perpendicular to the pattern's wave-vector, respectively. At finite distances beyond the pattern-forming instability, the amplitude equation is often reducible to a phase equation. This is because of the marginal translation mode which makes sufficiently long-wavelength phase modulations slow enough to dominate the long-term dynamics.

While the amplitude equation approach is limited to the close vicinity of instability points, the singular perturbation method does not share this limitation, and often becomes applicable far from instability points. Singular perturbation problems arise when a small parameter multiplies a derivative term of some state variable. In such cases the domain can be divided into outer regions, where the state variable changes gradually and the derivative term is negligible, and inner regions, where the state variable changes abruptly and the derivative term cannot be neglected. The method consists of solving the dynamic equations separately in the outer and in the inner regions, where different approximations can be made, and matching the solutions.

Chapter 7

Basic mechanisms of pattern formation

7.1 Non-uniform instabilities of uniform states

The growth of a spatially structured mode at an instability point of a uniform state is a fundamental mechanism of pattern formation. As discussed in Section 5.3.3, there are two general cases leading to the growth of spatially structured modes, a non-uniform stationary instability leading to stationary periodic patterns, and a non-uniform oscillatory instability leading to traveling-wave patterns. We use here the amplitude equation approach presented in the previous chapter to analyze these two types of patterns close to their onset. In Section 7.1.1 we consider a variant of the SH equation that breaks the inversion symmetry $u \to -u$ and allows for stationary hexagonal patterns in addition to stripe patterns. Using symmetry consideration we motivate the form of the amplitude equations and use them to study two types of hexagonal patterns and the interaction between hexagonal and stripe patterns.

In Section 7.1.2 we consider the amplitude equations for a pair of traveling waves propagating in opposite directions, and study the wave patterns that can result from the interaction between the two.

Farther from onset these periodic patterns may still persist, as Figure 6.6(a,b) shows for stationary periodic solutions of the FHN model. The patterns remain periodic, but lose their sinusoidal form by developing narrow transition layers. Farther yet from onset, the transition layers become sharper and the patterns are not necessarily periodic. In fact, they may lack any characteristic scale of the up-state or down-state domains and appear "scale free." In Section 7.1.3 we study such patterns, by analyzing the dynamics of the transition layers, and show that the lack of a characteristic scale is related to the development of a global constraint. The analysis sheds light on a similar global constraint that develops in the context of dryland vegetation and leads to wide patch-size distributions (see Section 10.2.2).

7.1.1 Stationary patterns

The appearance of stripe patterns beyond a stationary non-uniform instability and their stability properties have been studied in Sections 5.3.1, 6.1.3.1 and 6.2.1 by analyzing the SH equation. Numerical simulations of the SH equation with stripe-like initial conditions confirm the analytical results. However, starting with random initial conditions about the unstable uniform state reveals more intricate dynamics as Figure 5.1 shows. Because of the system's isotropy, there exists a continuous family of stripe solutions with wave-vectors, $\mathbf{k}$, spanning the circle $|\mathbf{k}|^2 = k_x^2 + k_y^2 = k_0^2$, as Figure 7.2a illustrates. Thus, random initial conditions initiate patterns consisting of small domains of parallel stripes that are oriented in different directions, as the left-most panel of Figure 5.1 shows. On longer time scales these domains grow in size to form ever larger domains of parallel stripes, until a stripe pattern with a well defined wave-vector is formed (Figure 5.1) [76]. This is not the only possible scenario, however [291]. Random initial conditions beyond a stationary non-uniform instability can also lead to asymptotic labyrinthine patterns as Figure 7.1 demonstrates in simulations of the symmetric FHN model.

The stripe patterns discussed above result from the local growth of a single mode[1]. As we have already pointed out, the instability of the uniform state involves the growth of a continuous family of modes ($|\mathbf{k}| = k_0$), which may lead to patterns involving several modes at each point in space rather than a single mode. Whether such multi-mode patterns are feasible or not depends on the nonlinear interactions between the growing modes. Because of the symmetry between the modes, each mode either acts to damp the other modes or acts to enhance the other modes. The latter possibility amounts to a positive feedback

[1]Since physical quantities are represented by real valued variables, any Fourier mode they contain must be accompanied by its complex conjugate. Thus, along with a mode of a wave-vector $\mathbf{k}$ there also exists a mode of a wave-vector $-\mathbf{k}$. We nevertheless refer to this pair of modes as representing a single mode.

FIGURE 7.1: A simulation of the FHN model in a range of fast inhibitor diffusion, showing the development of an asymptotic labyrinthine pattern from random initial conditions. The snapshots correspond (from left to right) to times $t = 0,\ 100,\ 500,\ 1500,\ 5000$, and indicate a fast convergence to a labyrinthine pattern with small slow adjustments as time proceeds. Parameters: $a_0 = 0$, $a_1 = 0.9$, $\epsilon = 1.5$, $\delta = 10$.

that may result in a multi-mode pattern. In the following we will show that any triad of modes on the circle $|\mathbf{k}| = k_0$ that satisfy the condition $\mathbf{k}_1 + \mathbf{k}_2 + \mathbf{k}_3 = 0$ (see Figure 7.2b) resonate in a way that enhance their simultaneous growth.

Consider the SH equation with a quadratic term which breaks its inversion symmetry $u \to -u$,

$$\partial_t u = \lambda u + \alpha u^2 - u^3 - \left(\nabla^2 + k_0^2\right)^2 u\,, \tag{7.1}$$

and an approximate solution of the form

$$u(\mathbf{r},t) \approx A_1 e^{i\mathbf{k}_1\cdot\mathbf{r}} + A_2 e^{i\mathbf{k}_2\cdot\mathbf{r}} + A_3 e^{i\mathbf{k}_3\cdot\mathbf{r}} + c.c.\,, \tag{7.2}$$

where $|\mathbf{k}_i| = k_0$, $i = 1,2,3$. Equations for the amplitudes A_i can be derived using a multiple-scale analysis as we did in the case of the NWS equation. Here, we will motivate the general form of the amplitude equations using symmetry considerations (see Section 6.1.2.1).

The symmetry of (7.1) to translations, $\mathbf{r} \to \mathbf{r} + \Delta\mathbf{r}$, implies that the amplitude equation for A_1 should be invariant under the transformations $A_1 \to A_1 \exp(i\mathbf{k}_1 \cdot \Delta\mathbf{r})$, $A_2 \to A_2 \exp(i\mathbf{k}_2 \cdot \Delta\mathbf{r})$, $A_3 \to A_3 \exp(i\mathbf{k}_3 \cdot \Delta\mathbf{r})$, and the same argument applies to the other two amplitude equations. As a consequence, the allowed terms on the right side of the amplitude equation can include the quadratic term $A_2^* A_3^*$ with the extra condition $\mathbf{k}_1 = -(\mathbf{k}_2 + \mathbf{k}_3)$, but no other quadratic terms. Since the three wave-vectors, $\mathbf{k}_1, \mathbf{k}_2, \mathbf{k}_3$, have the same magnitude, k_0, near the instability of the zero state, they must be equally spaced at angles $2\pi/3$, as Figure 7.2b illustrates. Such a combination of modes can give rise to hexagonal patterns as we will shortly see. The translational symmetry also implies that among all possible cubic terms only terms of the form $|A_j|^2 A_i$ in the equation for A_i are allowed. The symmetry of (7.1) to parity transformations, $\mathbf{r} \to -\mathbf{r}$, implies that the amplitude equations should be invariant under the transformation $A_i \to A_i^*,\quad i = 1,2,3$, which requires the coefficients of all terms in the amplitude equations to be real valued.

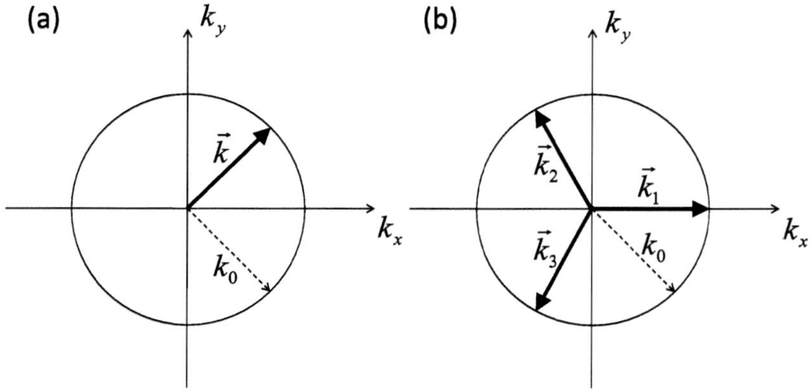

FIGURE 7.2: Wave-vectors of growing modes in the case of stripe patterns (a) and hexagonal patterns (b).

These symmetry considerations motivate the amplitude equations [55, 136]:

$$\dot{A}_1 = \lambda A_1 - a(|A_1|^2 + b|A_2|^2 + b|A_3|^2)A_1 + cA_2^*A_3^* \quad (7.3a)$$
$$\dot{A}_2 = \lambda A_2 - a(|A_2|^2 + b|A_1|^2 + b|A_3|^2)A_2 + cA_1^*A_3^* \quad (7.3b)$$
$$\dot{A}_3 = \lambda A_3 - a(|A_3|^2 + b|A_1|^2 + b|A_2|^2)A_3 + cA_1^*A_2^*\,, \quad (7.3c)$$

where a, b and c are real valued constants[2]. A multiple scale analysis of the SH equation (7.1) gives the particular values: $a = 3$, $b = 2$, $c = 2\alpha$. In deriving these equations we restricted our attention to solutions described by spatially uniform amplitudes [193]. Note that in the presence of an inversion symmetry $u \to -u$, which requires α in (7.1) to vanish, the coefficient c of the quadratic terms in (7.3) should vanish too. This is because of the required invariance of the amplitude equations under the transformation $A_i \to -A_i$, $i = 1, 2, 3$, implied by the inversion symmetry. We will soon see that the quadratic terms, and therefore the absence of an inversion symmetry, are essential for the stability of hexagonal patterns.

Equations (7.3) can be used to study the existence and stability ranges of both stripe and hexagonal patterns in the vicinity of the instability point of the zero state ($\lambda = 0$). Consider first stripe solutions for which only one amplitude is nonzero, e.g., $A_1 = A_{10}$ and $A_2 = A_3 = 0$, where $A_{10} = \sqrt{\lambda/a}\exp(i\phi)$ and ϕ is an arbitrary constant phase. This solution, and the other two solutions obtained by cyclic permutations of the indices, exist for $\lambda \geq 0$. To study the stability of these solutions we linearize (7.3) around $(A_1, A_2, A_3) = (A_{10}, 0, 0)$. Note that (7.3) represents a six-dimensional dynamical system, consisting of A_1, A_2, A_3 and their complex conjugates, and the linearization results in a

[2]The constant a can be eliminated by rescaling the amplitudes, but we choose to avoid this rescaling in order to leave the amplitudes in (7.2) unchanged.

6×6 Jacobian matrix $\mathbf{J}$ (see Section 2.2.2). Fortunately, the Jacobian matrix has a block diagonal form consisting of a 2×2 block, $\mathbf{J}_1$, related to A_1, A_1^*, and a 4×4 block, $\mathbf{J}_2$, related to A_2, A_2^* and A_3, A_3^*. Solving the characteristic equation

$$\det(\mathbf{J}_1 - \sigma_1 \mathbf{I}) = (\lambda - 2a|A_{10}|^2 - \sigma)^2 - a^2|A_{10}|^4 = 0\,, \tag{7.4}$$

we find the eigenvalues $\sigma_1^+ = 0$ and $\sigma_1^- = -2\lambda$, where the zero eigenvalue is related to the marginal stability of stripe patterns to phase modulations or space translations. The characteristic equation

$$\det(\mathbf{J}_2 - \sigma_2 \mathbf{I}) = \left[(\lambda - ab|A_{10}|^2 - \sigma)^2 - c^2|A_{10}|^2\right]^2 = 0\,, \tag{7.5}$$

gives the eigenvalues $\sigma_2^\pm = (1-b)\lambda \pm c\sqrt{\lambda/a}$, each of multiplicity 2. While $\mathbf{J}_1$ cannot contribute positive eigenvalues, $\mathbf{J}_2$ does contribute a positive eigenvalue (of multiplicity 2), σ_2^+, implying that the stripe solutions are unstable, whenever they appear, to the growth of the other two modes. Note that for $b > 1$ the sign of $\sigma_2^+ = (1-b)\lambda + c\sqrt{\lambda/a}$ changes to negative at sufficiently high λ values, implying the stabilization of stripe solutions. This is the case of the SH model for which $b = 2$; the stabilization of the stripe solutions occurs for $\lambda > \lambda_s$, where $\lambda_s = 4\alpha^2/3$.

Consider now solutions of (7.3) for which all three amplitudes are nonzero. We could proceed as before but it is more convenient in this case to express (7.3) in terms of the magnitude, $\rho_i = |A_i|$, and the phase, $\phi_i = \arg A_i$, of each amplitude. This leads to

$$\dot{\rho_1} = \lambda\rho_1 - a(\rho_1^2 + b\rho_2^2 + b\rho_3^2)\rho_1 + c\rho_2\rho_3\cos\Phi\,, \tag{7.6a}$$

$$\rho_1\dot{\phi_1} = -c\rho_2\rho_3\sin\Phi\,, \tag{7.6b}$$

where $\Phi = \phi_1 + \phi_2 + \phi_3$, and similar equations for $\rho_2,\ \phi_2$ and $\rho_3,\ \phi_3$, obtained from (7.6) by cyclic permutation of the indices. One advantage of this form is that the three phase equations are reducible to a single equation for Φ:

$$\dot{\Phi} = -cW\sin\Phi\,, \qquad W = \rho_1\rho_2\rho_3\left(\frac{1}{\rho_1^2} + \frac{1}{\rho_2^2} + \frac{1}{\rho_3^2}\right). \tag{7.7}$$

Another advantage is the decoupling of the phase and magnitude equations in the linear stability analysis, as we will shortly see. Equation (7.6a), its cyclic permutations, and equation (7.7) have constant solutions of the form

$$\rho_1 = \rho_2 = \rho_3 = \rho_0\,, \qquad \Phi = 0\,, \tag{7.8}$$

where ρ_0 satisfies

$$a(2b+1)\rho_0^2 - c\rho_0 - \lambda = 0\,, \tag{7.9}$$

and

$$\rho_1 = \rho_2 = \rho_3 = \rho_\pi\,, \qquad \Phi = \pi\,, \tag{7.10}$$

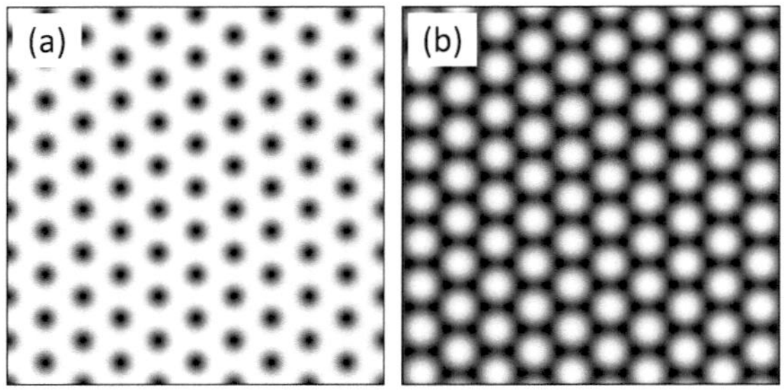

FIGURE 7.3: Two types of hexagonal patterns, 0-hexagons (a) and π-hexagons (b). Shown is the approximate solution (7.2) with $|A_i| = \rho_i$ and $\arg(A_i) = \phi_i$ satisfying (a) equation (7.8) and $\phi_1 = \phi_2 = \phi_3 = 0$, (b) equation (7.10) and $\phi_1 = \phi_2 = 0$, $\phi_3 = \pi$.

where ρ_π satisfies

$$a(2b+1)\rho_\pi^2 + c\rho_\pi - \lambda = 0\,. \tag{7.11}$$

These solutions describe two types of hexagonal patterns as Figure 7.3 shows. They are often referred to as 0-hexagon solutions ($\Phi = 0$) and π-hexagon solutions ($\Phi = \pi$). To study their stability we first observe that the linearization of (7.7) around any of them decouples the phase deviation, $\delta\Phi$ from the magnitude deviations, $\delta\rho_i$, and leads to

$$\delta\dot{\Phi} = \mp cW\delta\Phi\,, \tag{7.12}$$

where the minus and plus signs refer to deviations from $\Phi = 0$ and from $\Phi = \pi$, respectively. Since $W > 0$, the solution (7.10) ($\Phi = \pi$) is always unstable for $c > 0$, and the solution (7.8) ($\Phi = 0$) is always unstable for $c < 0$. Note that for both signs of c the term $c\cos\Phi$ in (7.6a) is positive for the stable solutions. This shows that for stable solutions any pair of the three modes act to enhance the amplitude of the third mode. For the sake of concreteness let us assume that $c > 0$. A similar analysis can be done for $c < 0$.

We now address the stability of (7.8) to magnitude perturbations. Insertion of the phase-stable solution $\Phi = 0$ in (7.6a) and its cyclic permutations results in a three-dimensional dynamical system for the magnitudes ρ_1, ρ_2, ρ_3. The Jacobian matrix, $\mathbf{J}$, with respect to the solution $\rho_1 = \rho_2 = \rho_3 = \rho_0$ has a simple form: all diagonal elements are equal and given by $J_{ii} = \lambda - 3a\rho_0^2 - 2ab\rho_0^2$, and all off-diagonal elements are equal and given by $J_{ij} = -2ab\rho_0^2 + c\rho_0$. Solving the characteristic equation, $\det(\mathbf{J} - \sigma\mathbf{I}) = 0$, for the eigenvalues σ we find an eigenvalue, $\sigma_1 = \lambda - c\rho_0 - 3a\rho_0^2$, of multiplicity two, and another eigenvalue, $\sigma_2 = \lambda + 2c\rho - 3a(1+2b)\rho_0^2$.

Let us apply these results to the SH equation for which $a = 3$, $b = 2$ and $c = 2\alpha$ ($\alpha > 0$). The two solutions of the quadratic equation (7.9) are

$$\rho_0^{\pm} = \frac{1}{15}\left(\alpha \pm \sqrt{\alpha^2 + 15\lambda}\right) . \tag{7.13}$$

The solution ρ_0^+ exists for $\lambda > \lambda_f$, where $\lambda_f = -\alpha^2/15$, while the solution ρ_0^- exists for $\lambda_f < \lambda < 0$. The eigenvalues associated with these solutions are

$$\sigma_1 = \lambda - 2\alpha\rho_0^{\pm} - 9\rho_0^{\pm 2} , \tag{7.14a}$$

$$\sigma_2 = \lambda + 4\alpha\rho_0^{\pm} - 45\rho_0^{\pm 2} . \tag{7.14b}$$

Consider first the solution ρ_0^-. It is straightforward to show that $\sigma_2 > 0$ in its entire existence range[3], implying that the solution is always unstable. For the solution ρ_0^+, the eigenvalue σ_2 is always negative, except at $\lambda = \lambda_f$ where $\sigma_2 = 0$. The eigenvalue σ_1, on the other hand, is negative for $\lambda < \lambda_h$, where $\lambda_h = 16\alpha^2/3$, but becomes positive as λ is increased past λ_h, implying an instability of the solution ρ_0^+. As discussed above, the π-hexagon solutions are always unstable to phase modulations for the choice $\alpha > 0$. Their analytical form is given by $\rho_i = \rho_\pi = (-\alpha + \sqrt{\alpha^2 + 15\lambda})/15$ ($i = 1, 2, 3$) and by the phase constraint $\phi_1 + \phi_2 + \phi_3 = \pi$, and they appear at $\lambda = 0$.

The results obtained above for stripe and hexagon solutions are summarized in the bifurcation diagram shown in Figure 7.4. The zero state is destabilized at $\lambda = 0$ to 0-hexagon solutions in a subcritical bifurcation, as stable hexagon solutions appear already at $\lambda_f = -\alpha^2/15$ in a fold bifurcation. Two other solutions appear at $\lambda = 0$, stripe solutions and π-hexagon solutions. Both are unstable when they appear, but the stripe solutions become stable at $\lambda = \lambda_s = 4\alpha^2/3$. The π-hexagons remain unstable in the entire validity range of the amplitude equations (7.3). The 0-hexagon solutions become unstable at $\lambda_h = 16\alpha^2/3$, in a range where the stripe solutions are stable.

The analysis of this section is based on universal amplitude equations for systems with broken inversion symmetry in which a zero state undergoes a non-uniform stationary instability. We may therefore expect such systems to go through a universal sequence of states as the control parameter is increased: zero state, hexagonal patterns, stripe patterns. Because of the bistability ranges, $\lambda_f < \lambda < 0$ of the zero state and hexagons, and $\lambda_s < \lambda < \lambda_h$ of hexagons and stripes, we also expect the transitions between these states to involve hysteresis (see Section 2.2.3). In Chapter 10.1.1 we will relate these findings to the sequence of vegetation states along environmental gradients.

It is instructive to discuss the limit $\alpha \to 0$ in which the system regains inversion symmetry. In this limit the range $\lambda_f < \lambda < \lambda_h$ of stable hexagon solutions and the range $0 < \lambda < \lambda_s$ of unstable stripe solutions both shrink to zero. The zero state then loses stability to stripe solutions in a supercritical

[3]The eigenvalue $\sigma_2(\lambda)$ for $\rho_0 = \rho_0^-$ has two zeros at $\lambda = \lambda_f$ and $\lambda = 0$ and a single maximum in between.

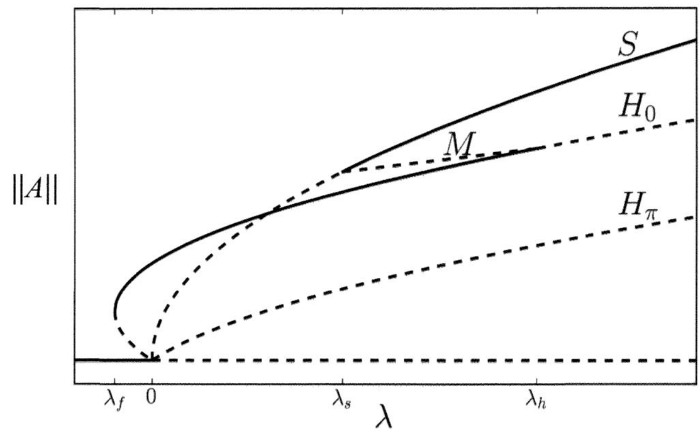

FIGURE 7.4: A bifurcation diagram for the amplitude equations (7.3), where $\|A\|$ represents a rescaled form of the $L2$ norm ($\sqrt{|A_1|^2 + |A_2|^2 + |A_3|^2}$) that graphically improves the presentation of the different solution branches. Solid (dashed) lines represent stable (unstable) solutions. The labels H_0 and H_π denote 0-hexagon and π-hexagon solutions, respectively (see Figure 7.3), S denotes a stripe solution, and M stands for an unstable mixed hexagonal-stripe solution. The diagram shows a universal sequence of stable states along the bifurcation parameter axis: uniform (zero) state ($\lambda < 0$), hexagonal pattern (H_0: $\lambda_f < \lambda < \lambda_h$), and stripe pattern ($S$: $\lambda > \lambda_s$). Note the bistability ranges of the uniform state and the hexagonal pattern ($\lambda_f < \lambda < 0$), and of the hexagonal pattern and the stripe pattern ($\lambda_s < \lambda < \lambda_h$).

bifurcation as discussed in Section 6.2.1. In many systems, broken inversion symmetry is the rule rather than the exception, which explains the common appearance of hexagonal patterns. In thermal convection (see Section 2.1.2) broken symmetry of this kind appears for fluids with temperature-dependent properties [25].

7.1.2 Traveling-wave patterns

In Section 7.1.1 we studied patterns that emerge beyond a non-uniform stationary instability. Here we consider the case of a non-uniform oscillatory instability (see Section 5.3.3) beyond which traveling waves appear, and confine ourselves to 1d. We assume a system with a parity symmetry, $x \to -x$, in which along with a wave that propagates to the right there also exists a symmetric wave that propagates to the left. Close to the instability point a typical state variable can be approximated by (6.4) or

$$u \approx A_R e^{ik_0x - i\omega_0 t} + A_L e^{-ik_0x - i\omega_0 t} + c.c.,$$

where A_R and A_L are the amplitudes of right and left traveling waves. The general form of the equations for these amplitudes can be deduced using symmetry considerations (see Section 6.1.2.1) and the actual form for a particular system can be derived using the method described in Section 6.1. The outcome, assuming space independent amplitudes, can be brought to the form

$$\dot{A_R} = \lambda A_R - (1+ib)|A_R|^2 A_R - g(1+ic)|A_L|^2 A_R\,, \tag{7.15a}$$
$$\dot{A_L} = \lambda A_L - (1+ib)|A_L|^2 A_L - g(1+ic)|A_R|^2 A_L\,, \tag{7.15b}$$

where λ, b, c and g are real valued parameters[4].

To study solutions of (7.15) that represent traveling waves, it is more convenient to transform these equations to equations for the magnitudes $\rho_{R,L} = |A_{R,L}|$ and for the arguments $\phi_{R,L} = \arg(A_{R,L})$ of the amplitudes. Inserting in (7.15) $A_{R,L} = \rho_{R,L}\exp(i\phi_{R,L})$ we obtain

$$\dot{\rho}_R = \lambda\rho_R - \rho_R^3 - g\rho_L^2\rho_R\,, \tag{7.16a}$$
$$\dot{\phi}_R = -b\rho_R^2 - cg\rho_L^2\,, \tag{7.16b}$$
$$\dot{\rho}_L = \lambda\rho_L - \rho_L^3 - g\rho_R^2\rho_L\,, \tag{7.16c}$$
$$\dot{\phi}_L = -b\rho_L^2 - cg\rho_R^2\,. \tag{7.16d}$$

The decoupling of the magnitude dynamics (7.16)(a,c) from the phase dynamics (7.16)(b,d) reduces the dimension of the system to two and simplifies the analysis.

Right traveling waves are represented by solutions of (7.16) of the form $\rho_R = \rho_T$, $\rho_L = 0$. Using this form in (7.16) we can find expressions for ρ_T and for ϕ_R. The results, expressed in terms of the amplitudes, are

$$A_R = \rho_T\exp(-i\Omega_T t)\,, \qquad A_L = 0\,, \tag{7.17a}$$
$$\rho_T = \sqrt{\lambda}\,, \qquad \Omega_T = b\lambda\,. \tag{7.17b}$$

A similar symmetric form describes left traveling waves. Equal magnitudes of both traveling waves describe standing waves. Inserting $\rho_R = \rho_L = \rho_S$ in (7.16) we obtain

$$A_R = A_L = \rho_S\exp(-i\Omega_S t)\,, \tag{7.18a}$$
$$\rho_S = \sqrt{\frac{\lambda}{1+g}}\,, \qquad \Omega_S = (b+cg)\rho_S^2\,. \tag{7.18b}$$

The stability of these traveling and standing wave solutions can be studied by calculating the eigenvalues of the Jacobian matrix associated with (7.16)

[4]Note that in bringing the amplitude equations to this compact form we generally have to make a few transformations such as rescaling the amplitudes. For simplicity we keep the same notations A_R and A_L for the transformed amplitudes.

For traveling waves we obtain the eigenvalues $\sigma_{T1} = -2\lambda$, $\sigma_{T2} = \lambda(1-g)$ and $\sigma_{T3} = \sigma_{T4} = 0$. For standing waves the eigenvalues are $\sigma_{S1} = -2\lambda$, $\sigma_{S2} = 2\lambda(g-1)/(g+1)$ and $\sigma_{S3} = \sigma_{S4} = 0$. The zero eigenvalues are associated with the phase variables and are related to the symmetry of the system under space and time translations. Assuming $\lambda > 0$, these results imply that traveling waves exist for all g values but are stable only for $g > 1$, while standing waves exist for $g > -1$ but are stable only in the range $-1 < g < 1$. We refer the reader to Ref. [55] for an additional account of traveling wave patterns.

7.1.3 Scale-free patterns

The general results of the analysis described in the Section 7.1.1 apply to the FHN model (5.37) too. In the symmetric case, $a_0 = 0$, for which the model has the inversion symmetry $(u, v) \to (-u, -v)$, the stationary patterns that emerge beyond the non-uniform stationary (Turing) instability are stripes or labyrinths. When this symmetry is broken hexagonal patterns appear. In both cases the patterns that develop have a characteristic length, the width of the stripes that make a labyrinthine pattern, or the diameter of the spots that make an hexagonal pattern. An interesting aspect of the FHN model is that it allows to study a parameter range, far beyond the Turing instability, where global spatial coupling develops. In that range the system lacks a characteristic length in a sense that will become clear in the following.

Consider case (b) of the FHN model (see Figure 5.2) in the limit of extremely fast inhibitor diffusion, $\epsilon/\delta \to 0$, and non-oscillatory behavior, $\epsilon > \epsilon_H \sim O(1)$. Assuming periodic or no-flux boundary conditions ($\hat{n} \cdot \nabla v = 0$ where $\hat{n}$ is a unit vector normal to the system boundary), the fast diffusion acts to homogenize the inhibitor v and therefore, apart of short initial transients, we can assume that $v = v(t)$ is independent of space. Averaging (5.37b) over space, the FHN model becomes

$$\partial_t u = u - u^3 - v + \nabla^2 u\,, \quad (7.19a)$$

$$\dot{v} = \epsilon(\langle u \rangle - a_0 - a_1 v)\,. \quad (7.19b)$$

Here $\langle u \rangle = A^{-1} \int_\Omega u d\mathbf{r}$ is the spatial average of u, where Ω is the system's domain and A is the domain's area. Note that $\delta\nabla^2 v$ is not necessarily small and the absence of this term in (7.19b) is due to the spatial averaging and the boundary conditions. The term $\langle u \rangle$ introduces global spatial coupling. Global coupling is an inherent property of systems satisfying conservation laws, such as two-phase mixtures with mass conservation [232, 201, 243]. It is also an inherent property of certain pattern forming systems, such as catalytic surface reactions that involve chemically reactive agents in the gas phase above the catalytic surface [86]. Here, global coupling is not an inherent property of the system; it is rather a property that develops in a limited parameter range that corresponds to fast inhibitor diffusion relative to inhibitor production. In Section 10.2.2 we will discuss physical and ecological conditions under which

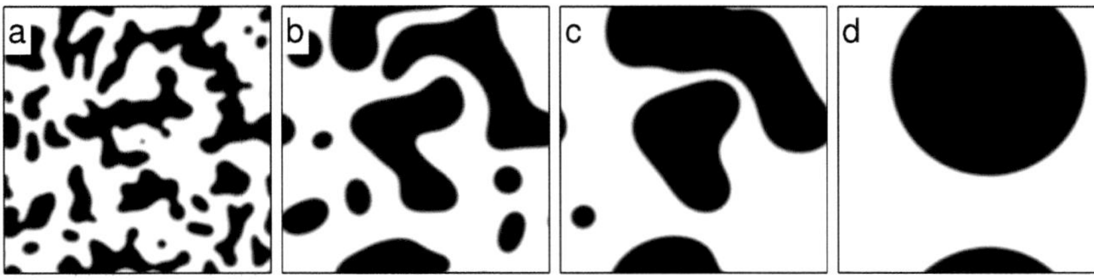

FIGURE 7.5: Development of scale free patterns (a) followed by phase coarsening (b,c) and the convergence to a single circular domain (d) in (7.19). Parameters: $a_0 = -0.15$, $a_1 = 0.5$, $\epsilon = 3.0$. Snapshots were taken at $t = 15$ (a), $t = 100$ (b), $t = 250$ (c) and $t = 9250$ (d). From [156].

water-limited vegetation becomes globally coupled, and draw an interesting analogy to the FHN model.

Numerical simulations of (7.19), starting with random initial conditions about the single equilibrium solution (u_0, v_0) reveal a remarkably different behavior than that described in Section 6.3.2 near and farther away from the Turing instability (Figure 6.6(a,b) and Figure 7.1). Instead of Turing patterns with a clear characteristic length, amorphous scale-free patterns with sharp transition layers appear, as Figure 7.5(a) shows. However, on a longer time scale a phase coarsening process begins (Figure 7.5(b-d)) whereby small spots shrink and disappear while big spots grow and merge together. During this process the spot boundaries smooth out and approach circular forms.

To gain a better understanding of the emergence of scale-free patterns and the phase-coarsening processes that follow the emergence, let us analyze (7.19) in 1d, focusing first on the dynamics of a single transition layer or front. We will assume that $|v| \ll 1$ and test the validity of this assumption a posteriori. Equation (7.19a) has front solutions that separate domains of high and low u values ("up-state" and "down-state" domains), as we have already discussed in Sec. 6.3.2 (see (6.95)). Let p be the position of a front solution. For $v = 0$ the front is stationary and p is constant. We may therefore expect the front to be slowly propagating for $|v| \ll 1$. We quantify the smallness of v by introducing a small auxiliary parameter $\mu \ll 1$, and describe the slow front propagation by letting the position to be time dependent, $p = p(T)$, where $T = \mu t$ is a slow time coordinate. Introducing a moving coordinate frame, $z = x - p(T)$, and expressing (7.19a) in terms of $u(z, T)$ and $v(T)$, we obtain

$$\mu \partial_T u - \mu \frac{dp}{dT} \partial_z u = u - u^3 - v + \partial_z^2 u \,. \tag{7.20}$$

For simplicity we use the same notations for u and v even though they are now expressed as functions of the new coordinates z and T.

We now expand u and v as

$$u(z,T) = u_0(z) + \mu u_1(z,T) + \dots\,, \qquad v(T) = \mu v_1(T) + \dots\,. \tag{7.21}$$

This form reflects the understanding that the main effect of v is to induce front propagation, with only a slight change (μu_1) in the spatial profile of the front. This is because of the existence of a marginal translation mode, which is "activated" by the asymmetry between the up-state and the down-state that v induces. Inserting the expansion (7.21) into (7.20) we obtain at order unity,

$$\frac{d^2 u_0}{dz^2} + u_0 - u_0^3 = 0\,. \tag{7.22}$$

Equation (7.22) has front solutions,

$$u_0 = \pm \tanh(z/\sqrt{2})\,, \tag{7.23}$$

which are biasymptotic to the up and down states, $u_+ = 1$ and $u_- = -1$, respectively.

At order μ we obtain

$$\mathcal{L} u_1 = v_1 - \frac{dp}{dT}\frac{du_0}{dz}\,, \tag{7.24}$$

where

$$\mathcal{L} \equiv \frac{\partial^2}{\partial z^2} + 1 - 3u_0^2\,. \tag{7.25}$$

The reader may recognize already that the right side of (7.24) contains a term proportional to the translation mode du_0/dz (see (5.61)). If the translation mode belongs to the null space of $\mathcal{L}^\dagger$, this term is secular and should be projected out in order for μu_1 to represent a small correction to u_0, as we expect (see Sec. 6.1.2.3). This is indeed the case, as we now show. Consider a vector space of complex-valued functions with space derivatives that asymptote to zero as $z \to \mp\infty$, and an inner product $\langle f, g\rangle = \int_{-\infty}^{\infty} f^* g dz$. The operator $\mathcal{L}$ defined in (7.25) is then self adjoint. Taking the derivative of equation (7.22) with respect to z we find

$$\mathcal{L}\left(\frac{du_0}{dz}\right) = \mathcal{L}^\dagger\left(\frac{du_0}{dz}\right) = 0\,. \tag{7.26}$$

Thus, the translation mode indeed belongs to the null space of $\mathcal{L}^\dagger$ and we therefore demand the right side of (7.24) to be orthogonal to du_0/dz. This leads to

$$\frac{dp}{dT} = b v_1\,, \qquad b = \frac{\int_{-\infty}^{\infty} u_0' dz}{\int_{-\infty}^{\infty} (u_0')^2 dz}\,, \tag{7.27}$$

where $u_0' = du_0/dz$, or in terms of the original time

$$\dot{p} = bv\,. \tag{7.28}$$

For a front that is biasymptotic to $u_\pm$ as $z \to \mp\infty$, b is negative. Using (7.23) we find for such a front that $b = -3/\sqrt{2}$. For a front that is biasymptotic to $u_\mp$ as $z \to \mp\infty$, $b = 3/\sqrt{2}$.

These results for the dynamics of a single front in 1d shed light on the properties of patterns consisting of alternate up and down state domains. In order for such patterns to be stationary we must require that the fronts that bound these domains are stationary too, or, using (7.19b) and (7.28), that

$$v = 0\,, \qquad \langle u \rangle = a_0\,. \tag{7.29}$$

Thus, stationary pattern solutions should satisfy the global constraint $\langle u \rangle = a_0$. To evaluate the stability of such solutions we first argue that $\langle u \rangle$ is a decreasing function of v. The dependence of $\langle u \rangle$ on v comes through the values of the up and down states, $u_\pm(v)$, and through the widths of the up and down-state domains. Solving $u - u^3 - v = 0$ for u assuming $|v| \ll 1$ we find $u_\pm = \pm 1 - v/2$, thus $u_\pm$ are decreasing functions of v. The widths of the up-state domains are also decreasing functions of v, because the larger v the faster the invasion of a down-state domain into an upstate domain ($\dot{p} < 0$ for a front biasymptotic to $u_\pm$ as $z \to \mp\infty$, and $\dot{p} > 0$ for a front biasymptotic to $u_\mp$ as $z \to \mp\infty$). For the same reason the widths of the down-state domains are increasing functions of v. Since decreasing the values of the up and down states, decreasing the widths of the up-state domains, and increasing the widths of the down-state domains all lower the value of $\langle u \rangle$, we conclude that $\langle u \rangle \equiv h(v)$ must be a decreasing function. Writing (7.19b) as $v_t = f(v)$ where $f(v) = \epsilon(h - a_1 v - a_0)$ we find that $f'(0) < 0$ which implies the linear stability of the stationary solutions (7.29) to small deviations of v. This observation justifies the assumption $|v| \ll 1$; if $|v|$ is sufficiently small it will remain small because it decays to zero.

The analysis described above is based on the dynamics of a single front and does not take into account front interactions. These interactions, however, are exponentially weak [148, 41], and become significant only for very small domains. As Figure 7.6 demonstrates, apart from transients during which the fronts that bound sufficiently small domains attract and lead to the disappearance of the domains, the dynamics freeze[5] on an apparently stationary pattern dictated by the initial conditions. Many frozen patterns of this kind are possible; the only requirement is that they share the same average value, $\langle u \rangle = a_0$. In fact, the range of possible frozen patterns also includes periodic patterns provided they satisfy this global constraint.

While long-lasting scale-free patterns can develop in 1d, the dynamics in two dimensions clearly show a relatively fast phase coarsening process (Figure 7.5) that results asymptotically in a single circular domain or a single

---

[5] Strictly speaking, stationary patterns are unstable; the attractive front interactions lead to a phase coarsening process in which nearby attracting fronts collide and annihilate one another until a single front is left. In practice, however, this process is extremely slow, as the average domain size grows like $\ln t$, and in the presence of small spatial noise can often be halted by front pinning.

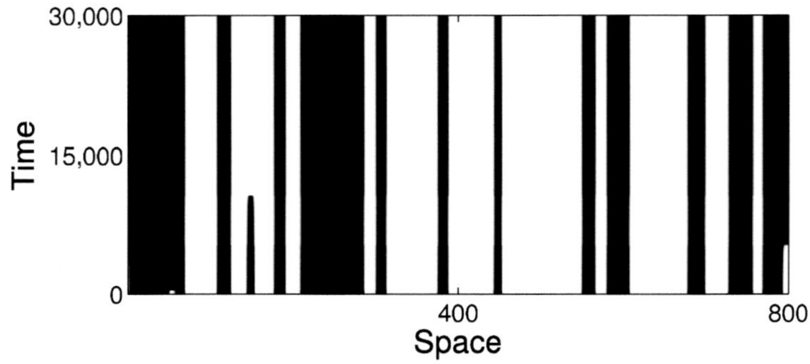

FIGURE 7.6: A space-time plot, obtained by numerical integration of (7.19a) and (7.19b), showing the development of a frozen scale-free pattern in the limit $\delta \to \infty$ of a one-dimensional system. The initial conditions consist of pink noise perturbations about the unstable equilibrium state (u_0, v_0). Parameters are as in Figure 7.5. From [156].

front. The major difference between the two cases is the curvature-induced front motion in 2d (see Sec. 7.3.1). Phase coarsening phenomena have extensively been studied in the context of two-phase mixtures, where they are commonly referred to as Ostwald ripening [326]. They are characterized by power-law growth forms of the average domain radius $\bar{R}$. A power law $\bar{R} \sim t^{1/2}$ is known to exist for (5.37a) with a global constraint, resulting, for example, from mass conservation [210]. In Section 10.2.2.3 we will discuss a similar Ostwald ripening process in vegetation patterns. In that context even this relatively fast process may be too slow to observe.

7.2 Multiplicity of stable states and localized structures

The multiplicity of stable uniform states, either stationary or oscillatory, is another fundamental mechanism of pattern formation; as local domains converge to different states, spatial patterns develop. Although multiplicity of three and more stable states can be realized in practice (see Section 8.2.1), more common are bistable systems, i.e., systems that can be realized in two alternative stable states over some parameter range. Three bistability types will be analyzed in Section 7.2.1. The first is the bistability of uniform states that gives rise to transient patterns, culminating in one of the two stable uniform states (Section 7.2.1.1). The second bistability type also involves two uniform states, but the interactions between fronts that separate domains of different states lead to stable asymptotic patterns (Section 7.2.1.2). The third bistability type involves a uniform state and a patterned state. In this case, the bistability range often contains a subrange where fronts are pinned in place and a multitude of stable hybrid states exists in addition to the original two stable states (Section 7.2.1.3).

Another class of multiplicity of stable uniform states is formed by uniform oscillatory instabilities. In this case the multiplicity is continuous and parameterized by the oscillation phase, which spans the circle. Initial phase gradients above the instability threshold give rise to long-wavelength traveling waves, spiral waves, and various localized structures as discussed in Section 7.2.2. Finally, there are circumstances in which a uniform state can go through two different instabilities, e.g., a non-uniform stationary instability and a uniform oscillatory instability. Mode interactions beyond both instabilities can lead to two alternative stable states, and to a variety of multi-mode localized structures as discussed in Section 7.2.3.

7.2.1 Bistable systems and fronts

7.2.1.1 Bistability of uniform states: Transient patterns

Bistability can arise in several ways, the most common of which are symmetry breaking bifurcations that lead to pairs of symmetric stable states (e.g., the pitchfork bifurcation), and subcritical bifurcations to bistability of asymmetric states. As an example, consider the equation

$$\partial_t u = \lambda u + \alpha u^2 - u^3 + \partial_x^2 u \,. \tag{7.30}$$

When $\alpha = 0$ the zero state goes through a pitchfork bifurcation at $\lambda = 0$ in which a pair of stable symmetric states appears, $u_\pm = \pm\sqrt{\lambda}$. When $\alpha \neq 0$ the instability of the zero state at $\lambda = 0$ is subcritical giving rise to bistability of the zero state with a nonzero state, $u_+ = \alpha/2 + \sqrt{(\alpha/2)^2 + \lambda} > 0$, in the range $-\alpha^2/4 < \lambda < 0$ (see Figure 2.8 with a replaced by $-\alpha$).

The building blocks of patterns in bistable systems are the transition layers or fronts[6] that separate domains of different states, as Figure 7.7(c) illustrates. Much of the overall pattern dynamics can be understood in terms of the dynamics of single fronts and the mutual interaction between pairs of adjacent fronts. These interactions determine, in particular, whether initial patterns are merely transients on the way toward one of the stable uniform states, or lead to asymptotic stable patterns.

To study the dynamics of a single front we first consider the variational equation,

$$\partial_t u = -\frac{dV}{du} + \partial_x^2 u = -\frac{\delta L}{\delta u}\,, \tag{7.31}$$

[6] A few other terms for "fronts" in bistable systems are in use in the literature, including "domain walls" and "kinks."

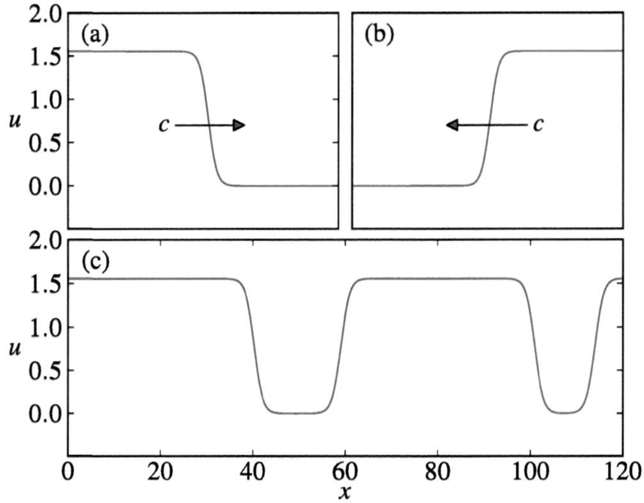

FIGURE 7.7: Fronts as building blocks of spatial patterns for the bistability of asymmetric uniform states. (a,b) Two symmetric front solutions of (7.30). Both fronts, which propagate in opposite directions at speed c, represent an invasion of the nonzero state into the zero state. (c) Alternating arrays of these fronts give rise to spatial patterns. From [16].

where the Lyapunov functional, L, is given by

$$L[u] = -\int_{\Omega} \left[V + \frac{1}{2} (\partial_x u)^2 \right] dx \,, \tag{7.32}$$

and $V = V(u; \lambda)$ is a double-well potential function having two minima, $u_{\pm}$, separated by a maximum u_0 in an appropriate parameter range. The two minima, hereafter the up (u_+) and down (u_-) states, correspond to the two stable uniform states of the bistable system. For example, the potential function associated with (7.30) is

$$V = -\frac{\lambda}{2} u^2 - \frac{\alpha}{3} u^3 + \frac{1}{4} u^4 + V_0 \,, \tag{7.33}$$

where V_0 is a constant reference potential, and has two minima, $u = u_{\pm}$, in the range $-\alpha^2/4 < \lambda < 0$, where

$$u_- = 0 \qquad u_+ = \frac{\alpha}{2} + \sqrt{\left(\frac{\alpha}{2}\right)^2 + \lambda} \,. \tag{7.34}$$

Consider front solutions of (7.31) that asymptote to the two stable states as $x \to \pm\infty$ and propagate at constant velocity c. In the moving frame $z = x - ct$ such solutions are time independent and satisfy the equation

$$\frac{d^2 u}{dz^2} + c\frac{du}{dz} - \frac{dV}{du} = 0 \,. \tag{7.35}$$

Note that (7.35) is invariant under the transformation $(z, c) \to (-z, -c)$. This implies that if $u(z)$ is a solution representing a front that asymptotes to $u = u_\pm$ as $z \to \mp\infty$ and propagates at velocity c (Figure 7.7(a)), then there also exists a symmetric solution $u(-z)$ representing a front that asymptotes to $u = u_\mp$ as $z \to \mp\infty$ and propagates at velocity $-c$ (Figure 7.7(b)). These front solutions are highly localized, tailing off to the two stable states exponentially fast.

The front velocity c is uniquely determined by the system's parameters. This can be readily understood by drawing an analogy to a particle of unit mass moving under the influence of a double-hump potential $-V$. In this analogy u represents the particle's position, z represents time, c represents a drag coefficient and $dV/du - cdu/dz$ is the total force (conservative plus drag) acting on the particle. A front solution that asymptotes to $u = u_\pm$ as $z \to \mp\infty$ and propagates at velocity c corresponds to a particle that has begun moving down the hump at u_+ in the far past and will be approaching the top of the lower hump at u_- in the far future. Obviously there is a unique value of the drag coefficient, c, for which the particle will reach the lower hump without passing it (i.e at zero velocity, $du/dz = 0$).

To find the unique velocity value we multiply equation (7.35) by du/dz and integrate over the whole line. Using integration by parts and the fact that $du/dz \to 0$ as $z \to \pm\infty$ we obtain

$$c = A^{-1} \int_{-\infty}^{\infty} \frac{dV}{du}\frac{du}{dz}dz = A^{-1} \int_{u_+}^{u_-} \frac{dV}{du}du = A^{-1}[V(u_-) - V(u_+)]\,, \qquad (7.36)$$

where

$$A = \int_{-\infty}^{\infty} u'^2 dz \qquad u' = du/dz\,. \qquad (7.37)$$

According to equations (7.36) and (7.37) the sign of the front velocity depends on the relative value of the potential V at the two stable states. When $V(u_+) < V(u_-)$ the velocity is positive, implying that the up state, u_+, propagates into the down state, u_-. Indeed, this process minimizes the Lyapunov functional, L, as it transforms high-potential areas into low-potential areas. When $V(u_+) > V(u_-)$ the down state propagates into the up state, and when $V(u_+) = V(u_-)$ the front is stationary ($c = 0$).

The potential values of the two states are affected by the control parameter λ in (7.33). The particular value, $\lambda = \lambda_m$, for which the potential values of the two states are equal is called the *Maxwell point* (see for example [243]). For $\lambda > \lambda_m$ the up state propagates into the down state and for $\lambda < \lambda_m$ the down state propagates into the up state. The Maxwell point can be readily obtained by solving $V(u_+) = V(u_-)$ for λ. For V given by (7.33) and $u_\pm$ by (7.34), such a calculation gives $\lambda_m = -2\alpha^2/9$. More generally, we will define the Maxwell point as the parameter value at which $c = 0$. This definition will be useful for more complex models of bistable systems, the potential functions of which are not known or do not exist.

Consider now an initial pattern consisting of a succession of up and down states as illustrated in Figure 7.7(c). As long as the domains are very wide compared to the front width the interactions between adjacent fronts are negligible and the dynamics can be inferred from the single front analysis: up-state domains will expand (contract) at the expense of down-state domains when $\lambda > \lambda_m$ ($\lambda < \lambda_m$). In the course of these dynamics, however, some of the fronts get closer to one another and the interactions between them may no longer be negligible. The subsequent dynamics depend on whether the interactions are attractive or repulsive. In the case of attractive interactions a uniform state will eventually be reached through the collapse of up-state or down-state domains, as Figure 7.8 demonstrates. For $\lambda > \lambda_m$ the collapse of down-state domains will lead to a uniform up state (Figure 7.8(a,d)), whereas for $\lambda < \lambda_m$ the collapse of up-state domains will lead to a uniform down state (Figure 7.8(c,f)). When the interactions are repulsive, fronts propagating toward one another slow down and may come to a stop if the interactions are strong enough. In this case stable asymptotic patterns can develop, as will be discussed in the next subsection[7].

For a single equation of the form (7.31) the front interactions are always attractive and the dynamics lead to uniform states. This becomes apparent from the Lyapunov functional (7.32); fronts are eliminated in the course of time because they cost "energy" (by contributing to the $(\partial_{x'} u)^2$ term). Note that at the Maxwell point, where $c = 0$, the dynamics is only due to front interactions and can be extremely slow. In fact, the time it takes for a domain of width w_d to collapse is exponentially long, $t \sim \exp(w_d/w_f)$, where w_f is the front width [148, 41].

7.2.1.2 Bistability of uniform states: Stable asymptotic patterns

Stable patterns can be obtained with repulsive front interactions which prevent the collapse of small domains. There are several mechanisms by which repulsive front interactions can arise [50, 83, 236]. We focus here on one generic mechanism—fast inhibitor diffusion in activator-inhibitor systems, and use the FHN model (5.37) (case (c) in Figure 5.2) to study it.

We begin by considering solutions of the FHN model that describe fronts propagating at a constant velocity c. We confine ourselves to the regime $\mu \equiv \epsilon/\delta \ll 1$, $\eta \equiv \sqrt{\epsilon\delta} \sim \mathcal{O}(1)$, and use the singular perturbation method presented in Section 6.3. In this section, however, we study the inner solutions only, postponing the full analysis to Section 7.3.1. Inserting $u = u(z)$ and $v = v(z)$, with $z = x - ct$, into (5.37) we obtain

$$\frac{d^2u}{dz^2} + c\frac{du}{dz} + u - u^3 - v = 0\,, \tag{7.38a}$$

$$\frac{d^2v}{dz^2} + \sqrt{\mu}\tilde{c}\frac{dv}{dz} + \mu(u - a_1 v - a_0) = 0\,, \tag{7.38b}$$

[7] Front interactions may also change from repulsive to attractive, or vice versa, as the distance between a pair of fronts increases. Such distance-dependent interactions can lead to bound front pairs [158].

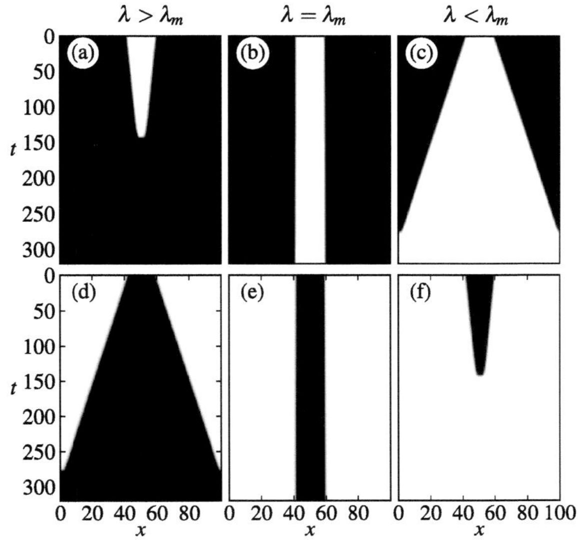

FIGURE 7.8: Domain collapse and the convergence to uniform states in the case of bistability of uniform states and attractive front interactions. Shown are space-time plots of the dynamics of initial down-state domains (light shade) (a,b,c) and up-state domains (dark shade) (d,e,f), obtained by solving (7.30) numerically. For $\lambda > \lambda_m$ an initial small down-state domain shrinks (a), while an initial small up-state domain expands (d). Both initial conditions culminate in the uniform up state. For $\lambda < \lambda_m$ an initial small down-state domain expands (c), while an initial small up-state domain shrinks (f). In this case both initial conditions culminate in the uniform zero state. At the Maxwell point, $\lambda = \lambda_m$, the initial domains are stationary, provided they are not too small for the attractive front interactions to be significant (b,e). From [16].

where $\tilde{c} = c/\eta$. In the limit $\mu \to 0$ the equation for v becomes $d^2v/dz^2 = 0$. Choosing a constant solution for v, the equation for u takes the the form of (7.35) with

$$V = -\frac{1}{2}u^2 + \frac{1}{4}u^4 + v_f u + V_0 \,, \tag{7.39}$$

where v_f is the constant value of v in the narrow inner region, i.e., the value of v at the front position, and V_0 is an arbitrary constant (reference potential). This equation has a front solution of the form (see Figure 6.7(b))

$$u(z) = \frac{u_-(v_f)e^{qz} + u_+(v_f)}{1 + e^{qz}} \,, \tag{7.40}$$

that asymptotes to $u_\pm$ as $z \to \mp\infty$, where

$$q = \frac{1}{\sqrt{2}}\left[u_+(v_f) - u_-(v_f)\right], \tag{7.41}$$

and $u_\pm(v_f)$ are the two minima of the potential V that is given by (7.39). Another symmetric front solution is obtained from (7.40) by the transformation $z \to -z$, $c \to -c$. The information these front solutions provide is not yet complete as both v_f and c are still unknown. The front velocity c can be calculated from (7.36). Let us calculate it for $|v_f| \ll 1$. A Taylor expansion of $u(z)$ around $v_f = 0$ gives the leading order form

$$u(z) = -\tanh\left(\frac{z}{\sqrt{2}}\right) - \frac{v_f}{\sqrt{2}}\cosh^{-2}\left(\frac{z}{\sqrt{2}}\right) + \mathcal{O}(v_f^2), \tag{7.42}$$

and linearizing the cubic nullcline $u - u^3 - v_f = 0$ around $v_f = 0$ gives $u_\pm(v_f) = \pm 1 - v_f/2$. Using these forms in (7.36), where V is given by (7.39), gives

$$c = -\frac{3}{\sqrt{2}}v_f + \mathcal{O}(v_f^2). \tag{7.43}$$

Note that this result has already been obtained in Section 7.1.3 using a different approach (compare (7.43) and (7.28)). According to (7.43) a front that asymptotes to $u_\pm$ as $z \to \mp\infty$ has a positive velocity if $v_f < 0$. Its symmetric counterpart, i.e., the front that asymptotes to $u_\mp$ as $z \to \mp\infty$, has a negative velocity. Both fronts describe an up state invading a down state. When $v_f > 0$ the pair of symmetric fronts describe a down state invading an up state. The value of v_f is still undetermined. In order to determine it we need to complement the analysis of the inner front region described above by an analysis of the outer regions on both sides of the front. This analysis will be described in Section 7.3.1. However, the dependence (7.43) of the front velocity on v_f can already help us understand the interaction between a pair of fronts.

Imagine a pair of symmetric fronts that bound a down state, as illustrated in Figure 7.9. Assume further that the value of the inhibitor at the front position is negative, $v_f < 0$, so that the up state invades the down state. Under these conditions the two fronts approach one another and the down-state domain contracts in time. During this process the dynamics of the two fronts are no longer independent of one another. When the diffusion of the inhibitor is sufficiently fast relative to the activator diffusion, the accumulation of v in the contracting down-state domain gradually increases v_f. As a result, the two approaching fronts slow down and eventually come to a stop when v_f increases to zero, forming a stationary down-state domain.

When the initial conditions consist of a number of front pairs, stationary patterns can result. Approximate solutions for periodic patterns of this kind have been derived in Section 6.3.2 using the singular perturbation method. A solution describing a single down-state domain is obtained by taking the limit $s_+ \to \infty$. Figure 7.10 shows the development of a stationary pattern

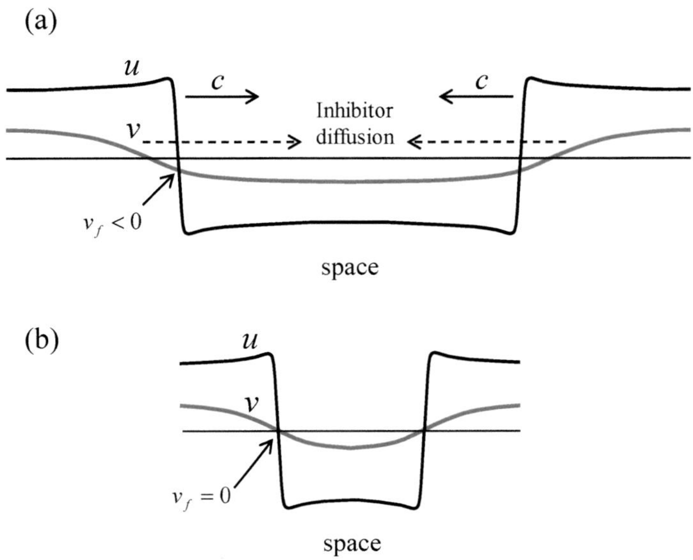

FIGURE 7.9: A schematic illustration of repulsive front interactions that lead to stable spatial structures. (a) Fast diffusion of the inhibitor (dashed arrows) into a contracting down-state domain increases the inhibitor value, v_f, at the front positions and slow down their velocities c. (b) A stable stationary down-state domain forms when v_f increases to zero (horizontal solid line).

from random initial conditions. Three time scales can be distinguished in this process. A short time scale at which persisting up-state domains are formed, a longer time scale at which domains of approximately fixed size are formed, and a yet longer time scale at which the domains slowly move as integral units. The short initial phase is dominated by the local dynamics that drive the system to one of the two stable states. The longer intermediate phase is dominated by the repulsive interactions of the front pairs that bound the domains. If the initial domain is too big the fronts that bound it propagate and the domain contracts until the repulsive front interactions become significant and slow them down. If the initial domains are too small the repulsive interactions are strong and the domains expand. In the final, very long phase the nearly stationary up-state domains slowly displace themselves by the weak repulsive interactions between the pairs of fronts that bound the down-state domains. During this phase each domain moves as a whole. This is because the up-state domains are smaller than the down-state domains and the "forces" that bind them are stronger than the those controlling their mutual interactions. These interactions may eventually lead to periodic patterns but the dynamics can be extremely slow and not observable in practice (see discussion in Section 7.1.3).

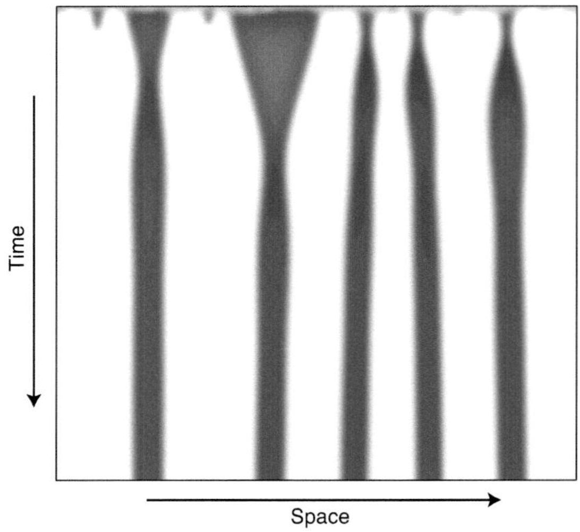

FIGURE 7.10: Development of stationary patterns in bistability of uniform states with repulsive front interactions. Shown is a space-time plot of a numerical solution of the bistable FHN model in the case of fast inhibitor diffusion.

7.2.1.3 Bistability of uniform and patterned states

So far we considered bistability of two uniform states. There are many circumstances, however, in which one of the two stable states is spatially patterned. In Chapter 10 we will encounter two examples related to water limited vegetation—bistability of bare soil and vegetation spot patterns, and of uniform vegetation and gap patterns (see also Sections 3.2.1 and 3.3.2). The coexistence of uniform and patterned stable states allows for a multitude of additional patterns and constitutes another mechanism of pattern formation.

To study this case let us consider the modified SH equation (7.1) in 1d. The amplitude of a stationary periodic solution satisfies, in this case, the amplitude equation

$$\partial_t A = \lambda A - \left(3 - \frac{38\alpha^2}{9k_0^4}\right)|A|^2 A + 4k_0^2\partial_x^2 A\,, \tag{7.44}$$

where α is the coefficient of the quadratic term in (7.1). Bistability of the zero and the periodic-pattern state is achieved when the coefficient of the cubic term in (7.44) is positive, or when

$$\alpha > \sqrt{\frac{27}{38}}k_0^2\,, \tag{7.45}$$

for then the instability of the zero state at $\lambda = 0$ to stationary periodic patterns is subcritical.

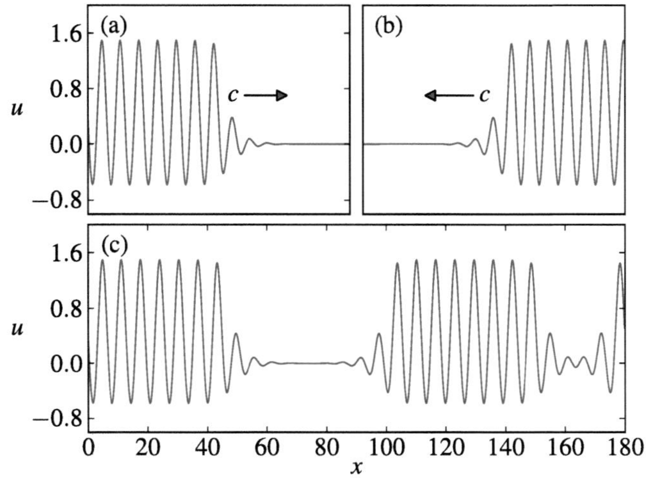

FIGURE 7.11: Fronts as building blocks of spatial patterns for bistability of uniform and patterned states. (a,b) Two symmetric front solutions of (7.1). Both fronts, which propagate in opposite directions at speed c, represent an invasion of the patterned state into the uniform zero state. (c) Alternating arrays of these fronts give rise to spatial patterns. From [16].

As in the case of two stable uniform states, much insight about the structure and dynamics of spatial patterns can be obtained by studying the dynamics of single fronts and their interactions. The fronts separate domains of uniform and patterned states as shown in Figure 7.11. Such fronts, however, behave differently in one important respect compared to fronts that separate two uniform states—they are stationary in a *range* of the control parameter, $\lambda_1 < \lambda < \lambda_2$, that includes the Maxwell point[8], rather than being stationary at the Maxwell point only [244]. This is a consequence of the Lyapunov functional associated with (7.1) having a periodic dependence on the phase of the patterned state; the front is pinned at phase values that minimize the Lyapunov functional [55].

The implications for pattern formation are illustrated in Figure 7.12(a-c) with numerical solutions of equation (7.1). The solutions describe the dynamics of a zero-state domain in an otherwise periodic pattern state within the bistability range, $\lambda_f < \lambda < 0$ (see Figure 7.13). This range can be divided into three parts where different behaviors take place: (i) $\lambda_2 < \lambda < 0$ where the zero-state domain diminishes in size and disappears, (ii) $\lambda_1 < \lambda < \lambda_2$ where the zero-state domain retains a fixed size because the fronts are pinned, and (iii) $\lambda_f < \lambda < \lambda_1$ where the zero-state domain expands until it occupies the whole system. In the first range the system converges to the stable

[8]The Maxwell point in this case is the value $\lambda = \lambda_m$ at which the values of the Lyapunov functional that correspond to the uniform and patterned states are equal.

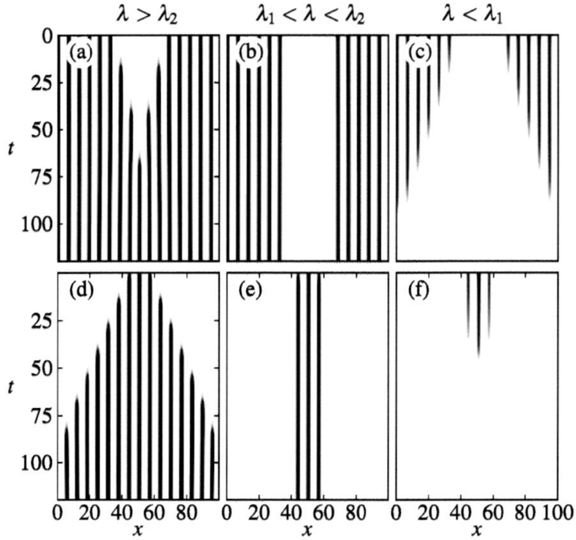

FIGURE 7.12: The emergence of stable localized structures in the case of bistability of uniform and patterned states. Space-time plots showing the dynamics of initial zero-state domains (a,b,c) and pattern-state domains (d,e,f). Unlike the case of two uniform states shown in Figure 7.8, a control-parameter range exists where the initial zero and pattern-state domains quickly converge to stationary hole solutions (b) and localized solutions (e), respectively. The dynamics resemble that of two uniform states only outside this control-parameter range as panels (a,c) and panels (d,f) show. From [16].

periodic-pattern state. In the third range it converges to the stable uniform zero state. In the range in between it converges to a stable *hole solution*, that is, a solution describing a periodic pattern with a confined domain or "hole" of the zero state. Analogous behavior is found for an initial pattern-state domain in an otherwise zero state as Figure 7.12(d-f) shows. The pattern-state domain in the intermediate range, $\lambda_1 < \lambda < \lambda_2$, retains a fixed size and forms a stable *local-pattern solution* (Figure 7.12(e)).

The local-pattern solution and the hole solution that the system converges to in panels (b) and (e) of Figure 7.12 are just examples of many more stable localized solutions that exist in the range $\lambda_1 < \lambda < \lambda_2$. Figure 7.13 shows a bifurcation diagram that displays a family of hole solutions, besides the zero and the periodic solutions. The family emerges from the periodic-pattern branch near the fold bifurcation at $\lambda = \lambda_f$, snakes down and terminates at the instability point of the zero state. The upper branches of this solution family describe small holes of the zero state in an otherwise periodic pattern, that widen in the descent toward the zero state. The full bifurcation

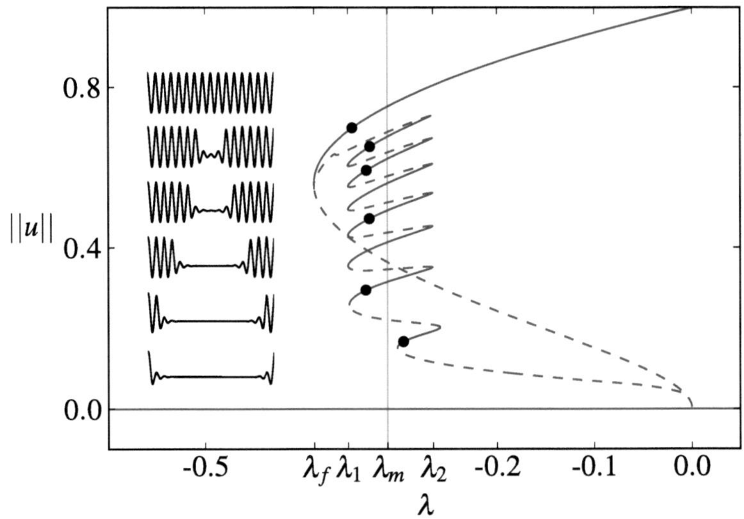

FIGURE 7.13: Bifurcation diagram for equation (7.1) showing a family of hole solutions in the bistability range, $\lambda_f < \lambda < 0$, of a uniform zero state and a periodic-pattern state. The vertical axis, $\|u\|$, represents the $L2$ norm of u, a global measure involving the square root of the spatial integral of u^2. Solid (dashed) lines denote stable (unstable) solutions. The family of hole solutions exists in a smaller range, $\lambda_1 < \lambda < \lambda_2$ (the snaking range), and describes zero-state domains of increasing sizes as the insets on the left show. The insets correspond to the solid circles on the stable solution branches. The vertical line denotes the Maxwell point $\lambda = \lambda_m$. From [16].

diagram of solutions is more involved and contains additional periodic-pattern branches and many more hybrid states involving mixtures of hole and local-pattern domains of various sizes. Since the hole and local-pattern solutions are homoclinic orbits in an appropriately defined phase space[9], this behavior is often called "homoclinic snaking" and the range $\lambda_1 < \lambda < \lambda_2$ is called the "snaking range." The reader is referred to Ref. [157] and references therein for additional information and for fuller descriptions of the bifurcation structure.

The solution structure described above for 1d extends to 2d with richer solution forms [179]. We illustrate some of these solutions in Figure 7.14. The leftmost and rightmost panels show the alternative stable states, periodic hexagonal pattern (left) and zero state (right). The panels in between show hybrid states consisting of holes (zero-state domains) of different sizes, combinations of holes and local patterns, and local-pattern domains of different sizes.

[9] A homoclinic orbit in the present context is a trajectory in phase space that emanates from a saddle point at $x \to -\infty$ and returns to it at $x \to +\infty$.

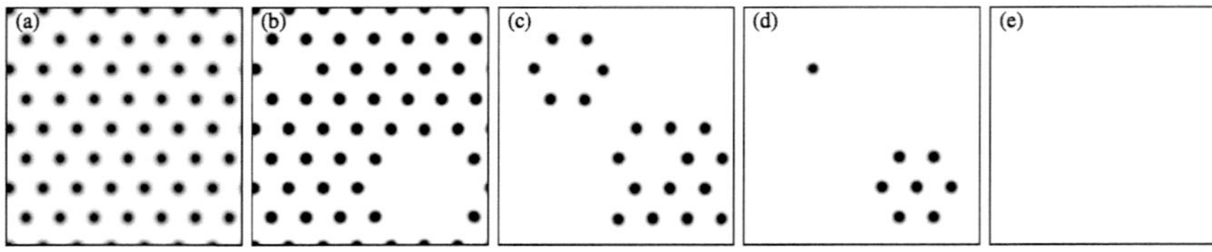

FIGURE 7.14: Examples of 2D patterns in the bistability range of a uniform zero state (e) and a hexagonal-pattern state (a) in equation (7.1). The examples include a spatial mixture of holes of different sizes (b), a spatial mixture of local-pattern domains of different sizes (d), and a combination of local patterns and holes (c). Dark shading indicates high u values. From [16].

It should be noted that the periodic phase modulation of the Lyapunov functional, which is responsible for the front pinning effect, is of order $\exp(-a/\sqrt{\lambda})$, where a is some constant [55], i.e., exponentially small and beyond all orders of the asymptotic expansion (6.22) in the small parameter $\sqrt{\lambda}$. As a consequence, the amplitude equation approach, pursued to any given order, cannot capture the pinning effect. The inapplicability of the amplitude equation approach is related to the coupling between the long length scale of the amplitude and the short length scale of the pattern through the locking of the front position to the phase of the periodic pattern.

Note also the difference between stationary localized structures in bistability of uniform states with fast inhibitor diffusion and bistability of uniform and patterned states. In the former case individual fronts propagate and the localized structures result from repulsive front interactions. In the latter case individual fronts are stationary.

7.2.2 Oscillatory systems and spiral waves

Uniform oscillatory instabilities constitute another class of pattern forming systems where pattern formation results from state multiplicity. This case is characterized by a *continuous* family of oscillating states with different phase values. Loosely speaking, spatial patterns in such systems can be regarded as spatial mixtures of oscillating states that differ by incremental phase changes[10]. In the following we present a brief overview of pattern formation in oscillating systems, and refer the reader to Ref. [6] for additional information.

The simplest type of patterns are long-wavelength traveling waves in 1d. Near the oscillatory instability the amplitudes of such waves are described by

[10]Consider, as an illustrative example, the FHN model beyond the Hopf bifurcation with the diffusion terms removed. In the absence of spatial coupling each point in space represents an independent oscillator. If these oscillators are assigned different phases, arbitrary oscillating patterns can be formed. Switching on diffusion will damp high wavenumber modes by reducing phase gradients, and smooth the wavy patterns to form long-wavelength traveling waves.

the CGL equation (6.53) and assume the form (6.54)–(6.56). These waves can go through secondary instabilities such as the Eckhaus and the Benjamin–Feir instabilities discussed in Section 6.2.2. When two traveling waves meet a localized sink or a shock structure forms, if the two waves have different frequencies the sink propagates toward the domain occupied by the lower frequency wave [27]. There are also localized source or hole solutions that emit waves. A source solution is characterized by a localized dip in the amplitude and can be stationary or propagating. Wave patterns may consist of several sink and source structures, in which case any pair of sinks is separated by a source and vice versa [27, 6, 287].

In 2d another type of source structure appears—the spiral vortex. At the vortex core the magnitude $\rho = |A|$ vanishes and the phase $\phi = \arg(A)$ becomes singular. Along any circular trajectory that surrounds the vortex core the phase varies by $2\pi n$,

$$\oint \nabla\phi \cdot \mathbf{dl} = 2\pi n \,. \tag{7.46}$$

where n is an integer number representing the topological charge of the spiral vortex. The vortex emits traveling waves that propagate outward to form a rigidly rotating spiral wave, a snapshot of which is shown in Figure 7.15. Spiral-wave solutions of the CGL equation, which we will call *Hopf spiral-wave solutions*, have the general form

$$A(r,\theta,t) = \rho(r)\exp\{i[-\nu t + n\theta + \psi(r)]\}\,, \tag{7.47}$$

where r, θ are polar coordinates and ν is the rotation frequency. Far from the

FIGURE 7.15: A snapshot of a Hopf spiral-wave solution of the CGL equation with a topological charge $n = 1$. As the spiral wave rotates (counterclockwise in this case) outwardly propagating waves are emitted. From [55].

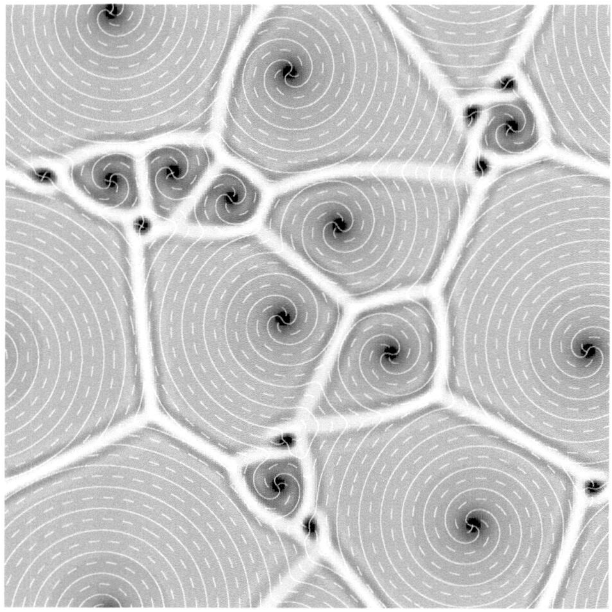

FIGURE 7.16: Quasi-stationary vortex patterns in the CGL equation. Shown are superimposed pictures of $|A|$ and zero contour lines of ReA (solid curves) and ImA (dashed curves). Light (dark) shades denote high (low) $|A|$ values. The light stripes that form the cellular pattern represent shock lines formed by incoming traveling waves. The dark dots denote zeros of $|A|$ that correspond either to spiral cores or to vertices of shock lines. The zero contour lines show the spiral waves emitted by the spiral cores within each cell. From [45].

spiral core ($r \to \infty$) the amplitude ρ and the frequency ν approach those of the traveling-wave solutions given by (6.55) and (6.56), respectively. The asymptotic wave number $q = d\psi/dr$, however, is uniquely determined by the parameters α and β in the CGL equation. Spiral waves with topological charges other than $n = \pm 1$ are unstable. Instabilities of spiral wave solutions with $n = \pm 1$ can originate from the core region or from the asymptotic traveling waves far from the core. The reader is referred to Refs. [45, 6] for a detailed account of the stability limits of these spiral wave solutions.

Wave patterns in oscillatory systems often contain many spiral vortices. These may form regular vortex lattices, quasi-stationary glass-like vortex patterns, or chaotic spatio-temporal patterns. An example of a glass-like pattern is shown in Figure 7.16. While the phase field, $\arg(A)$, is dynamic, consisting of outgoing spiral waves bounded by shock lines, the magnitude field, $|A|$, is quasi-stationary, consisting of nearly stationary spiral cores and shock lines. Chaotic vortex patterns are dynamic with respect to both phase and amplitude. They involve the spontaneous appearance and disappearance of vortex pairs and are characterized by exponentially decaying correlations in space and time [45, 6]. More about the chaotic patterns in Section 7.3.2.

7.2.3 Multimode systems and multimode localized structures

Quite often a uniform state can go through two types of instabilities at which different modes begin to grow. The two instabilities generally appear sequentially by varying some control parameter. When the thresholds of the instabilities are sufficiently far apart the first mode to grow often damps the growth of the second mode through nonlinear coupling. However, when the thresholds are made close enough, by tuning another control parameter, bistability of two pure-mode states or a stable mixed-mode state can result. The example of a Hopf–Turing bifurcation will help explain these behaviors.

The Hopf–Turing bifurcation refers to the coincidence of a non-uniform stationary instability and a uniform oscillatory instability. The occurrence of a Hopf–Turing bifurcation in the FHN model has been discussed briefly in Section 5.2.2. As Figure 5.3 shows, the bifurcation occurs at a codimension-2 point where the threshold lines of the two instabilities cross one another. In the vicinity of this point a typical dynamical variable can be approximated as

$$u(x,y,t) \approx u_0 + c_1 A e^{ikx} + c_2 B e^{i\omega t} + c.c.\,, \tag{7.48}$$

where u_0 is the value of u at the state that undergoes the Hopf–Turing bifurcation, k is the wavenumber of the Turing mode, ω is the frequency of the Hopf mode, A and B are the modes' amplitudes, which vary on space and time scales much longer than k^{-1} and ω^{-1}, and $c.c.$ stands for the complex conjugate. Other dynamical variables will have the same form but with different values for the constants c_1 and c_2. The amplitude equations for A and B are [61]

$$\begin{aligned}
\partial_t A &= \epsilon A + (2k_0\partial_x - i\partial_y^2)^2 A - (\lambda|A|^2 + \kappa|B|^2)A \\
\partial_t B &= (\mu + i\nu)B + \alpha\nabla^2 B - (\delta|A|^2 + \beta|B|^2)B\,,
\end{aligned} \tag{7.49}$$

where $\epsilon > 0$ and $\mu > 0$ are the distances from the Turing and the Hopf instabilities, λ and κ are real-valued constants, and α, β and δ, are complex-valued constants. In the following we will assume that both κ and $Re(\delta)$ are positive, that is, the modes act to damp one another or compete. We also assume that λ and $Re(\beta)$ are positive, otherwise higher order terms in the amplitude equations should be included.

Equations (7.49) have three types of stationary uniform solutions in addition to the zero solution $(A,B) = (0,0)$: pure Turing state $(A_0, 0)$, pure Hopf state $(0, B_0)$, and mixed-mode state (A_M, B_M). The two dynamical behaviors referred to above, bistability of two pure-mode states or a stable mixed-mode

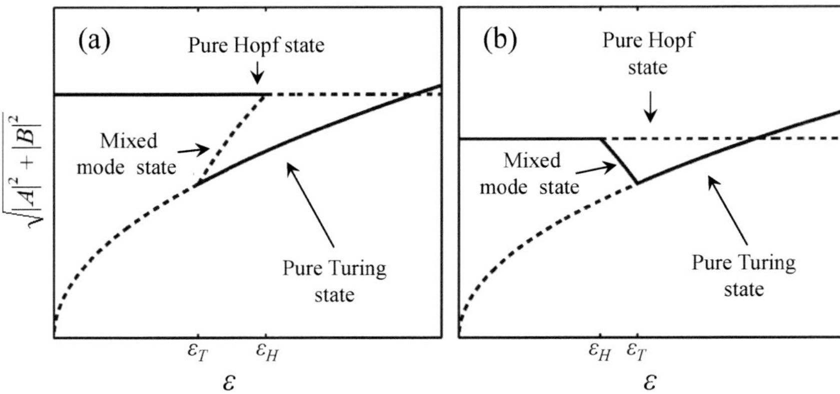

FIGURE 7.17: Bifurcation diagrams showing stationary uniform solutions of (7.49) for (a) $\Delta < 0$ (mixed mode is unstable) and (b) $\Delta > 0$ (mixed mode is stable). Solid (dashed) lines denote stable (unstable) solutions. The bifurcation parameter is the distance, ϵ, from the Turing bifurcation (the distance from the Hopf bifurcation, μ, is kept constant). From [167].

state, are obtained with opposite signs of the quantity

$$\Delta \equiv \lambda Re(\beta) - \kappa Re(\delta)\,.$$

The sign of Δ determines whether self-mode damping is weaker ($\Delta < 0$) or stronger ($\Delta > 0$) than mutual-mode damping. Weak self-mode damping favors the pure states whereas strong self-mode damping favors the mixed state. Thus, when $\Delta < 0$, the mixed-mode solution is unstable and a bistability range, $\epsilon_T < \epsilon < \epsilon_H$, exists where both the pure-Turing and the pure-Hopf solutions are stable (Figure 7.17a). When $\Delta > 0$, there is a range, $\epsilon_H < \epsilon < \epsilon_T$, where the mixed mode solution is stable, and no bistabilty range of pure mode solutions exist (Figure 7.17b).

The bistability range allows for front solutions that separate in space the two pure states. Front solutions of the amplitude equations (7.49) will generally discribe propagating fronts [235], but the solutions of the original system, from which the amplitude equations are derived, might describe fronts that are pinned by the phase of the stationary periodic pattern [61] (see Section 7.2.1.3). Figure 7.18(a) shows a Hopf–Turing front solution obtained by solving numerically a two-variable reaction diffusion model—the Brusselator [61]. The solution is stationary in a range of parameters (i.e., describes pinned fronts), in which a wide variety of stable localized solutions exist. A few of them are shown in Figure 7.18(b-d).

Multimode systems allow for an additional pattern formation mechanism that is not necessarily associated with a multiplicity of stable states. Consider

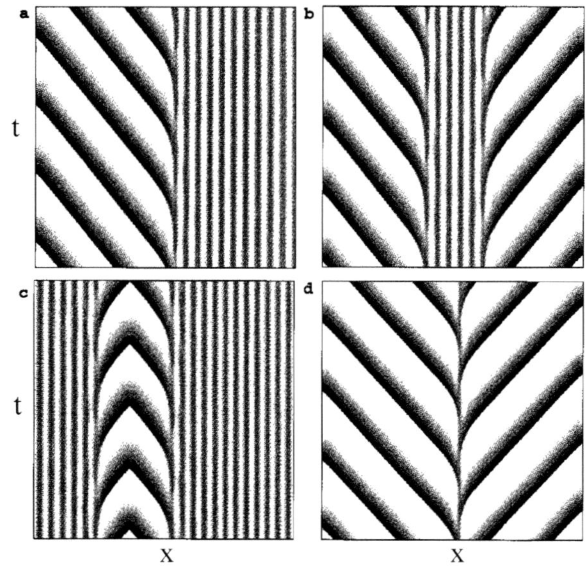

FIGURE 7.18: Space-time plots of various 1d patterns in the Hopf-Turing bistability range of the Brusselator model. (a) A front between a traveling wave state and a stationary periodic pattern. (b) A domain of a stationary pattern embedded in a traveling wave background. (c) A domain of traveling wave embedded in a stationary pattern background. (d) A localized structure acting as an asymmetric source of counter-propagating traveling waves. From [61].

the pure mode ranges in Figure 7.17, e.g., the pure Hopf range $\epsilon < \epsilon_T$ for $\Delta < 0$ or $\epsilon < \epsilon_H$ for $\Delta > 0$. In these ranges the Turing mode has a positive growth rate ($\epsilon > 0$), yet it is nonlinearly damped by the Hopf mode, which for sufficiently small ϵ values has a larger amplitude. This damping, however, may not be effective in localized structures of the Hopf mode where its amplitude vanishes or becomes very small [167]. An example of such a localized structure is the spiral vortex. The numerical studies of the FHN model presented in Figure 5.4 indeed demonstrate the possible growth of a Turing mode at the core of a Hopf spiral wave. This suggests the existence of an instability of a single-mode spiral vortex to a dual-mode vortex that hosts the Turing mode at its core. Actual bifurcation diagrams demonstrating instabilities of this type have been calculated for front solutions in Hopf–Turing systems that are subjected to spatial or temporal periodic forcing [167]. An experimental realization of such an instability has been found in the context of thermal convection where a spiral roll pattern is destabilized to the growth of an hexagonal mode at the spiral core [8, 4]. We will return to dual-mode spiral waves in Section 7.3.2.

7.3 Instabilities of localized structures

We have already discussed the role of localized structures as building blocks of global patterns that extend over the entire system (see Section7.2.1). Localized structures can induce global patterns also by means of instabilities that they go through. In Section 7.3.1 we will consider front structures and describe two types of front instabilities. The first is the so-called Nonequilibrium Ising–Bloch (NIB) bifurcation, a pitchfork front bifurcation that designates the onset of traveling waves, and the second is a transverse front instability that leads to labyrinthine patterns. We will also show that the two instabilities can act in concert to induce a state of spatio-temporal chaos. In Section 7.3.2 we will consider Hopf spiral waves (Section 7.2.2), and describe two types of spiral-core instabilities. In both cases the effects of the instabilities do not remain local, but rather extend over the entire system. As pointed out earlier, the amplitude of Hopf spirals is bounded away from zero except at the spiral core. The spiral-core instabilities to be described induce sharp amplitude variations throughout the whole system, including the formation of zeros and the consequent spontaneous nucleation of new spiral-vortex pairs. As the latter are subject to the same instabilities, a state of spatio-temporal chaos can result.

7.3.1 Front instabilities in bistable systems

We begin addressing the NIB bifurcation in 1d using the FHN model for $\mu = \epsilon/\delta \ll 1$. In Section 7.2.1.2 we obtained a relation (see (7.43)) between the velocity c of a front and the value v_f of the inhibitor field at the front position. This relation has been derived by considering the inner region in a stretched coordinate system, $z = x - ct$, where the inhibitor field hardly varies and can be assumed to be constant, $v = v_f$. The value of v_f, however, has remained undetermined. We now complement this analysis by considering the outer regions on both sides of the front. This will provide another relation between c and v_f and will allow solving the two relations for v_f and c.

Moving to the unstretched coordinate system, $\zeta = \sqrt{\mu} z$, Equation (7.38) becomes

$$\mu\frac{d^2u}{d\zeta^2} + c\sqrt{\mu}\frac{du}{d\zeta} + u - u^3 - v = 0\,, \tag{7.50a}$$

$$\frac{d^2v}{d\zeta^2} + \tilde{c}\frac{dv}{d\zeta} + u - a_1v - a_0 = 0\,. \tag{7.50b}$$

Taking into account that the derivatives of u with respect to ζ in the outer regions are of order unity, the limit $\mu \to 0$ gives $u - u^3 - v = 0$, which we can solve for u. Of the three solutions of this cubic equation we are interested in the two solutions, $u = u_\pm(v)$ (see Section 6.3.2), which for $|v| \ll 1$, can be approximated by $u_\pm = \pm 1 - v/2$. Considering now a front that asymptotes to $u_\pm$ as $\zeta \to \mp\infty$, and using the solutions $u_+(v)$ and $u_-(v)$ in the domains $\zeta < 0$ and $\zeta > 0$, respectively, we obtain the boundary value problems:

$$\frac{d^2v}{d\zeta^2} + \tilde{c}\frac{dv}{d\zeta} - q^2(v - v_+) = 0\,, \quad \zeta < 0 \tag{7.51a}$$
$$v(0) = v_f \quad v(-\infty) = v_+\,,$$
$$\frac{d^2v}{d\zeta^2} + \tilde{c}\frac{dv}{d\zeta} - q^2(v - v_-) = 0\,, \quad \zeta > 0 \tag{7.51b}$$
$$v(0) = v_f \quad v(\infty) = v_-\,,$$

where as before $\tilde{c} = c/\eta$, $q^2 = a_1 + 1/2$ and $v_\pm = q^{-1}(\pm 1 - a_0)$ (see Section 6.3.2). The solutions to these boundary value problems are

$$v(\zeta) = v_+ (v_f - v_+)\, e^{\sigma_+\zeta}\,, \quad \zeta < 0\,, \tag{7.52a}$$
$$v(\zeta) = v_+ (v_f - v_-)\, e^{\sigma_-\zeta}\,, \quad \zeta > 0\,, \tag{7.52b}$$

where

$$\sigma_\pm = -\frac{\tilde{c}}{2} \pm \sqrt{\frac{\tilde{c}^2}{4} + q^2}\,. \tag{7.53}$$

Demanding that the solution (7.52) is continuously differentiable at $\zeta = 0$ we obtain the second relation between v_f and c:

$$v_f = -\frac{\tilde{c}}{q^2\sqrt{\tilde{c}^2 + 4q^2}} - \frac{a_0}{q^2}\,. \tag{7.54}$$

Inserting $v_f = -\sqrt{2}c/3$ (from (7.36)) into (7.54) we obtain the following implicit relation for c [117]:

$$c = \frac{3c}{\sqrt{2}q^2\sqrt{c^2 + 4\eta^2 q^2}} + c_\infty, \tag{7.55}$$

where $c_\infty = 3a_0/\sqrt{2}q^2$.

Equation (7.55) can be readily solved in the symmetric case where $a_0 = 0$. It has a zero solution, $c = 0$, which exists for all η values and represents a stationary front. In addition, two new solutions, $c = \pm 2q\sqrt{\eta_c^2 - \eta^2}$ appear as η is decreased below $\eta_c = 3/2\sqrt{2}q^3$ [138, 118]. These solutions represent a symmetric pair of counter propagating fronts that appear in a pitchfork front bifurcation. This is the *nonequilibrium Ising–Bloch (NIB) bifurcation* [118]. The name of this front bifurcation was motivated by drawing an analogy to ferromagnets with weak anisotropy [52]. In this terminology the stationary front solution is called an *Ising front* and the two counter-propagating fronts solutions are called *Bloch fronts*.

The bifurcation diagram and the front structures associated with the different front solutions are shown in Figure 7.19. Note that for the stationary front $v_f = 0$ and the zero points of u and v, coincide, whereas for the propagating

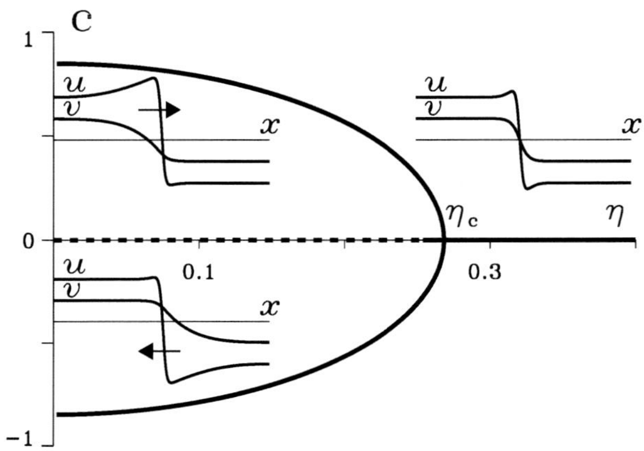

FIGURE 7.19: The NIB bifurcation and the internal structure of front solutions. For $\eta > \eta_c$ a single stationary Ising-front solution exists. The value v_f of the inhibitor v at the front position (where $u = 0$) vanishes for this solution. Beyond the bifurcation, $\eta < \eta_c$, a pair of counter-propagating Bloch-front solutions appears. The value of v_f is negative (positive) for fronts representing an up (down) state invading a down (up) state. From [119]

fronts $v_f \neq 0$ and the zero points do not coincide–the v field lags behind the u field for both fronts. Figure 7.20 shows the front solutions as phase trajectories in the (u, v) plane, parameterized by the spatial coordinate. For the case $a_0 \neq 0$ it is necessary to solve (7.55) numerically. The effect of a_0 is to unfold the symmetric pitchfork bifurcation to one of the two imperfect bifurcation forms shown in Figure 2.9(a,c), depending on the sign of a_0 [117].

Expressed in terms of ϵ and δ the front bifurcation threshold (for $a_0 = 0$) is given by $\delta = \delta_F(\epsilon) = 9/(8q^6\epsilon)$ and has the form of a hyperbola as Figure 7.21 shows. This result, however, is valid only for $\epsilon/\delta \ll 1$. A different perturbation analysis for $\delta \ll 1$ shows that the bifurcation line in the $\delta - \epsilon$ plane intersects the ϵ axis at $\epsilon_F = a_1^{-2}$, rather than approaching it asymptotically [118]. Since $a_1 > 1$ (bistability case) $\epsilon_F < a_1^{-1}$, that is, the front bifurcation occurs outside the range where the FHN model is variational (see Section 5.2.2). Indeed the appearance of counter-propagating front solutions is a non-variational effect.

The NIB bifurcation has significant implications for pattern formation in bistable systems. First and foremost it designates the onset of traveling waves. Up-state and down-state domains can travel if they are bounded by a pair of fronts corresponding to an up-state invading a down-state and a down-state invading an up-state. These front types are only available in the Bloch side of the NIB bifurcation.

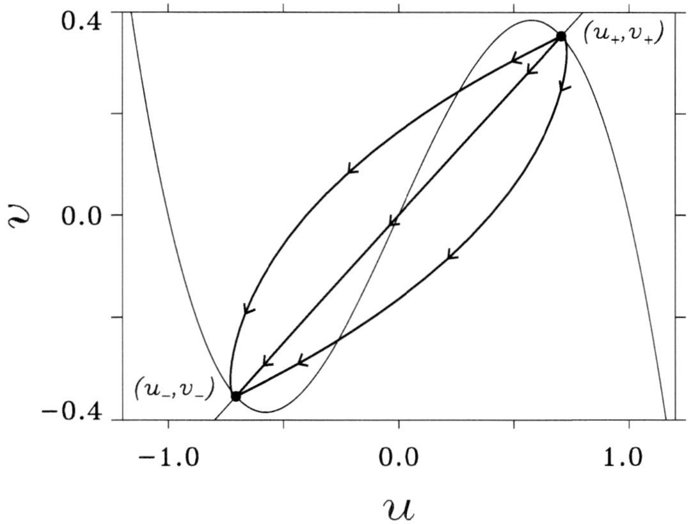

FIGURE 7.20: Phase trajectories of Ising and Bloch front solutions in the (u, v) plane, parameterized by the spatial coordinate x. All trajectories connect the up state at $x \to -\infty$ to the down state at $x \to \infty$. The Ising trajectory goes through the origin, whereas the two Bloch trajectories avoid the origin from above or below. From [118]

In 2d spiral waves can appear as Figure 7.22 demonstrates. These waves can be regarded as two front lines, corresponding to the two different Bloch fronts, that meet at the spiral core. Across the core the inhibitor field v changes from v_f to $-v_f$ (or vice versa) along the front line, and from v_+ to v_- (or vice versa) in a direction normal to the front line. We refer to this type of spiral wave as *Bloch spiral waves* to distinguish them from Hopf spiral waves in oscillatory systems (see Section 7.2.2). Although the two types of spiral waves look similar, there are essential differences between the two that become sharp near the Hopf and NIB bifurcations. The Hopf spiral is a phase wave with a slowly varying amplitude that is bounded away from zero, except for the core region where large phase gradients develop and the amplitude sharply goes to zero. A Bloch spiral, on the other hand, experiences sharp variations across the Bloch fronts it consists of, even far from the core. Another difference pertains to the spiral-wave frequency; the frequency of a Hopf spiral wave is finite even at the onset of the Hopf bifurcation, whereas the frequency of a Bloch spiral wave goes to zero as the NIB bifurcation is approached, along with the propagation speed of the Bloch fronts.

Fronts in 2d can also become unstable to the growth of transverse perturbations, i.e., perturbations that modulate front properties along the front line, such as curvature. This is unlike the NIB bifurcation, which involves the

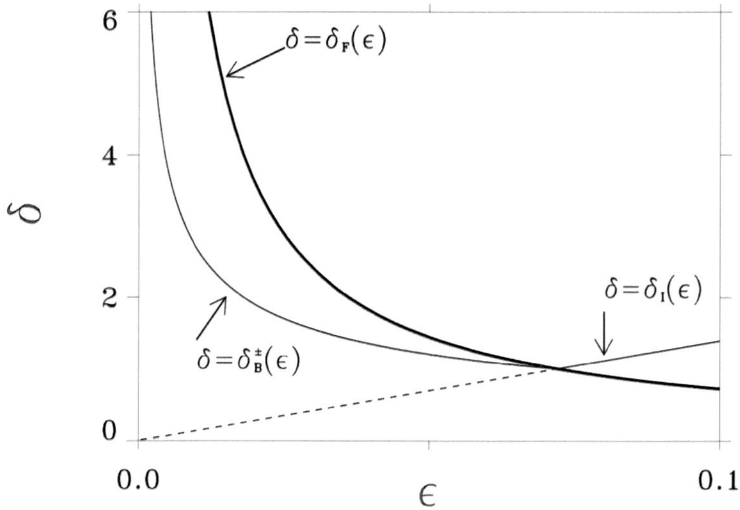

FIGURE 7.21: Front instability lines in the $\epsilon - \delta$ plane for the symmetric FHN model. The thick line is the NIB threshold $\delta_F = 9/(8q^6\epsilon)$. The thin lines represent the transverse instability thresholds of Ising fronts, $\delta_I = \frac{8}{9}q^6\epsilon$ and of Bloch fronts, $\delta_B^{\pm} = \frac{3}{2\sqrt{2}q^3\sqrt{\epsilon}}$. From [117].

displacement of one field relative to the other in a direction normal to the front line. In activator-inhibitor systems transverse instabilities occur when the inhibitor diffusion is fast enough; diffusion of the inhibitor to the sides of a protruding front segment will accelerate its growth and, at the same time, will inhibit the growth of the side front segments. Transverse instabilities occurring in the vicinity of the NIB bifurcation may lead to complex spatiotemporal behavior, including spontaneous nucleation of spiral-vortex pairs and the onset of spatiotemporal chaos. We refer the reader to Refs. [117, 109, 119, 121] for detailed analytical and numerical studies of 2d front dynamics near and far from the NIB bifurcation. Here, we briefly describe the main results of these studies in a rather qualitative way.

The dynamics of a front in the x, y plane can be described by following the contour line $u = 0$ in time, as Figure 7.23 illustrates. The front line can be described by the Cartesian coordinates, $x = X(s,t)$ and $y = Y(s,t)$, of points along it, where s is a coordinate along the front line, measuring the arclength from some reference point (origin). The front shape is given by the front curvature

$$\kappa(s,t) = -\partial_s\theta(s,t) = \partial_s Y \partial_s^2 X - \partial_s X \partial_s^2 Y \,, \tag{7.56}$$

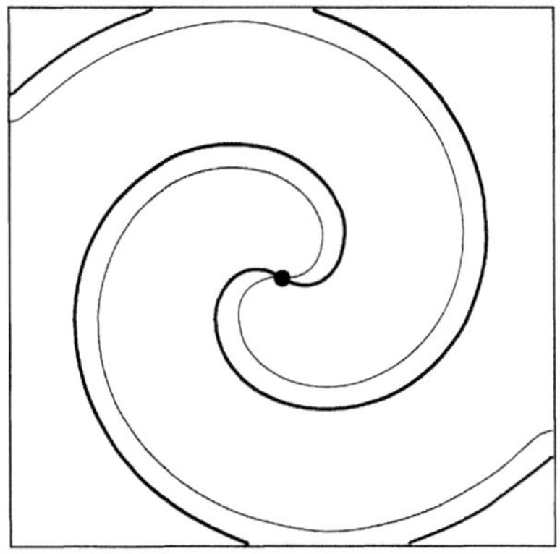

FIGURE 7.22: A Bloch spiral wave obtained by numerical integration of the bistable FHN model for $\eta < \eta_c$. The thick line is the $u = 0$ contour while the thin line is the $v = 0$ contour. Since the v field lags behind the u field the spiral-wave rotation in this example is clockwise. The crossing point of the two lines represents the stationary spiral core.

where $\theta(s,t)$ is the angle between the tangent to the front at a point s to a fixed direction, e.g., the x axis (with the choice illustrated in Figure 7.23 the curvature is positive for convex up domains). The front motion is determined by the normal front velocity $c_n(s,t)$, i.e., the velocity of the front in a direction perpendicular to the front tangent at point s, as Figure 7.23 illustrates. In general both κ and c_n are time dependent and satisfy kinematic equations that take the form of coupled nonlinear PDEs [119, 121].

Away from the NIB bifurcation, and for weakly curved fronts, c_n varies on a time scale much shorter than that of κ, and can be assumed to follow the curvature adiabatically. Under these conditions the relation between c_n and κ becomes algebraic. A relatively simple calculation [117, 121], based on expressing the FHN equations (5.37) in a coordinate system that moves with the front and assuming a radius of curvature much larger than the front width, i.e., $\kappa^{-1} \gg q^{-1}$, leads to the implicit velocity-curvature relation

$$c_n + \kappa = \frac{3(c_n + \delta\kappa)}{\sqrt{2}q^2\sqrt{(c_n + \delta\kappa)^2 + 4\eta^2 q^2}} + c_\infty\,, \tag{7.57}$$

which generalizes (7.55) for weakly curved fronts. The form of the velocity-curvature relation provides information about the number of front solutions that exist for a given curvature value; a single-valued relation implies the existence of an Ising front, while a multi-valued relation implies the existence of a pair of counter-propagating Bloch. It also provides information about the stability of fronts to transverse perturbations—a positive slope, $dc_n/d\kappa > 0$,

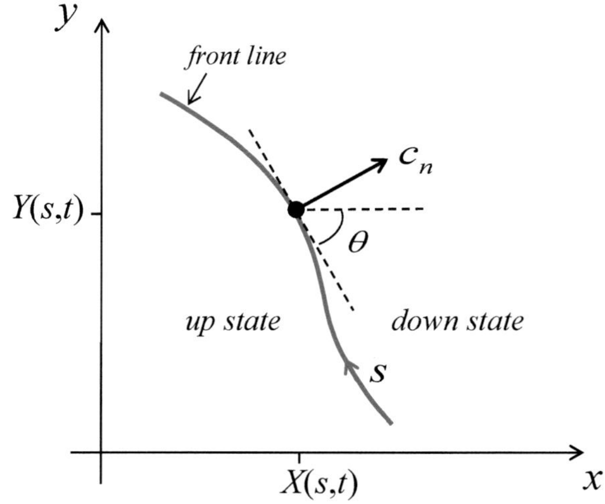

FIGURE 7.23: A schematic illustration of a front line in a 2d bistable system.

implies a transverse front instability. This can easily be inferred by considering a bulge-like perturbation of a front line, which creates a front segment with higher curvature than its neighbor front segments. Because of the positive slope of the velocity-curvature relation the more curved front segment propagates faster than its neighbor segments, and thereby accelerates the perturbation growth (see Figure 7.25 below).

Transverse instability thresholds for linear (zero curvature) Ising and Bloch fronts can be derived by linearizing the velocity-curvature relation (7.57) with respect to κ to obtain

$$c_n = c_0 - D\kappa + \mathcal{O}(\kappa^2)\,, \tag{7.58}$$

where

$$\begin{aligned} D &= \alpha^{-1} + (1-\alpha^{-1})\delta\,, \\ \alpha &= 1 - \frac{c_0 - c_\infty}{c_0}\left(1 - \frac{2q^4}{9}(c_0 - c_\infty)^2\right). \end{aligned} \tag{7.59}$$

The instability thresholds for linear Ising and Bloch fronts can then be derived by using in (7.59) the corresponding solution c_0 of (7.55) and setting $D = 0$. This gives the instability thresholds shown in Figure 7.21: $\delta = \delta_I(\epsilon)$ for an Ising front and $\delta = \delta_B^{\pm}(\epsilon)$ for Bloch fronts [117].

Figure 7.24 shows typical velocity-curvature relations at various points in the $\epsilon - \delta$ plane relative to the front bifurcation line, obtained by solving (7.57). Deep in the Bloch regime, at sufficiently small epsilon and delta values,

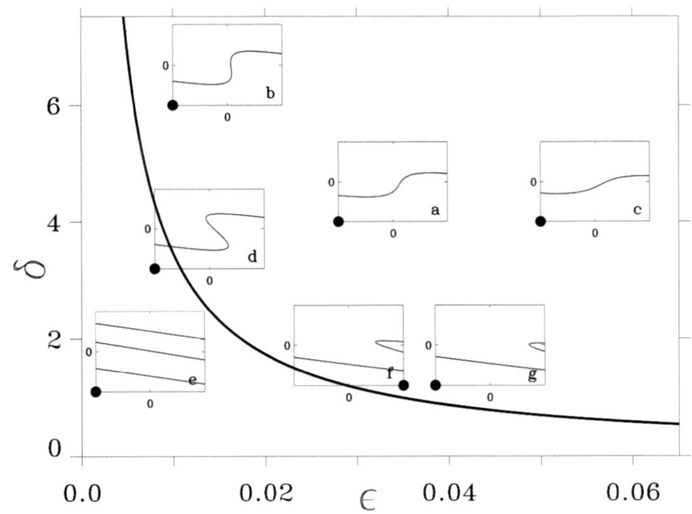

FIGURE 7.24: Velocity–curvature relations for front solutions of the (asymmetric) FHN model at various points in the $\epsilon - \delta$ parameter plane. The thick line represents the NIB bifurcation.

the velocity-curvature relation is multi-valued and the slopes of the solution branches are negative (e.g., point e in Figure 7.24). In this range stable spiral waves are found. Deep in the Ising regime, at sufficiently large epsilon and delta values, the relation is single valued with a positive slope (e.g., point e in Figure 7.24). In this range a transverse instability develops and, as Figure 7.25 shows, drives the formation of a stationary labyrinthine pattern. During this process initial bulge-like perturbations grow into fingers that do not merge with one another because of the inhibitor diffusion into the spaces that separate them[11]. The tips of the fingers often split because of the positive slop of the velocity-curvature relation even at the highly curved tips and the persistence of the transverse instability. The repeating processes of fingering and tip splitting lead eventually to the formation of labyrinthine patterns.

Close to the NIB bifurcation (e.g., points b,d) the velocity-curvature relation assumes the typical S shape associated with the cusp singularity (see Figure 2.9f). Although such relations are no longer valid as the front velocity is no longer a fast variable relative to the curvature, they are still instrumental in predicting possible dynamical behaviors associated with transitions from one Bloch front to another [117, 80]. When such a transition is uniform along the front line it simply involves a reversal in the direction of front propagation.

[11]The inhibitor diffusion can induce repulsive front interactions between front segments that are far apart from one another along the front line. A kinematic description of the front dynamics that takes into account this effect, leads to nonlocal kinematic equations [109].

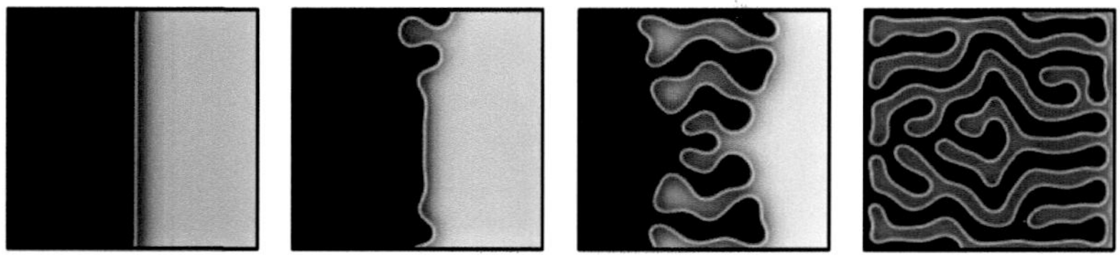

FIGURE 7.25: Solutions of the FHN model at successive times (from left to right), showing a transverse front instability deep in the Ising regime (e.g., point c in Figure 7.24) and the subsequent formation of a labyrinthine pattern by fingering and tip splitting.

When it is limited to a front segment it involves the nucleation of a spiral-vortex pair at the segment edges, where two counter-propagating Bloch fronts meet. Such local front transitions can take place as a result of a transverse instability that modulates the front curvature as Figures 7.26 and 7.27 show; the low (often negative) curvature that develops in a modulated front segment can drive a Bloch front branch beyond its end point where it no longer exists (see e.g., point d in Figure 7.24) and induce a transition to the other Bloch front that propagates in the opposite direction. As Figure 7.27 shows, such local front transitions involve the nucleation of spiral-vortex pairs. The occurrence of a transverse front instability in the vicinity of a NIB bifurcation can lead to complex spatio-temporal dynamics, as the subsequent dynamics in Figure 7.26 indicate. At the initial stage of these dynamics the number of vortices sharply increases, but as the vortex density becomes high, pair-annihilation events act to balance this growth. The outcome of these opposing processes is

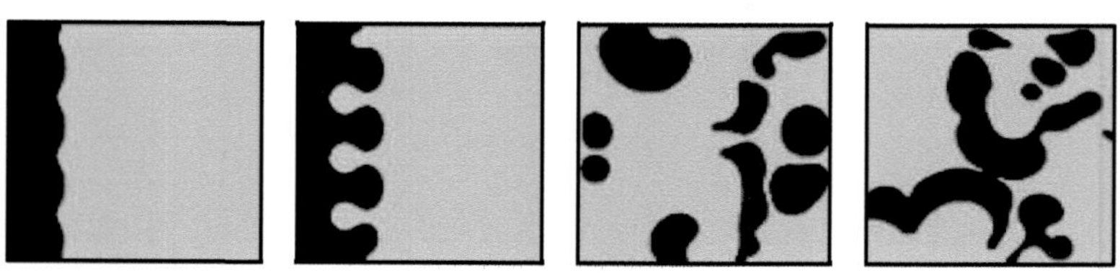

FIGURE 7.26: Solutions of the FHN model at successive times (from left to right), showing a transverse front instability near the NIB bifurcation (e.g., points b,d in Figure 7.24) and the development of a chaotic spatio-temporal state. The dynamics involve local front transitions accompanied by spiral-vortex nucleation.

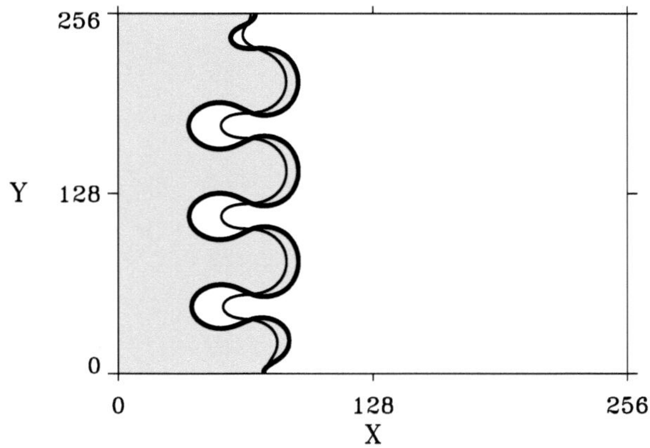

FIGURE 7.27: Zero contour lines of the u (thick line) and v (thin line) fields for the second snapshot shown in Figure 7.26. The intersection points represent vortices that have nucleated as a result of a transverse front instability.

a disordered state involving repeated events of vortex-pair nucleation[12] and annihilation in which the number of vortices fluctuates about a mean value that does not decay in time.

Transitions from one Bloch front to another, which become feasible near the NIB bifurcation, can lead to a variety of additional interesting dynamical behaviors. We have already mentioned spiral break up in Chapter 5 (Figure 5.5), a behavior found at point g in Figure 7.24 where the $C_n - \kappa$ relation is multi-valued only at sufficiently high curvature. This form implies that spiral waves can persist only near the spiral core where the curvature is high enough to allow for Bloch fronts. Additional behaviors include spiral breakup by advective fields, spot splitting and spot breathing [80, 122].

7.3.2 Spiral-core instabilities in oscillatory systems

We consider here two examples of spiral-core instabilities of Hopf spiral waves and their pattern formation effects. The first example is the CGL equation (6.53) in the high-dispersion range, that is, with large values of the coefficient α. Rescaling the spatial coordinate by $\sqrt{\epsilon}$, where $\epsilon = 1/|\alpha| \ll 1$, the

[12]At this stage additional processes contribute to local front reversal and spiral-vortex nucleation e.g., repulsive front interactions.

CGL equation can be rewritten as

$$\partial_t A = \lambda A + (\epsilon + i)\nabla^2 A - (1 + i\beta)|A|^2 A\,. \tag{7.60}$$

For $\epsilon = 0$ this equation is invariant under the Galilean transformation [84, 320]

$$\mathbf{x}' = \mathbf{x} + \mathbf{v}t\,, \qquad t' = t\,, \qquad A(\mathbf{x}', t') = e^{\frac{i}{2}\mathbf{v}\cdot(\mathbf{x}+\frac{1}{2}\mathbf{v}t)} A(x, t)\,, \tag{7.61}$$

where $\mathbf{v}$ is an arbitrary constant velocity vector. This implies in particular that if $A(\mathbf{x}, t)$ is a spiral wave solution of the form (7.47) (with $n = 1$), then the following continuous family of moving spiral-wave solutions exist too:

$$A(\mathbf{x}, t) = \rho(r')e^{i[-\nu' t + \theta + \psi(r') - \mathbf{x}'\cdot\mathbf{v}/2]}\,, \tag{7.62}$$

where $\mathbf{x}' = \mathbf{x} + \mathbf{v}t$, $\nu' = \nu - v^2/4$, and the functions ρ and ψ are the same as those in (7.47). When $\epsilon \neq 0$ the diffusion term $\epsilon\nabla^2 A$ destroys the family and leads to acceleration or deceleration of the spiral wave that is proportional to ϵv; for ϵ larger than a critical value, ϵ_c, the spiral wave decelerates and converges to a stable stationary solution, whereas for $\epsilon < \epsilon_c$ it accelerates [5]. When this spiral acceleration is combined with an Eckhaus instability of the emitted plane waves, perturbations produced by the accelerating spiral core are amplified as they move away from the core, due to the convective character of the instability (see Section 6.2.2), and can lead to spontaneous spiral-vortex nucleation and to intermittent spatio-temporal chaos [5, 6].

As a second example of a spiral core instability that projects on the spatio-temporal behavior of the entire system, let us consider the vicinity of a codimenson-2 point where a Hopf bifurcation coincides with another instability of the same uniform state. Specifically, we consider a Hopf-pitchfork bifurcation and assume that the Hopf bifurcation is the first to be traversed as some control parameter is increased. Imagine now a stable spiral-wave solution beyond the Hopf bifurcation. Increasing further the control parameter past the pitchfork bifurcation may not immediately affect the stability of the Hopf spiral; a parameter range generally exists where the Hopf mode is still effective in damping the pitchfork mode even at the spiral core. However, beyond another threshold the single-mode spiral wave solution can lose stability to the growth of the pitchfork mode at the spiral core, leading to a dual-mode spiral-wave solution, similar to that shown in Figure 5.4 for the case of a Hopf–Turing codimension-2 point except that the hosted mode is pitchfork rather than Turing.

In order to study the dynamics of such dual-mode spiral waves we consider the vicinity of a Hopf-pitchfork bifurcation where a typical dynamical variable can be written as

$$u = c_1 A e^{i\omega t} + c_2 v + c.c. + \dots\,. \tag{7.63}$$

Here, A and v are the amplitudes of the Hopf and pitchfork modes, respectively, c_1 and c_2 are constants, ω is the Hopf frequency, $c.c.$ stands for the complex conjugate, and the dots denote higher order terms. The amplitude equations for A and v can be brought to the following simplified form (after rescaling time, space and the two amplitudes, rotating the phase of A, and using the same notations for the transformed quantities) [203]:

$$\partial_t A = A - (1+i\beta)|A|^2 A + (1+i\alpha)\nabla^2 A - \gamma A v - \delta A v^2 , \quad (7.64a)$$
$$\partial_t v = \lambda v - v^3 + d\nabla^2 v + \sigma |A|^2 - \eta |A|^2 v . \quad (7.64b)$$

In these equations all parameters are real valued except for γ and δ. Note that in the absence of the terms that couple the two amplitudes these equations reduce to the normal form equations for the pitchfork bifurcation (2.6) and for the Hopf bifurcation (6.53) (the CGL equation). The coupling terms are the resonant terms that appear up to third order in the small amplitudes.

Equations (7.64) have a single mode spiral-wave solution, $(A, v) = (A_s, 0)$, where A_s has the form (7.47). To better clarify the dynamics that can be induced by instabilities of dual-mode spiral waves let us consider the simple case where $\alpha = \beta = 0$. In this case $\nu = 0$ (see (6.56)) and the spiral-wave solution becomes a stationary vortex solution with linear equiphase lines emanating from the core in radial directions as Figure 7.28(a) shows (the phase ϕ is independent of the radial coordinate r). This solution remains stable even for $\lambda > 0$, i.e., beyond the pitchfork bifurcation, because of the nonlinear damping by the Hopf mode represented by the term $-\eta|A|^2 v$ in (7.64b). Beyond some threshold, however, the single-mode solution loses stability to a pair of dual-mode spiral waves hosting the pitchfork mode at the spiral cores with

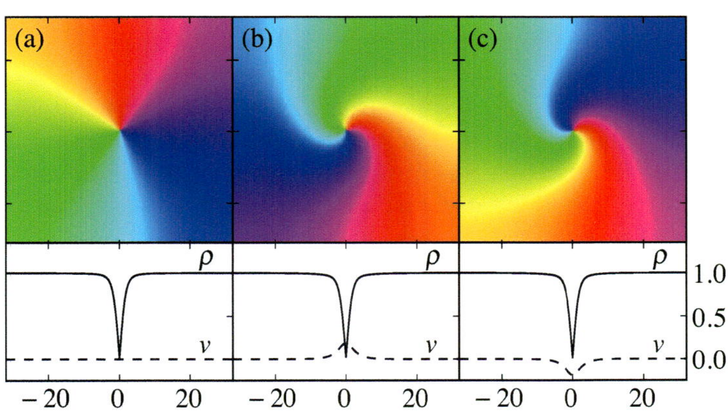

FIGURE 7.28: Vortex solutions of the Hopf–pitchfork system (7.64). (a) A single-mode vortex with $v = 0$ at the core, obtained with $\lambda = 0.62$. (b,c) Dual-mode vortices with (b) $v > 0$ and (c) $v < 0$ at the core, obtained with $\lambda = 0.635$. The top frames show the phase $\varphi = \arg A$ in the $x - y$ plane and the bottom frames show the amplitude $\rho = |A|$ and v at cross section at $y = 0$. Parameters: $\eta = 0.8$, $\kappa = 0$, $\mu = 0$, $\delta = 0.8$, $\gamma = -i$, $\alpha = 0$, $\beta = 0$, $d = 1$ $x = y = [-32, 32]$ From [203].

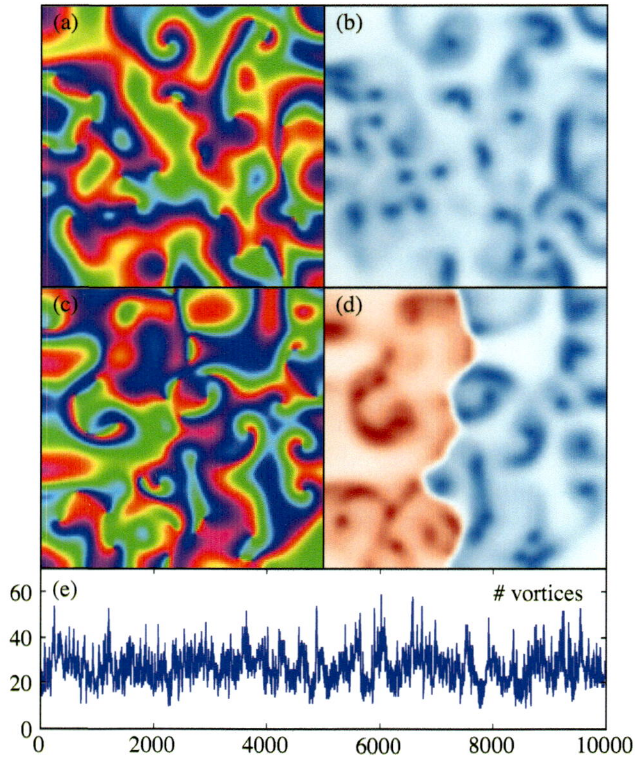

FIGURE 7.29: Spatio-temporal chaos induced by spontaneous hosting. Shown are the phase $\varphi = \arg A$ (a,c) and the mode v (b,d) far from the pitchfork bifurcation. Strong hosting events lead to steep phase gradients that induce zeros of the oscillation amplitude and spontaneous vortex-pair nucleation. This results in a chaotic state with a fluctuating number of vortices (e). Bistability of chaotic states with v positive (blue) and negative (red) allow for spatial mixture of the two chaotic states and for fronts separating them (c,d). The parameters are the same as in Figure 7.28 with $\lambda = 0.75$ and $x = y = [-64, 64]$. From [203].

positive and negative v values, as Figures 7.28(b,c) show. Note that the hosted pitchfork mode twists the phase of the solution and creates phase gradients in the radial direction. This phase-twist effect has important implications for the stability of dual-mode vortex solutions; the phase gradient it creates reduces the Hopf amplitude, $\rho = |A|$, in a circle around the core where it attains a maximal absolute value, and when steep enough, can induce *spontaneous hosting* of the pitchfork mode.

At moderate distances from the instability to dual-mode spiral waves, the phase gradients induced by the hosted pitchfork mode can reduce the Hopf amplitude, ρ, to values that enable hosting of v but are bounded away from zero. This can result in periodic oscillatory dynamics [203]. Farther away from the instability, zeros of ρ can form, resulting in vortex-pair nucleation and the onset of spatio-temporal chaos, as Figure 7.29 shows. Note that there are two symmetric chaotic states, corresponding to positive and negative v values at the vortex cores. This bistability of states allows for spatial co-existence of the chaotic states with a well defined front separating them, as Figures 7.29(c,d) show. Note also that this form of spatio-temporal chaos is unrelated to Benjamin–Feir turbulence (see Section 6.2.2) because of the choice $\alpha = \beta = 0$ in (7.64a). In fact, without the coupling to the pitchfork mode, this parameter choice amounts to the variational Ginzburg–Landau equation that cannot support spatio-temporal chaos. Hosting events of a pitchfork mode therefore provide a mechanism by which a spiral core instability develops into a chaotic state encompassing the entire system.

Chapter summary

Two main types of pattern-formation mechanisms can be distinguished; patterns formed by non-uniform instabilities of uniform states that involve monotonic or oscillatory growth of spatially structured modes, and patterns formed by local convergence to different states in multi-stable systems. The latter involve localized structures that constitute the building blocks of extended patterns. Combinations of these mechanisms are possible too, e.g., bistability of a uniform state and a patterned state that has been created in a non-uniform instability of a uniform state. Extended patterns, encompassing the entire system, can also be induced by instabilities of localized structures.

The patterns that appear beyond non-uniform instabilities can be analyzed using the amplitude-equations approach. In the case of stationary instabilities (monotonic growth) such an analysis predicts the onset of stationary periodic patterns. Systems with an inversion symmetry, $\mathbf{u} \to -\mathbf{u}$, form stripe or labyrinthine patterns, and systems that lack this symmetry first form hexagonal patterns. In the latter case a universal sequence of stable stationary states can be identified as the control parameter is varied: uniform state, hexagonal patterns and stripe (labyrinthine) patterns. Moreover, the stability ranges of these states generally overlap, leading to bistability of a uniform state and hexagonal patterns, and of hexagonal and stripe patterns. Non-uniform oscillatory instabilities lead to periodic traveling waves. An amplitude equation analysis of left and right traveling waves gives conditions for the existence and stability of standing wave patterns.

Under certain conditions scale-free patterns may appear. In the FHN model this behavior is obtained far from the onset of stationary periodic patterns, in the limit of very fast inhibitor diffusion. This limit imposes a global constraint on the spatial distribution of the activator, which results in transient patterns that lack a characteristic length. In the course of time these patterns go through a slow phase coarsening process whereby small patches of a high activator shrink and disappear while the big patches grow and merge. A similar behavior will be encountered in the context of vegetation patterns under conditions of fast water transport.

Multiplicity of stable states can be discrete, as in bistable systems, or continuous, as in oscillatory systems. Bistability of uniform states can lead to transient patterns or to stable asymptotic patterns, depending on whether the interactions between approaching fronts are attractive or repulsive. Bistability of uniform and patterned states can give rise to a multitude of stable hybrid states consisting of confined domains of one state in a background of the other state. The appearance of hybrid states is related to front pinning in a range of the control parameter. The continuous family of oscillatory states that appears beyond a uniform oscillatory instability gives rise to long-wavelength traveling waves, including Hopf spiral waves. The cores of spiral waves are localized structures that constitute building blocks for extended patterns or vortex lattices. Such patterns also contain shock lines, formed by incoming traveling waves, that separate adjacent spiral waves.

Quite often a uniform state can go through two different instabilities, e.g., a non-uniform stationary instability and a uniform oscillatory instability. Mode interactions beyond both instabilities can lead to two alternative stable states and patterns consisting of alternating domains of the two states. More intricate mode interactions can lead to a variety of dual-mode localized structures, e.g., Hopf spiral waves with Turing spots at the spiral cores. Such structures appear as instabilities of localized structures of primary modes where their low amplitudes are not effective in damping secondary modes.

Single-mode and dual-mode localized structures may go through instabilities that affect the entire system. Fronts that are bi-asymptotic to a symmetric pair of uniform stationary states can go through a pitchfork front bifurcation (the NIB bifurcation) that renders a stationary front unstable and gives rise to a pair of counter-propagating fronts. Such a bifurcation designates the onset of traveling wave patterns including Bloch spiral waves. Under suitable conditions (e.g., fast inhibitor diffusion) the stationary front can go through a transverse instability that leads to labyrinthine patterns. The combination of the two front instabilities can induce spatio-temporal chaos, a state that involves repeated events of spiral-vortex nucleation and annihilation. Similar chaotic behaviors can also be induced by spiral-core instabilities in oscillating systems, both of single-mode and dual-mode spiral waves, although the mechanisms for the onset of spatio-temporal chaos are different.

Chapter 8

External modulations of pattern forming systems

8.1 The interplay between intrinsic and extrinsic periodicities

Natural systems are often subjected to external periodic forcing. The entrainment of cardiac cells in the atrioventricular node to signals generated at the sinoatrial node, circadian rhythms entrained by the 24h day-night periodicity, and phytoplankton oscillations under seasonal forcing are examples of temporal periodic forcing of oscillatory systems [106, 110, 137]. There also exists a spatial counterpart for this class of systems, namely, periodic spatial structures, such as crystals or self-organized patterns, that are subjected to periodic spatial modulation. The modulation can be induced by an intrinsic instability, e.g., charge density waves in a 1d crystal [306, 10], or externally applied. Examples of the latter case include periodic spatial modulations of an electric field in nematic liquid-crystal convection [182, 127], periodic temperature modulation of the lower boundary of a Rayleigh–Bénard system [279, 96], and periodic illumination of a photo-sensitive chemical reaction [69, 68, 217]. An ecological example, discussed earlier in Section 3.3.3, is the restoration of degraded vegetation bands by means of parallel embankments. The ability of periodically forced systems to yield to the forcing in a synchronized manner is the subject of this Section. Yielding to temporal forcing by frequency locking will be analyzed in section 8.1.1, while yielding to spatial forcing by wavenumber locking will be analyzed in Section 8.1.2.

8.1.1 Frequency locking

An oscillatory system can respond to periodic temporal forcing by locking its oscillation frequency to the forcing frequency. By this we mean that there exists a range of the forcing frequency, ω_f, within which the system deviates from its natural frequency, ω_0, and oscillates at a frequency which is a rational fraction of the forcing frequency, i.e., at a frequency $\omega = (m/n)\omega_f$, where m and n are integers such that ω is close to ω_0. In general, this range becomes wider as the forcing amplitude increases, forming a tongue-like domain in the plane spanned by the forcing frequency and the forcing amplitude—the so-called $n : m$ resonance tongue. Thus, anywhere within the 1:1 tongue the system oscillates at exactly the forcing frequency, anywhere within the 2:1 tongue the system oscillates at exactly half the forcing frequency and so on. Outside a resonance tongue, the oscillations are no longer periodic; they are rather quasi-periodic, characterized by two rationally independent frequencies, ω_0 and ω_f.

Frequency locking and other general aspects of periodically-forced spatially extended oscillatory systems [218] have been studied experimentally, using the BZ reaction [176, 177] and catalytic surface reactions [147, 161], and theoretically, using reaction-diffusion models [177, 199, 328] and the normal-form or amplitude equation for forced oscillations [52, 51, 79, 338, 336, 11]. Forcing of the BZ reaction has been achieved by illuminating the reaction cell uniformly in space and periodically in time. Figure 8.1 shows experimental observation of resonance tongues in that reaction. Besides $n : 1$ tongues for $n = 1, ..., 5$, two additional tongues have been observed, 3:2 between 1:1 and 2:1, and 4:3 between 1:1 and 3:2. These tongues reflect a Farey-series order [125], according to which any two resonances $n : m$ and $n' : m'$ give rise to a daughter resonance $(n + n') : (m + m')$ in between [74, 108].

To get a better understanding of frequency locking let us consider the amplitude equation for a periodically forced system in the vicinity of a Hopf bifurcation. In the absence of forcing the amplitude equation coincides with the CGL equation (6.53). The modification that the forcing introduces can be deduced using symmetry considerations (see section 6.1.2.1). Consider a system with a natural (Hopf) frequency ω_0 subjected to periodic forcing with a frequency $\omega_f \approx n\omega_0$, where n is an integer. In the vicinity of the Hopf bifurcation a typical dynamical variable can be written as

$$u = Ae^{i\omega_0 t} + c.c. + \cdots = Be^{i\omega_f t/n} + c.c. + \dots , \tag{8.1}$$

where the dots represent higher order terms. The amplitude B is introduced for the sake of convenience; easily derivable constant solutions for B represent resonant or frequency-locked oscillations. As we will shortly see, existence conditions for such solutions provide the boundaries of the resonance tongues.

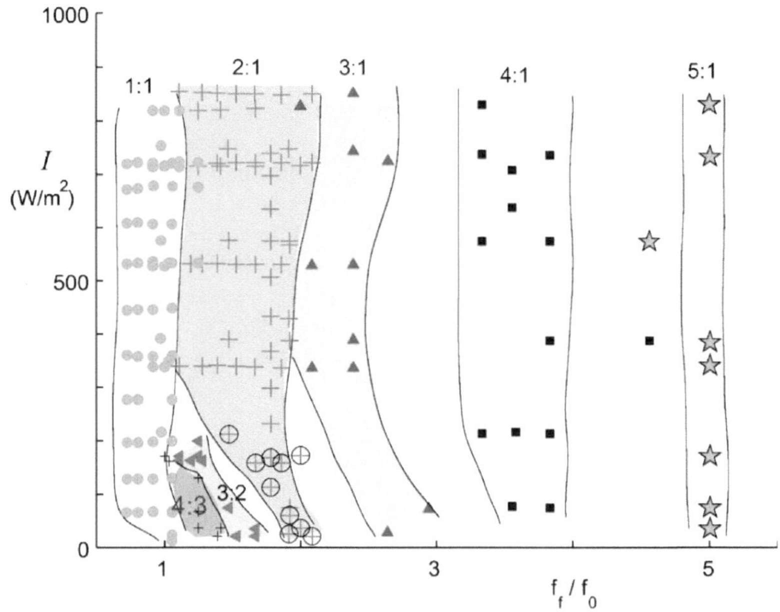

FIGURE 8.1: Experimental observation of resonance tongues for a light-sensitive BZ reaction forced by a spatially uniform time-periodic illumination. The horizontal axis represents the forcing frequency f_f (ω_f), scaled by the natural frequency f_0 (ω_0). The vertical axis represents the forcing amplitude. From [177].

The relation between the two amplitudes is $A = B\exp(-i\nu t)$, where ν is the detuning between the natural frequency ω_0 and the actual locked frequency $\omega = \omega_f/n$,

$$\nu = \omega_0 - \omega_f/n\,. \tag{8.2}$$

Note that the detuning must be regarded as small, $|\nu|/\omega_0 \ll 1$, in order for B to represent a slowly varying amplitude. The amplitude equation for B differs from the CGL equation (6.53) in that it includes a detuning term, which can be inferred by inserting $A = B\exp(-i\nu t)$ into the CGL equation, and an additional term proportional to the forcing strength. The form of the latter can be deduced from the following symmetry argument. The periodic forcing breaks the continuous time-translation symmetry of the system but still leaves a discrete symmetry, $t \to t + 2\pi/\omega_f$. This implies that the amplitude equation for B should be invariant under the transformation $B \to B\exp(2\pi i/n)$. This is a weaker invariance requirement compared to the requirement $B \to B\exp i\phi$, with ϕ being an arbitrary constant, that holds in the absence of the forcing and leads to the CGL equation. As a consequence additional terms in the amplitude equation are allowed, including in particular terms of the form B^{*n-1}.

We will mostly be concerned here with the 2:1 resonance ($n = 2$) for which the amplitude equation is

$$\partial_t B = (\lambda + i\nu)B + (1 + i\alpha)\nabla^2 B - (1 + i\beta)|B|^2 B + \gamma B^* \,, \tag{8.3}$$

where $\nu = \omega_0 - \omega_f/2$ and the constant γ is proportional to the forcing amplitude, which we assume to be small. The equation is obviously invariant under the transformation $B \to B \exp(2\pi i/n)$, which for $n = 2$ reduces to $B \to -B$. Equations of this form (hereafter the forced CGL equation) can be derived systematically for specific models and for various forms of forcing, using the multiple-scale method described in Section 6. We note that not all forcing forms necessarily contribute resonant terms at this low order of the amplitude equation. The resonant term B^* generally appears in models that include parametric forcing of linear terms.

According to (8.1) with $n = 2$, stable stationary solutions of (8.3) describe frequency-locked or resonant oscillations. In this section we confine ourselves to uniform solutions of this type and study them by writing the complex amplitude in polar form, $B = \rho e^{i\phi}$. Equation (8.3) then gives the system of equations

$$\begin{aligned} \lambda - \rho^2 + \gamma \cos(2\phi) &= 0 \\ \nu - \beta\rho^2 - \gamma \sin(2\phi) &= 0, \end{aligned} \tag{8.4}$$

which can be solved for ρ^2:

$$\rho_\pm^2 = (1 + \beta^2)^{-1} \left[\lambda + \nu\beta \pm \sqrt{(1 + \beta^2)\gamma^2 - (\nu - \lambda\beta)^2} \right] . \tag{8.5}$$

There are two conditions for the existence of these solutions. The first is that the argument in the square root is positive, and the second is that the expression for $\rho_\pm^2$ is positive. Within the detuning range $\nu > -\lambda/\beta$ for $\beta > 0$ and $\nu < -\lambda/\beta$ for $\beta < 0$ these conditions give the existence threshold:

$$\gamma_b = \frac{|\nu - \lambda\beta|}{\sqrt{1 + \beta^2}} \,. \tag{8.6}$$

Outside this detuning range the threshold is $\gamma_b = \sqrt{\lambda^2 + \nu^2}$. Solving (8.4) for the phase ϕ too, we find a pair of solutions of (8.3) that appear in a saddle-node bifurcation at $\gamma = \gamma_b$. A linear stability analysis of (8.3) shows that the solution obtained with ρ_+ is stable while the solution obtained with ρ_- is unstable. In addition to this solution pair there exists a symmetric solution pair associated with the invariance of (8.3) to the transformation $B \to B \exp(i\pi)$, i.e., a solution pair with a phase shifted by ϕ. The appearance of the four solutions in a pair of saddle-node bifurcations at $\gamma = \gamma_b$ is shown in Figure 8.2. The phases of the stable solutions are given by

$$\phi_0 = \frac{1}{2} \arcsin\left(\frac{\tilde{\nu} - \beta\tilde{\gamma}}{\gamma} \right), \qquad \phi_\pi = \phi_0 + \pi \,, \tag{8.7}$$

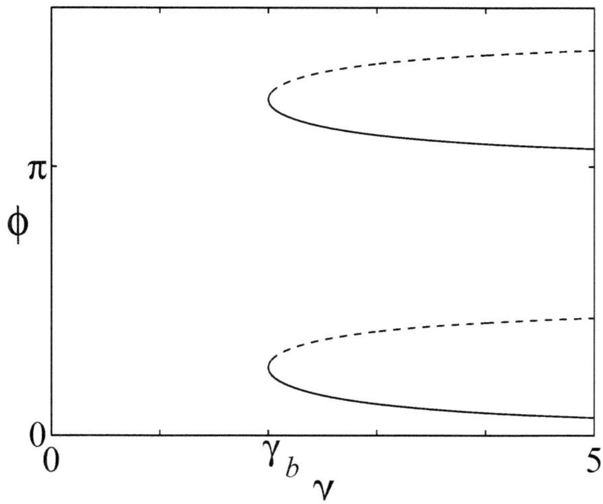

FIGURE 8.2: Bistability of phase states. Two stable phase states appear in a symmetric pair of saddle-node bifurcations as γ is increased past the boundary, γ_b, of the 2:1 resonance tongue. From [336].

where

$$\tilde{\nu} = \frac{\nu - \beta\lambda}{1+\beta^2}, \qquad \tilde{\gamma} = \sqrt{\frac{\gamma^2}{1+\beta^2} - \tilde{\nu}^2}\,. \tag{8.8}$$

To summarize this analysis, two stable stationary uniform solutions, $B = \rho_+ e^{i\phi_0}$ and $B = \rho_+ e^{i\phi_\pi}$, of the amplitude equation (8.3) appear as γ is increased past γ_b. They have the same amplitude, ρ_+ but differ in their phases; the phase of one solution, ϕ_π, is shifted by π with respect to the phase, ϕ_0, of the other solution. The existence range of these solutions, shown in grey in Figure 8.3, defines the 2:1 resonance tongue. Within the tongue the system adjusts its oscillation frequency to exactly half the forcing frequency, even when $\omega_0 \neq \omega_f/2$. Thus, the 2:1 forcing has created a bistable system characterized by two uniform states that oscillate out of phase at exactly half the forcing frequency anywhere within the 2:1 resonance tongue.

Two additional general aspects of periodically forced oscillatory systems are front structures, associated with multi-stability of phase states (bistability in the case of the 2:1 resonance), and symmetry-breaking spatial instabilities of the uniform (zero amplitude) state that are induced by the forcing. These two aspects will be studied in Sections 8.2.1 and 8.3.1.

8.1.2 Wavenumber locking

In analogy to frequency locking in periodically forced oscillatory systems, spatial periodic forcing of pattern-forming systems can result in wavenumber

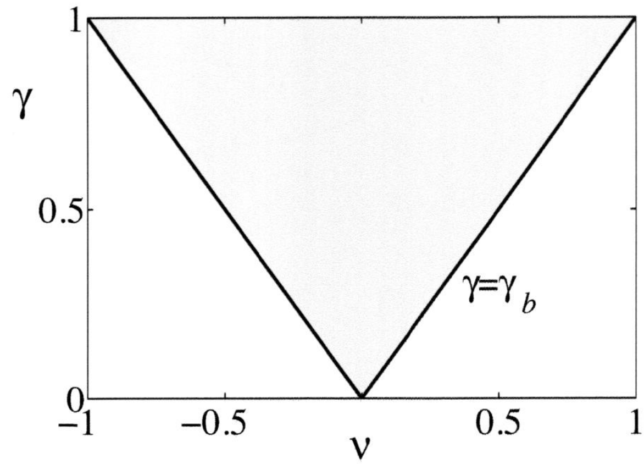

FIGURE 8.3: The 2:1 resonance tongue according to equation (8.6) with $\beta = 0$. Anywhere within the tongue (shaded area) the system oscillates at exactly half the forcing frequency. From [336].

locking. The picture, however, is more intricate because of the freedom of 2d spatial systems to respond in a direction orthogonal to the forcing direction. We will study this spatial resonance problem in greater detail using amplitude equations derived from a periodically forced SH equation. Motivated by the vegetation restoration problem (see Section 3.3.3), we consider a parametric spatial forcing of the SH equation rather than an additive forcing. The reason is that the redistribution of soil water by the parallel embankments modulates the biomass growth rate, a parameter that multiplies the biomass variable in the biomass equation (see Section 9.2.1). The forced SH equation reads

$$\partial_t u = \lambda u - \left(\nabla^2 + k_0^2\right)^2 u - u^3 + \gamma u \cos\left(k_f x\right), \tag{8.9}$$

where k_f is the forcing wavenumber, and γ is the forcing amplitude. Notice that (8.9) is invariant under the transformation $\gamma \to -\gamma$ and $x \to x + \pi/k_f$. The discussion can therefore be restricted to positive γ values without losing generality. We first look for 1d wavenumber-locked stripe solutions whose existence ranges define the resonance tongues. In analogy to (8.1) we approximate a stripe solution within or in the vicinity of an $n:1$ resonance as

$$u = Ae^{ik_0 x} + c.c. + \cdots = Be^{ik_f x/n} + c.c. + \ldots . \tag{8.10}$$

We consider the vicinity of the instability to stripe patterns, $0 < \lambda \ll 1$ (see Section 5.3.1), and use λ as the small parameter in a multiple-scale analysis (see section 6.1.3.1). Specifically, we assume a small, slowly varying amplitude that satisfies $|B| \sim O(\sqrt{|\lambda|})$, $|\partial_x B| \sim O(|\lambda|)$, $|\partial_t B| \sim O(|\lambda|^{3/2})$. We further consider the vicinity of an exact resonance $n:1$ by taking the detuning

$$\nu_n = k_0 - k_f/n\,, \tag{8.11}$$

to be small, $\nu_n \sim O(\sqrt{|\lambda|})$. Finally, we assume that the forcing is weak, $\gamma \sim O(\sqrt{|\lambda|})$. A multiple-scale analysis to order $|\lambda|^{3/2}$ then leads to the following amplitude equation [204, 205]

$$\begin{aligned}\partial_t B &= \lambda B - 3|B|^2 B - (2ik_0\partial_x + 2k_0\nu_n)^2 B \\ &\quad + \left(\tfrac{\gamma}{2}\right)^2 \left[(d_+ + d_-)B + \delta_{n,1} d_- B^*\right] ,\end{aligned} \tag{8.12}$$

where

$$d_\pm = \frac{1}{k_f^2 (k_f \pm 2k_0)^2} , \tag{8.13}$$

and $\delta_{n,1}$ is the Kronecker delta. Note that $d_\pm$ diverges for $k_f = \mp 2k_0$, that is, for the exact 2:1 resonance. This resonance requires a different scaling of the forcing, $\gamma \sim O(|\lambda|)$, and is described by a slightly different amplitude equation [195, 196]

$$\partial_t B = \lambda B - 3|B|^2 B - (2ik_0\partial_x - 2k_0\nu_2)^2 B + \frac{\gamma}{2} B^* . \tag{8.14}$$

Notice that a resonant term of the form ${B^*}^{n-1}$ appears only in the amplitude equation (8.14) for the 2:1 resonance ($n = 2$). For all other resonances such terms do not appear up to the order of calculation, $|\lambda|^{3/2}$. The amplitude equations for these resonances (8.12) contain other resonant terms proportional to γ^2. Terms of the form ${B^*}^{n-1}$ with $n > 2$ can appear at relatively low orders when the parametric forcing includes nonlinear terms, or when the inversion symmetry $u \to -u$ of (8.9) is broken [123].

Constant solutions of the amplitude equations (8.12) and (8.14) represent $n:1$ wavenumber-locked (resonant) stationary stripe patterns. For $n \neq 2$ they have the form [204, 205]

$$B = \rho_n \exp(i\phi) , \qquad \rho_n = \frac{1}{\sqrt{3}} \sqrt{\lambda - (2k_0\nu_n)^2 + d\gamma^2/4}, \tag{8.15}$$

with $d = d_+ + (1 + \delta_{n,1})d_-$. The phase ϕ is equal to zero or π for $n = 1$, but undetermined for higher resonances at the order, $|\lambda|^{3/2}$, of the calculation. Both the zero and π phase values for $n = 1$ correspond to solutions of (8.12) that are stable to uniform perturbations. The resonant stripe solutions exist for $\gamma > \gamma_n$ where

$$\gamma_n = 2\sqrt{\frac{(2k_0\nu_n)^2 - \lambda}{d}} . \tag{8.16}$$

Wavenumber-locked patterns corresponding to the 2:1 resonance have the form [196]

$$B = \rho_2 \exp(i\phi) , \qquad \rho_2 = \frac{1}{\sqrt{3}} \sqrt{\lambda - (2k_0\nu_2)^2 + \tfrac{\gamma}{2}\cos 2\phi} . \tag{8.17}$$

with $\phi = 0, \pi/2, \pi, 3\pi/2$. Of the four different phase values only $\phi = 0, \pi$ represent stable solutions. These solutions exist for $\gamma > \gamma_2$ where

$$\gamma_2 = 2\left[(2k_0\nu_2)^2 - \lambda\right] . \tag{8.18}$$

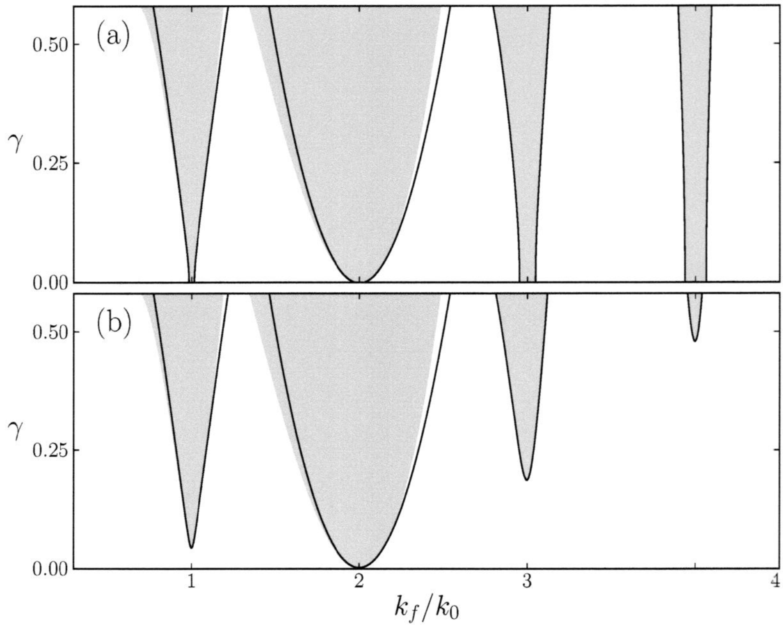

FIGURE 8.4: Existence domains of resonant stripe solutions of (8.9), (a) above the pattern forming instability ($\lambda > 0$), and (b) below it ($\lambda < 0$). The shaded regions indicate the range of resonant solutions computed from stationary solutions of (8.9), and the solid curves show the region boundary approximations based on the amplitude equation approach. From [204]. Parameters: (a) $\lambda = 0.001$, (b) $\lambda = -0.001$.

Figure 8.4 shows the tongue-shaped existence ranges of $n{:}1$ resonant stripe patterns with $n = 1, ..., 4$, for parameters above ($\lambda > 0$) and below ($\lambda < 0$) the pattern forming instability. The solid lines in the figure are the resonance boundaries obtained from equations (8.15) and (8.17) and the shaded regions are the resonance ranges obtained by solving the forced SH equation (8.9) numerically for stationary solutions, using a continuation method [67]. We emphasize that the resonance tongues depicted in Figure 8.4 represent ranges of existence and stability to uniform amplitude perturbations. Non-uniform instabilities, Eckhaus in 1d and 2d and zigzag in 2d, reduce the stability ranges [196].

Of all resonances shown in Figure 8.4, the 2:1 resonance region stands out in its robustness. It is wider and, for $\lambda < 0$, i.e., below the pattern-forming instability, it appears at lower forcing amplitude γ. This is because the forcing is parametric, involving the linear term in the SH equation. As in the oscillatory case, the 2:1 tongue contains a bistability range of two phase states, shifted by π with respect to one another, except that the states correspond to stationary stripe patterns rather than to uniform oscillatory states. In the next section we will study the implications of bistability of phase states for pattern formation.

8.2 Multistability of phase states and fronts

Resonant response to periodic forcing involves not only frequency or wavenumber locking but also a multiplicity of stable phase states. We have discussed already the bistability of phase states within the 2:1 resonance tongue of uniform oscillations and pointed out that an analogous bistability of phase states exists within the 2:1 resonance tongue of stationary periodic patterns. The local convergence to different phase states gives rise to spatial patterns that involve *phase fronts*. These fronts shift the oscillation phase or the spatial-pattern phase by $2\pi/n$, where n represents the multiplicity of stable phase states. We will mostly be concerned here with bistability of phase states, i.e., with $n = 2$, and study the structure and stability properties of fronts that shift the phase by π. In Section 7.3.1 we used the FHN model to study two front instabilities in bistable systems, a longitudinal instability—the NIB bifurcation, and a transverse instability. These are generic instabilities, which we may expect to find in periodically forced systems too. We will show here that this is indeed the case, addressing temporal forcing in Section 8.2.1 and spatial forcing in Section 8.2.2. We will also briefly consider resonances higher than $n = 2$ in the context of temporally forced oscillatory systems, and the multi-phase patterns that can arise in such cases (Section 8.2.1).

8.2.1 Temporally forced oscillatory systems

The amplitude equation for temporally forced oscillatory systems with $\omega_f \approx 2\omega_0$ is given by (8.3). Consider first the special case $\nu = \alpha = \beta = 0$ in which the equation is variational, that is, derivable from a Lyapunov functional. The two stable phase states in this case are given by $B = \pm\kappa_1$ where $\kappa_1 = \sqrt{\lambda + \gamma}$. Two types of front solutions, biasymptotic to $\kappa_1 = \sqrt{\lambda + \gamma}$ as $x \to -\infty$ and to $-\kappa_1$ as $x \to \infty$, are known to exist [52]. The first is a real-valued solution that has the characteristics of an Ising front (see Section 7.3.1), as we will shortly see, and is given by

$$B_{Ising}(x) = -\kappa_1 \tanh\left(\frac{1}{\sqrt{2}}\kappa_1 x\right). \tag{8.19}$$

The second type of front solution is complex-valued and has the characteristics of a Bloch front. There are two front solutions of this type, describing a pair of Bloch fronts:

$$B_{Bloch}{}^{\pm}(x) = -\kappa_1 \tanh(\kappa_2 x) \pm i\eta \,\mathrm{sech}(\kappa_2 x)\,, \tag{8.20}$$

where $\kappa_2 = \sqrt{2\gamma}$ and $\eta = \sqrt{\lambda - 3\gamma}$. The Ising front (8.19) exists for all γ values, while Bloch fronts (8.20) exist only below the threshold $\gamma_{NIB} = \lambda/3$. In this variational limit the Bloch fronts are stationary, like the Ising front, but their phase trajectories in the complex B plane are different as Figure 8.5 illustrates; while Ising-front trajectories go through the origin, Bloch-front trajectories bypass the origin, similar to the trajectories shown in Figure 7.20 for front solutions of the FHN model. In terms of the phase $\arg(B)$, Ising fronts are characterized by a constant phase that suffers a discontinuous jump by $\pm\pi$ at the front core, where B vanishes, while Bloch fronts are characterized by smoothly rotating phases, either clockwise or anti-clockwise.

In the non-variational case the pair of Bloch fronts propagate in opposite directions. Their velocities can be calculated analytically in the case of a weakly non-variational system by approximating a Bloch front solution that propagates at constant speed c as

$$B_{Bloch}(z) = B_0(z) + \epsilon B_1(z) + \ldots\,, \tag{8.21}$$

where B_0 is given by (8.20), B_1 is the leading order correction, $z = x - ct$ and ϵ is a small parameter representing the small deviation from the variational limit, and therefore the sizes of ν, α, β and c. Inserting this form into (8.3) leads at order ϵ to a linear equation for the correction B_1. A solvability condition, associated with the marginal translation mode, then provides the following expression for the front speed [52]:

$$c = \pm\frac{3\pi\eta\kappa_1}{2\kappa_2(3\lambda - \gamma)}\left[\beta\lambda - \nu + (\alpha - \beta)\gamma\right]. \tag{8.22}$$

A similar calculation carried out for B_0 representing an Ising front (8.19) shows that Ising fronts remain stationary, both below and above the threshold γ_{NIB}, also in the non-variational case. These results suggest the existence of a NIB bifurcation, that is, a pitchfork front bifurcation that renders a stationary Ising front unstable and gives rise to a pair of counter-propagating Bloch fronts, very much like the NIB bifurcation found for the FHN model. Numerical studies of (8.3) indeed confirm this expectation, as the bifurcation diagram in Figure 8.6 shows (compare with Figure 7.19). Note that while in the FHN model the fronts separate two uniform stationary states, in the context of forced oscillations they separate two uniform states that oscillate out of phase, i.e., with a phase shift of π. Figure 8.7 shows the threshold curve $\gamma = \gamma_{NIB}$ within the 2:1 resonance tongue, which split the tongue into a lower Bloch-front part and an upper Ising front part.

In Section 7.3.1 we argued that the NIB bifurcation designates a transition from stationary patterns in the Ising regime to traveling-wave patterns in the Bloch regime, and demonstrated that using the bistable FHN model.

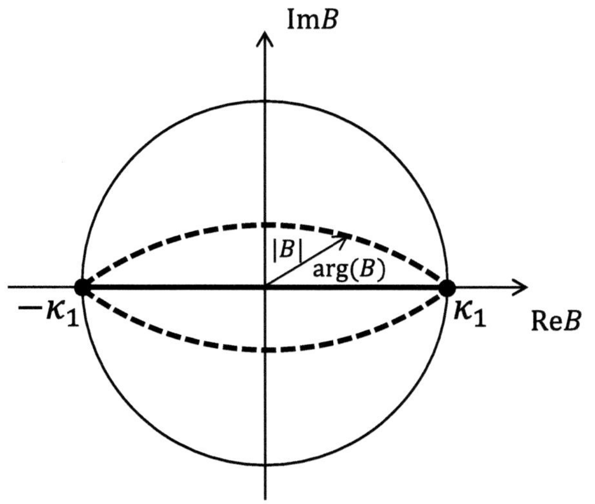

FIGURE 8.5: A schematic illustration of Ising and Bloch fronts as phase trajectories in the complex amplitude plane, parameterized by the spatial coordinate x. The solid horizontal line represents an Ising front while the dashed lines represent the two types of Bloch fronts. Along an Ising trajectory the phase, $\arg(B)$, is constant except at the origin, $B = 0$, where it jumps by $\pm\pi$. Along Bloch fronts the phase rotates either clockwise or anti-clockwise and $|B| > 0$.

The NIB bifurcation in the forced CGL equation plays a similar role in designating a transition from standing-wave patterns (stationary solutions of the amplitude equation) to traveling-wave patterns. However, in the context of forced oscillations it has an additional significance—it also designates a transition from resonant standing-wave patterns to non-resonant traveling-wave patterns. This is because stationary solutions of (8.3) for the oscillation amplitude B correspond to resonant oscillations at $\omega_f/2$ (see (8.1)), while time-dependent solutions do not. The NIB threshold $\gamma = \gamma_{NIB}$ therefore restricts the resonance range of non-uniform oscillations to the Ising regime, $\gamma > \gamma_{NIB}$. In the Bloch regime $\gamma_b < \gamma < \gamma_{NIB}$ uniform oscillations are still resonant but non-uniform oscillations involve traveling waves, which shift the oscillation frequency and therefore are not resonant. These theoretical predictions have been confirmed in numerical studies of a forced oscillatory FHN model and in experiments on a forced BZ reaction as Figure 8.8 shows [177].

In two-space dimensions fronts can go through transverse instabilities too [338, 336]. In the bistable FHN model we found that a transverse instability of an Ising front far from the NIB bifurcation can lead to fingering, tip splitting, and eventually to stationary labyrinthine patterns (see Section 7.3.1).

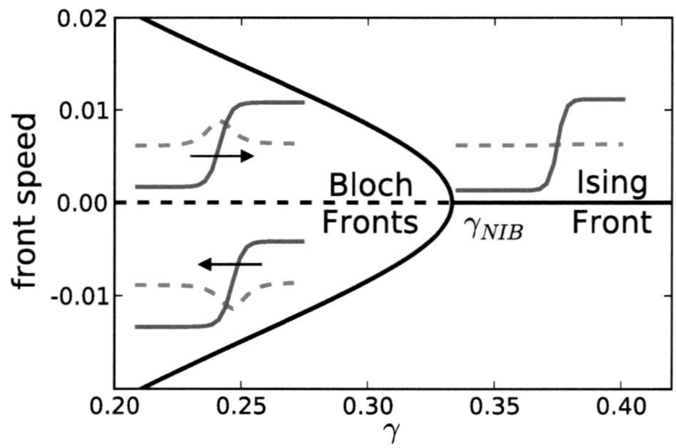

FIGURE 8.6: The nonequilibrium Ising–Bloch bifurcation diagram for equation (8.3). For $\gamma > \gamma_{\rm NIB}$ there is a single stable Ising front with zero speed. For $\gamma < \gamma_{\rm NIB}$ the Ising front is unstable and there are a pair of stable counter-propagating Bloch fronts. The insets show the $Re(B)$ curves (solid blue) and $Im(B)$ curves (dashed green) across the front position. Parameters: $\lambda = 1.0$, $\nu = 0.01$ and $\alpha = \beta = 0$. From [199].

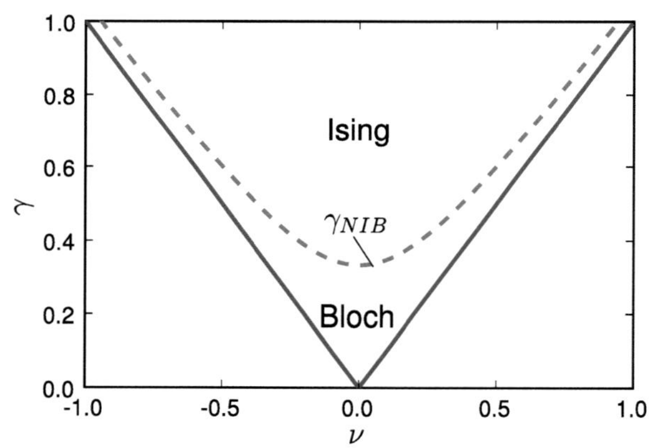

FIGURE 8.7: The NIB threshold within the 2:1 resonance tongue for the forced CGL equation (8.3). Inside the tongue-shaped region bounded by the solid lines $\gamma = \gamma_b$ uniform solutions are frequency-locked (resonant). Non-uniform oscillations are frequency-locked only above the dashed curve $\gamma = \gamma_{\rm NIB}$. Parameters $\lambda = 1.0$, $\nu = 0.01$ and $\alpha = \beta = 0$. From [199].

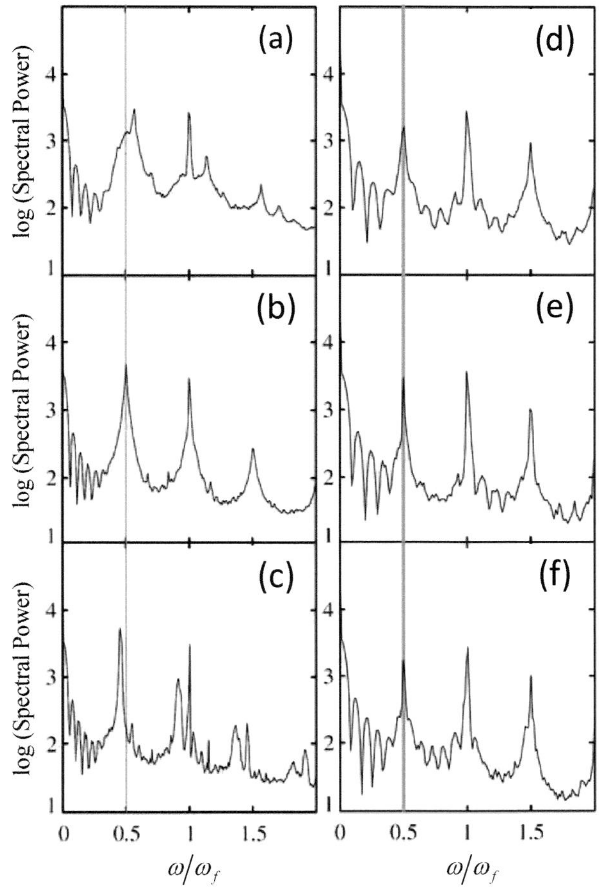

FIGURE 8.8: Power spectra for non-resonant Bloch spiral waves (a-c) and resonant Ising standing-waves (d-f) within the 2:1 resonance tongue of a forced BZ reaction. The three panels (top to bottom) correspond to negative, zero and positive detuning $\nu = \omega_f(\omega_0/\omega_f - 0.5)$. In panels (a–c) the power spectra, normalized to the forcing frequency f_f, show that the largest response at non-zero detuning (panels a,c) is near, but not exactly at half the forcing frequency as indicated by the vertical line. The spiral waves therefore do not lock to the forcing frequency. In panels (d–f) the largest response is at exactly half the forcing frequency also for non-zero detuning (panels d,f), indicating that the standing waves are resonant. Adopted from [177].

A similar behavior is found also in 2:1 forced oscillatory systems [338, 199]. In addition, approaching the NIB bifurcation can lead to spatio-temporal chaos as found in the bistable FHN model (see Section 7.3.1). Figure 8.9 shows such processes in numerical simulations of the FCGL equation (8.3) and in

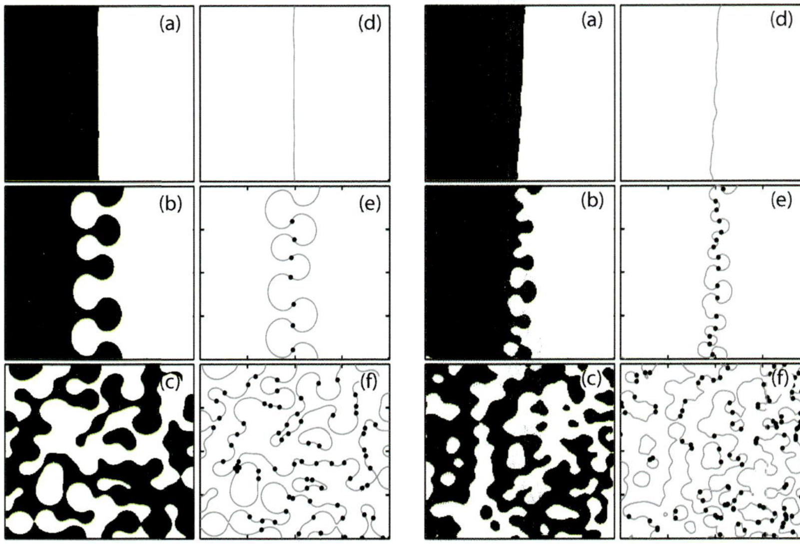

FIGURE 8.9: Spiral-vortex nucleation and formation of spatio-temporal chaos in a numerical solution of equation (8.3) (left) and in the forced BZ reaction (right). Panels (a)–(c) show the oscillation phase at three different times while panels (d)–(f) show the front line and the position of vortices along it (shown as solid circles) at the corresponding times. Perturbations on an unstable front grow and pairs of vortices form along the front (panels a,b and d,e). The process continues until an asymptotic state of spatio-temporal chaos is achieved with repeating events of vortex-pair nucleation and annihilation across the whole system (panels c and f). From [198].

experiments on the periodically forced BZ reaction. In both cases the initial condition is a nearly planar front separating the two stable phase states. The transverse instability that develops induces front-reversal events along the front line and the nucleation of spiral-vortex pairs. The process appears to continue indefinitely, leading to a state of spatio-temporal chaos with vortices spontaneously nucleating and annihilating in the whole spatial domain [198, 199].

Interesting front dynamics are also obtained within the 4:1 resonance, which gives rise to four stable uniform phase states. In all states the oscillating frequency is locked to $\omega_f/4$ but the oscillation phase shifts from one state to another by $\pi/2$. This state multiplicity allows for two types of fronts, traveling fronts that shift the phase by $\pi/2$ ($\pi/2$-fronts) and stationary fronts that shift the phase by π (π-fronts). Analytical and numerical studies [81] of the interaction between a pair of $\pi/2$-fronts using the amplitude equation for this resonance (the CGL equation with a forcing term γB^{*4}) reveal a change

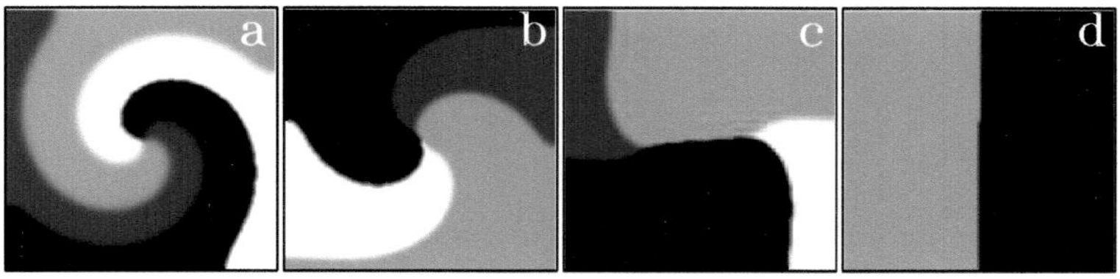

FIGURE 8.10: Numerical solution of the amplitude equation for the 4:1 resonance, showing the collapse of a rotating four-phase spiral-wave into a stationary two-phase pattern when the forcing is strong enough. The panels show the oscillation phase in the x-y plane. (a) The initial four-phase spiral wave computed as a stable solution at weak forcing. (b) The spiral core, a 4-point vertex, splits into two 3-point vertices connected by a π-front. (c) A two-phase pattern develops as the 3-point vertices further separate. (d) The final stationary two-phase pattern. From [81].

from attractive to repulsive interactions as the forcing strength decreases below a critical value. Associated with this front-interaction change is a front instability from a stationary π-front at strong forcing to a pair of propagating $\pi/2$-fronts at weak forcing. This π-front instability designates a transition from stationary two-phase patterns (representing standing two-phase waves) at strong forcing to traveling four-phase patterns at low forcing. An example of the reverse transition is shown in Figure 8.10; a stable four-phase spiral wave at weak forcing destabilizes into a stationary two-phase wave upon increasing the forcing strength. A Similar instability have been found for higher even resonances, e.g., a stationary π-front decomposing into three traveling $\pi/3$-fronts within the 6:1 resonance [82]. The π-front instability, predicted by the amplitude-equation analysis, has been found in direct numerical studies of a periodically forced FHN model but has not been confirmed yet in experiments [175].

Temporally forced oscillatory systems provide excellent examples of pattern formation that results from a multiplicity of stable uniform states (see Section 7.2). Besides two-phase patterns within the 2:1 resonance tongue and four-phase patterns within the 4:1 tongue of the forced BZ reaction, multiphase patterns have also been observed within the 3:1, 5:1 and 6:1 tongues, as Figure 8.11 shows [177]. The 2:1 resonance, however, also gives rise to pattern formation by a non-uniform instability of a uniform state (see Section 7.1). An instability of this kind will be discussed in Section 8.3.1.

8.2.2 Spatially forced stripe-forming systems

The 2:1 resonance of spatially forced stripe-forming systems gives rise to a pair of stable, wavenumber-locked stripe patterns that differ from one another by a phase shift of π (see (8.17)). Bistability of this kind allows for front

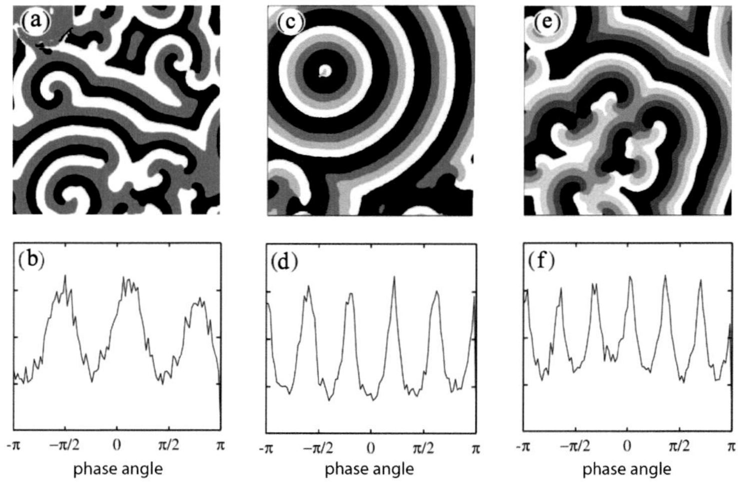

FIGURE 8.11: Multi-phase traveling waves in the forced BZ reaction within the 3:1 (a,b), 5:1 (c,d) and 6:1 (e,f) resonances. Shown for each resonance are the phase distribution in the x, y plane (top) and a phase histogram (bottom) that clearly shows the prevalence of domains with 3 (b), 5 (d) and 6 (f) distinct phase angles equally distributed along the circle. From [177].

solutions that are biasymptotic to the two periodic stripe solutions, with a sharp phase shift of π across their cores. Two types of fronts are possible, (i) longitudinal fronts aligned parallel to the stripe direction, and (ii) transverse fronts aligned perpendicular to the stripe direction. Both are solutions of the amplitude equation

$$\partial_t B = \lambda B - 3|B|^2 B - (2ik_0\partial_x - 2k_0\nu_2 + \partial_y^2)^2 B + \frac{\gamma}{2} B^* , \tag{8.23}$$

which is the 2d extension of (8.14) (see also Section 6.1.3.1). The longitudinal fronts are solutions $B = B(x,t)$ that are independent of y. Under conditions of exact 2:1 resonance, i.e., $\nu_2 = 0$, they satisfy an amplitude equation that can be brought to the form

$$\partial_t B = \lambda B - |B|^2 B + \partial_x^2 B + \gamma B^* , \tag{8.24}$$

with appropriate rescaling of B, x and γ (for simplicity we keep the same notation for the rescaled quantities). Equation (8.24) is the variational limit of the forced CGL equation (8.3) where all imaginary constants are set to zero. We have already considered this limit in Section 8.2.1, and presented analytical solutions, (8.19) and (8.20), describing stationary Ising and Bloch fronts. Both the Ising and the Bloch fronts shift the phase of the stripe pattern by π, but while the Ising front shifts the phase discontinuously and vanishes

FIGURE 8.12: Stationary transverse fronts in 2:1 spatially forced stripe patterns. An Ising (left panel) and Bloch (middle panel) fronts, shifting the phase of the stripe pattern along the vertical (y) axis by π. In an Ising front the phase shift is discontinuous and the amplitude vanish at the front core. In a Bloch front the phase shift is continuous; in the complex amplitude plane the phase rotates smoothly either clockwise or counter-clockwise, as in Figure 8.5. When the front line comprises the two Bloch front types, a vortex structure (dislocation defect) forms where two Bloch fronts meet. At the core of the vortex the phase is singular and the amplitude vanishes (right panel). From [166].

at the front core, the Bloch fronts shift the phase smoothly, clockwise and counter-clockwise in the complex B plane, and do not vanish at the core.

The transverse fronts are solutions $B = B(y,t)$ of (8.23) that are independent of x and satisfy an amplitude equation that can be brought to the form

$$\partial_t B = \lambda B - |B|^2 B - \partial_y^4 B + \gamma B^* \,. \tag{8.25}$$

This equation also has a stationary Ising front solution that goes through a NIB bifurcation as γ is decreased below a critical value γ_{NIB} [158]. Here too, all fronts are stationary and shift the phase of the stripe pattern by π except that now they are aligned perpendicular to the stripes. Figure 8.12 shows an Ising front, representing a discontinuous phase jump by π, and a Bloch front, representing a smooth transverse phase shift. The amplitude B vanishes along the core line of the Ising front but is bounded away from zero in the case of Bloch fronts. Also shown in Figure 8.12 is a Bloch vortex solution representing two different Bloch fronts that meet at a point defect, where the amplitude $B(x,y)$ vanishes.

Both Ising and Bloch fronts can act as buiding blocks for more complex spatial patterns. These patterns may involve stable line defects associated with longitudinal and transverse Ising and Bloch fronts, point defects associated with Bloch vortices, bounded pairs of transverse fronts [158] and possibly more complex clusters of fronts and vortices. The rich spatio-temporal dynamics found in forced oscillations, however, is absent here because of the variational nature of the amplitude equations.

8.3 Symmetry breaking instabilities

There is a third type of response to periodic forcing, besides frequency or wavenumber locking and the appearance of multiphase patterns, that involves the creation of a new state. The new state appears as a result of an additional instability of the zero state, i.e., the state that went through the basic instability to periodic oscillations or periodic patterns. Temporal forcing (periodic in time and uniform in space) of a system that goes through a Hopf bifurcation to uniform oscillations, can induce a non-uniform stationary instability (i.e., Turing-like) of the zero state, in addition to the Hopf instability that leads to standing waves [338, 336]. Spatial forcing (periodic in space and constant in time) can break the remaining symmetry along the stripes and lead to 2d oblique or rectangular patterns [195]. We will study these instabilities and their effects on resonant system response and pattern formation in Sections 8.3.1 (temporal forcing) and 8.3.2 (spatial forcing).

8.3.1 Instabilities induced by temporal forcing

We focus on the 2:1 resonance because forcing in higher resonances generally introduces to the amplitude equations terms of the form B^{*n-1}, which are nonlinear for $n > 2$ and therefore do not affect the linear stability of the zero state. In order to identify the Turing-like instability that the 2:1 temporal forcing induces, we study the amplitude equation (8.3). Writing the 1d version of (8.3) in terms of the real and imaginary parts U and V of the complex valued amplitude $B = U + iV$, we study the stability of $U = V = 0$ to infinitesimal perturbations of the form

$$\begin{bmatrix} \delta U(x,t) \\ \delta V(x,t) \end{bmatrix} = \begin{bmatrix} u_{0k} \\ v_{0k} \end{bmatrix} e^{\sigma t + ikx} + c.c.\,. \tag{8.26}$$

Insertion of this form into the coupled equations for U and V and linearization in u_{0k} and v_{0k} (see Section 5.3.2) lead to the dispersion relation [336]

$$\sigma(k) = \lambda - k^2 \pm \sqrt{\gamma^2 - (\nu - \alpha k^2)^2}\,. \tag{8.27}$$

Note that in the absence of the forcing ($\gamma = 0$) this dispersion relation gives a uniform oscillatory instability at $\lambda = 0$ with a frequency ν. This is the original Hopf bifurcation as can be seen by inserting $B = A\exp(i\nu)$ into (8.1) for $n = 2$.

The conditions for a non-uniform stationary instability at a critical forcing value γ_c are $\sigma(k_0) = 0$, $d\sigma/dk|_{k=k_0} = 0$ and $d^2\sigma/dk^2|_{k=k_0} < 0$, where k_0 is the wavenumber of the mode that begins to grow at $\gamma = \gamma_c$. Applying these conditions to (8.27) we obtain

$$\gamma_c = \frac{\nu - \alpha\lambda}{\sqrt{1+\alpha^2}} \qquad\qquad k_0^2 = \frac{\lambda + \nu\alpha}{1+\alpha^2}\,. \tag{8.28}$$

Above γ_c stationary periodic solutions $B = B_s(x)$ of (8.3) appear. These solutions describe resonant standing waves $A_s(x,t) = B_s(x)\exp(i\omega_f t/2)$ of the original problem. Recall that constant solutions of (8.3) represent resonant uniform oscillations, which appear above γ_b (see (8.6)). Comparing the expressions for γ_b and γ_c, a forcing range, $\gamma_c < \gamma < \gamma_b$, can be identified where the system is outside the 2:1 resonant tongue of uniform oscillations but above the threshold for resonant standing waves. This raises the following interesting question: can non-uniform oscillations extend the resonant response to forcing-frequency ranges outside the tongue of resonant uniform oscillations? In order to address this question we need to study the interplay between non-resonant uniform oscillations outside the 2:1 tongue and resonant standing waves.

Non-resonant oscillations are described by oscillatory solutions of (8.3). The appearance of such solutions can be studied using the dispersion relation (8.27), viewing λ as the bifurcation parameter. The conditions for a uniform oscillatory instability are $\sigma = i\omega_0$, $k = 0$, $d\sigma_R/dk|_{k=0} = 0$ and $d^2\sigma_R/dk^2|_{k=0} < 0$, where σ_R is the real part of σ. Applying these conditions to (8.27) we obtain the following results for the instability threshold and the oscillation frequency:

$$\lambda_c = 0 \qquad \omega_0^2 = \nu^2 - \gamma^2\,. \tag{8.29}$$

Setting $\lambda = \lambda_c$ in (8.28) we obtain the Hopf–Turing codimension-2 point,

$$\lambda_c = 0 \qquad \gamma_c = \frac{\nu}{\sqrt{1+\alpha^2}}\,,$$

at which the non-uniform stationary instability and the uniform oscillatory instability of the zero state coincide. At this point both the Turing mode and the Hopf mode have zero growth rates as Figure 8.13 shows. The frequency of the Hopf mode, ω_0, and the wavenumber of the Turing mode, k_0, at the codimension-2 point are given by

$$\omega_0 = \frac{\nu\alpha}{\sqrt{1+\alpha^2}} \qquad k_0^2 = \frac{\nu\alpha}{1+\alpha^2}\,.$$

The Hopf–Turing bifurcation has been discussed earlier in Section 7.2.3. As Figure 7.17(a) indicates a pure Turing state may exist beyond the Hopf bifurcation, either as a monostable state or as a stable state in a bistability range with the pure Hopf state. Both cases correspond to resonant standing waves outside the 2:1 resonance tongue of uniform oscillations. Figure 8.14 shows numerical solutions of (8.3) that demonstrate the prevalence of stationary Turing patterns in a bistability range with uniform Hopf oscillations; small initial pattern domains grow and expand into uniformly oscillating domains until a stationary labyrinthine pattern occupying the whole system is obtained. This behavior occurs outside the 2:1 resonance tongue and corresponds, in the original system, to the invasion of resonant labyrinthine standing waves into

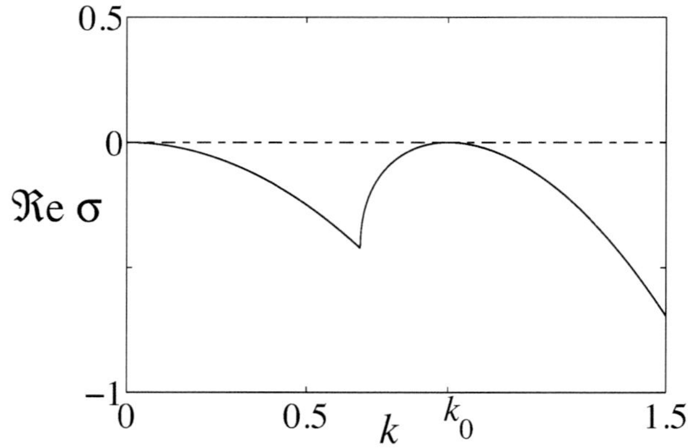

FIGURE 8.13: Hopf–Turing bifurcation in 2:1 temporally forced oscillatory systems. Shown are growth rates curves, obtained from the dispersion relation (8.27), showcasing the simultaneous growth of a uniform oscillatory mode ($Re[\sigma(0)] = 0$, $Im[\sigma(0)] = \omega_0 \neq 0$) and of a non-uniform stationary mode ($(\sigma(k_0)] = 0$). From [336].

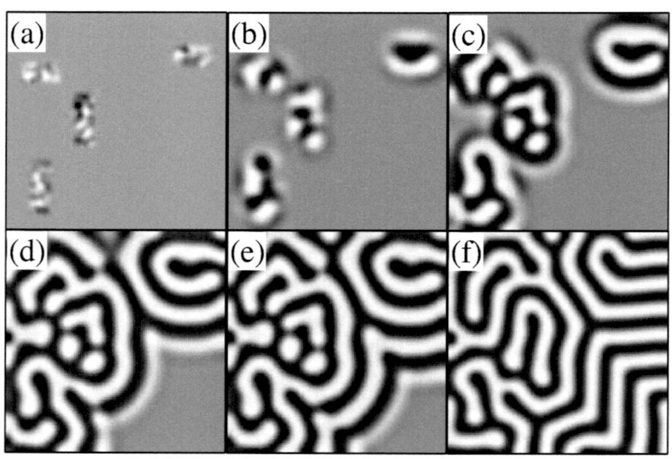

FIGURE 8.14: A numerical solution of (8.3) in the range $\gamma < \gamma_b$ (but sufficiently close to γ_b), showing the gradual development of a stationary pattern from uniform oscillations. Panels (a)–(f) show the U field at successive time steps. The uniform oscillations correspond, in the original system, to unlocked oscillations while the asymptotic stationary pattern corresponds to resonant labyrinthine standing waves. From [336].

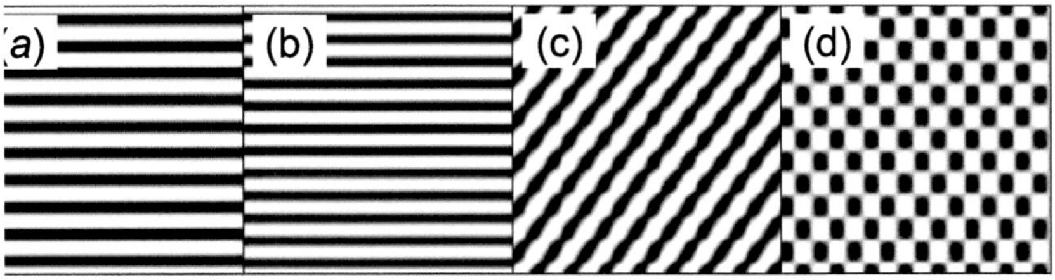

FIGURE 8.15: Two-dimensional response to a one-dimensional spatially-periodic parametric forcing in the forced SH equation (8.9). A system that supports a stripe pattern with wave-number k_0 (a), responds to a stripe-like forcing with a wave-number $k_f > k_0$ (b), by forming an oblique pattern when the forcing is sufficiently weak (c), or a rectangular pattern with stronger forcing (d). Both two-dimensional patterns are resonant since their wave-vector components in the forcing direction are exactly half the forcing wave-number.

non-resonant uniform oscillations. We conclude that the symmetry-breaking Turing instability induced by the temporal forcing can extend the range of resonant oscillations by forming standing wave patterns [337]. This range extension can be viewed as a dispersive effect that acts to reduce the detuning ν to an effective detuning $\nu - \alpha k^2$ within the 2:1 resonance tongue.

8.3.2 Instabilities induced by spatial forcing

In the previous section we discussed the effects of spatially uniform time-periodic parametric forcing of a system whose zero state goes through a uniform oscillatory instability. We found the surprising results that a uniform forcing can induced a symmetry-breaking instability of the zero state to resonant non-uniform standing-wave patterns and that these resonant patterns can persist outside the 2:1 resonance tongue of uniform oscillations. In this section we study the spatial counter part of this problem, i.e., stationary stripe-forming systems subjected to stripe-like parametric forcing, and find analogous behaviors: as Figure 8.15 shows, the forcing can break the remaining symmetry along the stripes and lead to 2d oblique or rectangular patterns. These patterns resonate with the forcing in a range much wider than the 2:1 resonance of stripe patterns, as we will shortly see [195].

We derive these results using the forced SH equation (8.9), which corresponds to a forcing wave-vector $\mathbf{k}_f = k_f\hat{\mathbf{x}}$, where $k_f \in (0, 2k_0)$. Unlike the 1d analysis in Section 8.1.2, we will not restrict ourselves here to small detuning from exact 2:1 resonance, and consider ν_2 to be of order unity. We recall that the zero state in the unforced system goes through a stationary non-uniform instability as λ becomes positive. Stable stripe patterns for $0 < \lambda \ll 1$ occupy a narrow band of wavenumbers bounded by the zigzag and Eckhaus instability thresholds as Figure 6.4 illustrates. The application of a near resonant ($|\nu_2| \ll 1$) weak forcing extends the existence range of 1d stable stripe patterns to λ values as low as $\lambda = -\gamma/2$ (for $\nu_2 = 0$), and stabilizes them by extending the boundaries of the zigzag and Eckhaus instabilities to lower and higher wavenumbers, respectively [196]. In these extended ranges the stripe patterns are locked to the forcing in a 2:1 resonance. Outside these ranges stripe patterns are mostly quasi-periodic, resembling the quasi-periodic oscillations in the temporal counter-part problem.

The two-dimensional extent of the system, however, allows for an additional response form; the growing spatial modes can still lock their wave-vector components in the forcing direction, k_x, to the forcing in a 2:1 resonant, $k_x = k_f/2$, and compensate for the unfavorable wavenumber by creating a wave-vector component, k_y, in the orthogonal direction, such that the total wavenumber, $k = \sqrt{k_x^2 + k_y^2}$, is equal to k_0, the preferred wavenumber that minimizes the Lyapunov functional of (8.9). Such *oblique modes* have wave-vectors $\mathbf{k}_\pm = k_x\hat{\mathbf{x}} \pm k_y\hat{\mathbf{y}}$, where $k_x = k_f/2$ and $k_y = \sqrt{k_0^2 - k_x^2}$, that lie on the circle $k = k_0$. To test this proposition we approximate a solution of (8.9) as

$$u \approx ae^{i(k_x x + k_y y)} + be^{i(k_x x - k_y y)} \; + \; c.c., \tag{8.30}$$

with k_x and k_y satisfying the above requirements, and assume λ and γ are small. Using the method of multiple scales, assuming a and b vary weakly in time, the following equations for the amplitudes a and b can be derived [196]:

$$\begin{aligned} \dot{a} &= \lambda a - 3(|a|^2 + 2|b|^2)a + \frac{\gamma}{2}b^*, \\ \dot{b} &= \lambda b - 3(|b|^2 + 2|a|^2)b + \frac{\gamma}{2}a^*. \end{aligned} \tag{8.31}$$

A linear stability analysis of the zero solution $a = b = 0$ (which corresponds to the zero state, $u = 0$, of (8.9)) shows that it becomes unstable as λ exceeds $\lambda_c = -\gamma/2$. To find out which solutions the zero state evolves towards beyond the instability point we write the amplitudes in polar forms, $a = \rho_a \exp(i\alpha)$ and $b = \rho_b \exp(i\beta)$, and express (8.31) as

$$\begin{aligned} \dot{\rho_a} &= \lambda\rho_a - 3(\rho_a^2 + 2\rho_b^2)\rho_a + \frac{\gamma}{2}\rho_b\cos\varphi, \\ \dot{\rho_b} &= \lambda\rho_b - 3(\rho_b^2 + 2\rho_a^2)\rho_b + \frac{\gamma}{2}\rho_a\cos\varphi, \end{aligned} \tag{8.32}$$

and

$$\dot{\varphi} = -\frac{\gamma}{2}\left(\frac{\rho_b}{\rho_a} + \frac{\rho_a}{\rho_b}\right)\sin\varphi, \tag{8.33}$$

where $\varphi = \alpha + \beta$. Constant solutions of (8.32) and (8.33) give the following solutions to (8.31)

$$a_0 = \rho_0 e^{i\alpha}, \qquad b_0 = \rho_0 e^{-i\alpha}, \tag{8.34}$$

where

$$\rho_0 = \frac{1}{3}\sqrt{\lambda + \frac{\gamma}{2}}, \tag{8.35}$$

and α is an arbitrary constant associated with the continuous symmetry of translations in the y direction. These solutions describe rectangular patterns. Their existence range is given by $\lambda > -\gamma/2$ and $0 < \nu_2 < k_0$ (in order for k_y to be real and $k_f > 0$). Linear stability analysis of these solutions yields the stability condition, $-\gamma/2 < \lambda < \gamma$. Since the solutions (8.34) are independent of x they describe *resonant* patterns; the x component of the pattern's wave-vector is locked to $k_f/2$. A significant outcome of this analysis is that rectangular patterns are resonant over a wide and continuous detuning range $|\nu_2| \sim O(1)$, despite the fact that the forcing intensity γ can be very small. This is unlike the resonance-tongues of stripe solutions, where the locking range is small and scales with the forcing intensity (see Figure 8.4).

What type of pattern solutions appear beyond the instability threshold $\lambda = \gamma$ of rectangular patterns? Looking for stationary solutions of (8.32) and (8.33) in the range $\lambda > \gamma$ we find two new solution families that bifurcate from the rectangular-pattern solutions:

$$a_{\pm} = \rho_{\pm}e^{i\alpha}, \qquad b_{\mp} = \rho_{\mp}e^{-i\alpha}, \tag{8.36}$$

where

$$\rho_{\pm} = \sqrt{\frac{\lambda \pm \sqrt{\lambda^2 - \gamma^2}}{6}}, \tag{8.37}$$

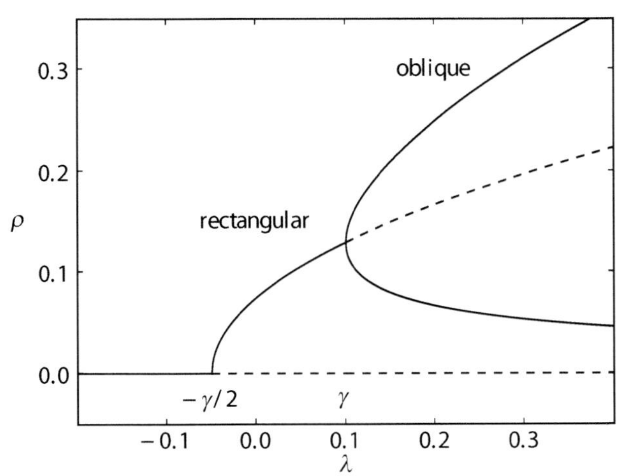

FIGURE 8.16: Bifurcation diagram showing the appearance of rectangular patterns (a_0, b_0) at $\lambda = -\gamma/2$ [see (8.34) and (8.35)], and their destabilization to oblique patterns (a_+, b_-) and (a_-, b_+) at $\lambda = \gamma$ [see (8.36) and (8.37)]. From [195].

and α is an arbitrary constant. These solutions break the symmetry between the two oblique modes, as the bifurcation diagram in Figure 8.16 shows, and give rise to oblique patterns. They are linearly stable in the range $\lambda > \gamma$, and therefore prevail when the forcing is weak. Like the rectangular patterns, the oblique patterns are also resonant, because the wave-vector component in the forcing direction x is exactly half the forcing wave-number, $k_x = k_f/2$.

The stability ranges of two-dimensional rectangular and oblique patterns in the $(k_f/k_0, \gamma)$ plane are shown in Figure 8.17, along with the existence range of one-dimensional stripe patterns in the 1:1 and 2:1 resonance. Also shown are actual patterns that exist at selected points in this plane. Both rectangular and oblique patterns change from stripe patterns along the x direction, in the limit $\nu_2 \to 0$, to stripe patterns along the y direction, in the limit $\nu_2 \to k_0$. Accordingly, the pattern's wave-vector component in the x direction is continuously controllable by the forcing from $k_x = k_0$ $(\nu = 0)$ to $k_x = 0$ $(\nu_2 = k_0)$. At $\nu_2 = k_0/4$ the rectangular patterns become square patterns and the oblique stripes are exactly diagonal.

According to the analysis presented above, stripe patterns associated with resonance tongues in the range $k_f < 2k_0$ coexist with 2d rectangular and oblique patterns, which raises the question: are resonant stripe patterns stable to the growth of the 2d patterns? We study this question by approximating a solution of (8.9) as a superposition of a stripe mode and two oblique modes

$$u \approx Be^{ik_f x} + a\, e^{i(k_x x + k_y y)} + b\, e^{i(k_x x - k_y y)} + c.c.\,, \tag{8.38}$$

where $k_x = k_f/2$ and $k_y = \sqrt{k_0^2 - k_x^2}$, and deriving coupled equations for the three amplitudes A, a and b in the vicinity of the 1:1 resonance. We present below the amplitude equations for the more general case of an asymmetric forced SH equation

$$\partial_t u = \lambda u + \alpha u^2 - u^3 - \left(\nabla^2 + k_0^2\right)^2 u + \gamma u \cos\left(k_f x\right), \tag{8.39}$$

where the additional quadratic term breaks the inversion symmetry $u \to -u$.

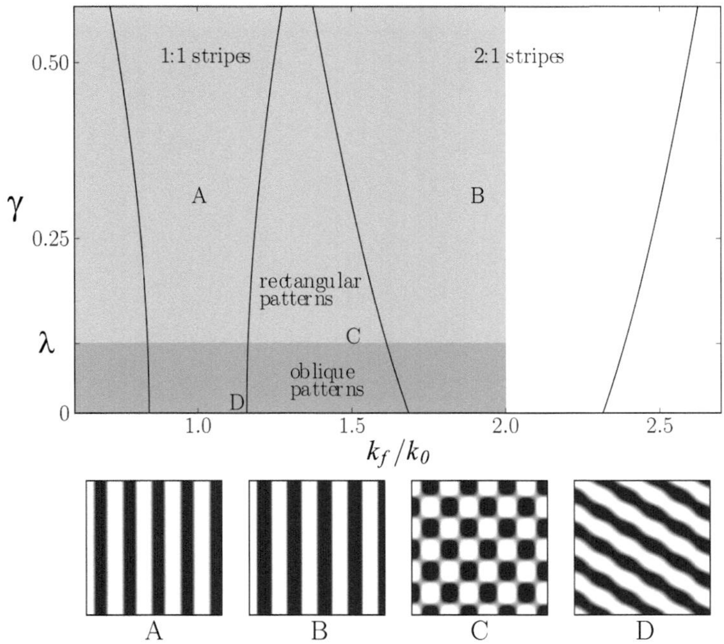

FIGURE 8.17: Stability domains of resonant rectangular and oblique patterns obtained from linear stability analysis of (8.32), and the existence domains of 1:1 and 2:1 resonant stripe solutions. Rectangular patterns destabilize to oblique patterns as γ decreases below λ. The patterns in the panels A,B,C,D below the diagram show representative numerical solutions of (8.9) with the parameters indicated by the corresponding letter in the diagram above. From [204].

A multiple-scale analysis, assuming $k_f \approx k_0$, $\gamma \sim \mathcal{O}(\sqrt{\lambda})$ and $\alpha \sim \mathcal{O}(\sqrt{\lambda})$, and considering for simplicity spatially uniform amplitudes, leads to

$$\begin{aligned} \dot{B} &= (\lambda - 4k_0^2\nu^2)B - 3\left(|B|^2 + 2|a|^2 + 2|b|^2\right)B + 2\alpha a^* b^* \\ &+ \tfrac{1}{4}\gamma^2(d_+ + d_-)B + \tfrac{1}{4}\gamma^2 d_- B^\star + \dots, \qquad (8.40\text{a}) \\ \dot{a} &= \lambda a - 3\left(|a|^2 + 2|b|^2 + 2|B|^2\right)a + \tfrac{1}{2}\gamma b^\star + 2\alpha B^* b^* + \dots, \qquad (8.40\text{b}) \\ \dot{b} &= \lambda b - 3\left(|b|^2 + 2|a|^2 + 2|B|^2\right)b + \tfrac{1}{2}\gamma a^\star + 2\alpha B^* a^* + \dots, \qquad (8.40\text{c}) \end{aligned}$$

where $d_\pm$ are given by (8.13) and the ellipses denote higher order terms [204, 205].

Using (8.40) with $\alpha = 0$ we can study the stability of resonant stripe solutions of the forced SH equation (8.9), which correspond to constant solutions of (8.40) of the form $(B, 0, 0)$. A linear stability analysis of such solutions indeed shows that 1:1 resonant stripe solutions can lose stability to the growth

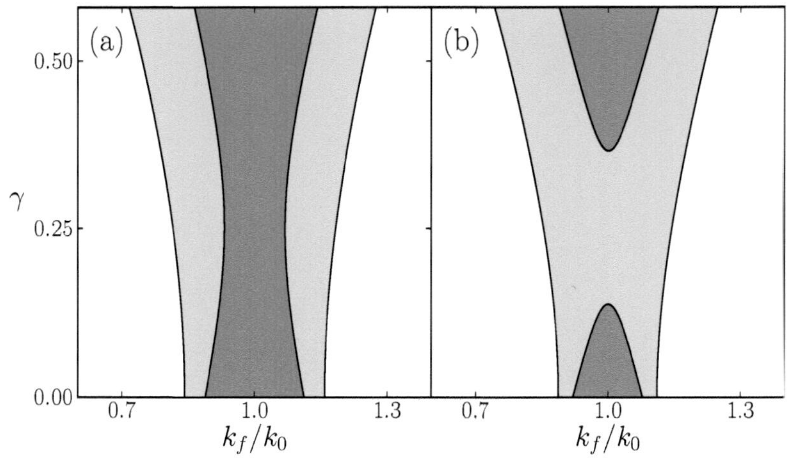

FIGURE 8.18: Existence and stability domains of $1:1$ resonant stripe solutions of (8.40) with $\alpha = 0$ (symmetric forced SH equation). The shaded areas indicate the existence domains and the dark shaded areas are the stability regions. (a) Above a critical λ value the stability region is contiguous. (b) Below that critical value, 1:1 stripe solutions are unstable to the growth of oblique modes in a range of the forcing amplitude, γ, even at exact resonance $k_f = k_0$. From [204]

of the two oblique modes. This instability reduces the stability range of these solutions, as Figure 8.18(a) shows, and may even completely destabilize them for some range of λ as Figure 8.18(b) shows. This is in contrast to the behavior of 1d systems for which a 1d forcing always acts to stabilize stripe patterns. It reflects the advantage of 2d patterns in avoiding a compromise on the preferred wavenumber k_0 by forming 2d patterns that are still locked to the forcing. These intriguing results have been confirmed in direct studies of (8.9), but are valid for weak forcing ($\gamma \ll 1$) and close to the pattern forming instability of the unforced system ($0 < \lambda \ll 1$).

Consider now equations (8.40) for $\alpha \neq 0$. Resonant stripe solutions of the form $(B, 0, 0)$ still exist, as Figure 8.19(a) shows, but solutions that describe pure rectangular or oblique patterns, i.e., solutions of the form $(0, a, b)$, no longer exist. Instead, equations (8.40) have mixed solutions, (B, a, b), involving both the stripe mode and the two oblique modes. As Figure 8.19(b,c) shows, these solutions still describe resonant patterns for which $k_x = k_f/2$. Moreover, we can still distinguish between patterns for which $|a| = |b|$, and patterns for which $|a| \neq |b|$. The presence of the stripe mode, however, imparts to the former patterns hexagonal-like forms and leads to rhombic patterns [239]. In fact, at exact 1:1 resonance ($k_f = k_0$) the three modes have wave-vectors that are oriented $2\pi/3$ degrees apart with moduli all equal to k_0, as is the

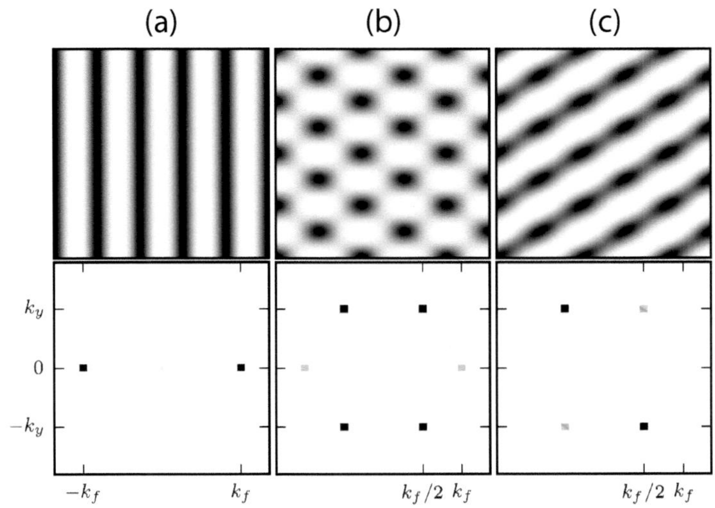

FIGURE 8.19: Resonant patterns in the forced asymmetric SH equation (8.39). The top panels show the patterns in the x, y plane (x is the horizontal axis). The bottom panels show the corresponding power spectra. The small squares denote power peaks and their positions relative to the circle of radius k_0 (darker shades of the squares correspond to higher power). (a) Resonant 1:1 stripes, $\mathbf{k} = \pm k_f \hat{\mathbf{x}}$. (b) Resonant 2:1 rhombic patterns with oblique modes $\mathbf{k} = \pm[k_x \hat{\mathbf{x}} \pm k_y \hat{\mathbf{y}}]$ of equal power, $|a| = |b|$, where $k_x = k_f/2$ and $|\mathbf{k}| = k_0$, and stripe modes $\mathbf{k} = \pm k_f \hat{\mathbf{x}}$. (c) Resonant 2:1 oblique patterns with the same oblique modes but of different power, $|a| \neq |b|$, and the same stripe modes. Parameters: $\gamma = 0.05$, $\alpha = 0.1$, $k_0 = 1$, $k_f = 1.15$, $\lambda = 0.05$ for the rhombic solution and $\lambda = 0.15$ for the stripe and oblique solutions.

case with hexagonal patterns (see Figure 7.2). We will reconsider the general structure of the amplitude equations (8.40) in Section 11.3.2 where we discuss the restoration of degraded areas by periodic landscape modulations.

Chapter summary

Pattern forming systems are often subjected to external periodic forcing, which can be an inherent aspect of the system's environment or intentionally applied as a means of pattern control. In model equations, the forcing often appears as modulations of parameters that multiply state variables (parametric forcing). There is much in common in the response of an oscillating system to a temporal periodic forcing and the response of a pattern-forming system to a spatial periodic forcing. There are also important differences that stem from the higher dimensionality of the physical space as compared with the one-dimensional time axis.

The first common aspect pertains to the interplay between intrinsic and extrinsic periodicities. A temporally forced oscillatory system can adjust its oscillation frequency to a rational fraction of the forcing frequency, provided the forcing is strong enough. Such a response, which is called frequency locking, occurs within tongue-like domains in the plane spanned by the forcing frequency and the forcing amplitude. Anywhere within a tongue the oscillation frequency of the system, ω, is determined by the forcing frequency, ω_f, through a relation of the form $\omega = (m/n)\omega_f$, where m and n are integers. For a given forcing frequency the integers n and m are determined such that ω is sufficiently close to the oscillation frequency, ω_0, of the unforced system. These integers define the $n : m$ resonance tongue. A similar behavior is found in spatially forced 1d pattern-forming systems with ω_f, ω_0 and ω replaced, respectively, by the forcing wavenumber, k_f, the pattern wavenumber of the unforced system, k_0, and the actual wavenumber, k, that is locked to the forcing. Amplitude equations derived for the forced systems can be used to determine the tongue boundaries and the resonant solutions that describe frequency-locked oscillations or wavenumber-locked patterns within the tongues.

The second common aspect is the appearance of multiple stable phase states inside resonance tongues. The local convergence to different phase states gives rise to spatial patterns that involve phase fronts. Special attention has been given in this chapter to the 2:1 resonance, which gives rise to bistability of states with a phase difference of π, and to phase fronts that locally shift the phase by π. These fronts can go through nonequilibriun Ising-Bloch (NIB) bifurcations, very much like fronts in other bistable systems (see Section 7.3.1). In the case of temporally forced oscillatory systems, the NIB bifurcation designates a transition from standing-wave patterns to traveling-wave patterns, and, when acting in concert with a transverse front instability, can lead to spontaneous spiral-wave nucleation and spatio-temporal chaos. In the case of spatially forced pattern-forming systems, two types of fronts can be distinguished; longitudinal fronts that shift the pattern phase in the wavevector direction, and transverse fronts that shift the phase in a direction normal to the pattern's wavevector. Both types of fronts can go through NIB bifurcations, but in this case all fronts are stationary. Higher resonance tongues of forced oscillations can give rise to multi-phase traveling waves, e.g., four-phase spiral waves within the 4:1 resonance tongue, and to phase-front instabilities involving a change in the phase shifts that the fronts make.

The third aspect that is common to temporally forced oscillatory systems and to spatially forced pattern-forming systems is the induction of a symmetry breaking instability by the forcing. In both cases this is an additional instability that the original stationary uniform state goes through. In the case of oscillatory systems the forcing leads to a Turing-like instability, in addition to the Hopf bifurcation, that results in standing-wave patterns. In the case of pattern-forming systems the forcing induces an instability to 2d stationary oblique or rectangular patterns, in addition to the instability to stripe patterns. An interesting consequence of these instabilities is the possible extension of the ranges of frequency-locked oscillations and wavenumber locked patterns.

The range extension of resonant response is also the main aspect by which wavenumber locking differs from frequency locking. While resonant oscillations occupy relatively small locking ranges—the narrow resonance tongues—which reflect their limited capacity to adjust their frequency to the forcing, resonant 2d patterns can occupy large locking ranges. This is because of the freedom of spatial patterns in 2d systems to respond in a direction normal to the forcing; they lock their wavevector component in the forcing direction to half the forcing wavenumber and compensate for unfavorable wavenumbers by building wavevector components in the orthogonal direction. This mechanism leads to resonant 2d patterns that occupy the whole range $0 < k_f < 2k_0$. As this wide range includes the 1:1 resonance tongue, the question of which pattern dominates the other, resonant stripes or resonant 2d patterns, arises. Analysis of coupled amplitude equations for a stripe and two oblique modes shows that resonant stripe patterns may become unstable to the growth of oblique modes. As a result, the range of stable resonant stripe patterns narrows down and may even vanish.

Part III

Applications to Ecology

Chapter 9

Modeling water-limited vegetation

9.1 Basic biomass-water feedbacks

In order to apply the concepts and tools of pattern-formation theory to specific ecological contexts we need mathematical models for these contexts. As pointed out in Section 4.1.1 PDE models are more suitable for pattern-formation studies than individual-based models, because they are more amenable to mathematical analysis. The application of continuum PDE models to discrete small populations of life forms, however, requires some attention. This question is especially relevant to strongly water-limited vegetation, which often consists of sparse patches comprising a single or a few plant individuals. In Section 9.1.1 we will argue that despite the small size of plant populations in dryland vegetation, the above-ground biomass can justifiably be regarded as a continuous variable. The next question we address in this section is how detailed should continuous-biomass models be? The discussion of this question will lead us to the next section (9.1.2) in which we describe three biomass-water feedback mechanisms capable of forming vegetation patterns, and the ways by which they can be modeled.

9.1.1 Continuum modeling of discrete plant populations

The biomass of a plant population in a water limited system can often be regarded as a continuous deterministic variable for two main reasons. The first reason pertains to the level of a single plant and is related to the modular design that dryland plants typically have. Rather than having a single stem that acts as an integrated hydraulic system, and is vulnerable to hydraulic failures caused by droughts, dryland plants often develop hydraulically independent multiple stems as Figure 9.1 illustrates. The stems can remain physically connected by nonconducting heartwood or split axially along woody stems to form physically separate plants. The resulting redundancy of independent conduits increases the plant's resistance to drought; failure of a single or a group of conduits can lead to partial plant mortality but still leaves the plant viable [273]. As a consequence, the response of a plant individual to water stress often involves a gradual biomass decrease rather than a sharp mortality event. Partial mortality may also result from other forms of stress such as insufficient light [233].

The second reason pertains to the level of a plant population occupying a small area element. Even when the mortality of individual plants is complete, the availability of long-lived seeds and their non-vanishing probability to germinate whenever the biotic and abiotic conditions allow, compensate for the mortality, reduce strong population fluctuations, and prevent the extinction of small populations. Although demographic noise [240, 28] generated by the mortality of plant individuals in small populations is strongly reduced, it may not always be ruled out [251] and further studies, highlighting conditions under which demographic noise is not negligible, are needed.

These considerations suggest the description of a plant population in terms of a deterministic continuous biomass variable, representing the above-ground biomass per unit area, irrespective of the number or identity of the plant individuals contributing to it. The next question is how detailed PDE models involving continuous biomass variables should be. Obviously, in order to account for self-organized vegetation patchiness (see Section 3.3.1), the models should capture pattern-forming feedbacks, i.e., feedbacks that can induce nonuniform instabilities of uniform vegetation. This can already be achieved with a single-variable model for the population biomass that does not take into account the associated water dynamics [169]. More detailed models include an additional water variable [155, 325, 290], or two water variables representing soil-water content and overland water flow [128, 104]. The more detailed models are advantageous for several reasons; they capture additional pattern-forming feedbacks, introduce better defined and measurable parameters, provide a better account of the roles that various physical factors play in pattern-formation processes, and can also exclude pattern-formation behaviors that require parameter values that are not realizable in particular contexts. On the other hand, detailed models are less amenable to mathematical analysis.

FIGURE 9.1: Integrated (a) and modular (b) hydraulic systems in plants. Dark lines illustrate pathways of water transport. From [273].

To benefit from both aspects we will consider a relatively detailed model for water-limited vegetation that captures three different pattern-forming feedbacks, but also study specific contexts that allow simplifications of the model by eliminating one or two feedbacks. We begin with a description of the pattern-forming feedbacks to be modeled.

9.1.2 Pattern-forming feedbacks

The three pattern-forming feedbacks to be modeled can all be viewed as different realizations of the same general positive feedback. As Figures 9.2 and 9.3 illustrate, this is a positive feedback between local vegetation growth and water transport toward the growing vegetation, where the three feedbacks differ in the mechanism of water transport–overland water flow, water conduction by laterally extended root systems, and soil-water diffusion. While the transport of water toward the growing vegetation helps further vegetation growth it inhibits the growth in the vicinity of the growth location, thereby favoring non-uniform vegetation growth and pattern formation. Why should local vegetation growth enhance the transport of water to the growth location? The answer depends on the water-transport mechanism, as explained in the following sections.

On a slope or in the presence of wind, another pattern-forming feedback, not related to the general feedback illustrated in Figure 9.2, becomes feasible. It is associated with water advection, such as fog advection by wind [2] or overland water flow on a slope [155], and involves the interception of advected water by vegetation. This produces a shadowing effect on water transport

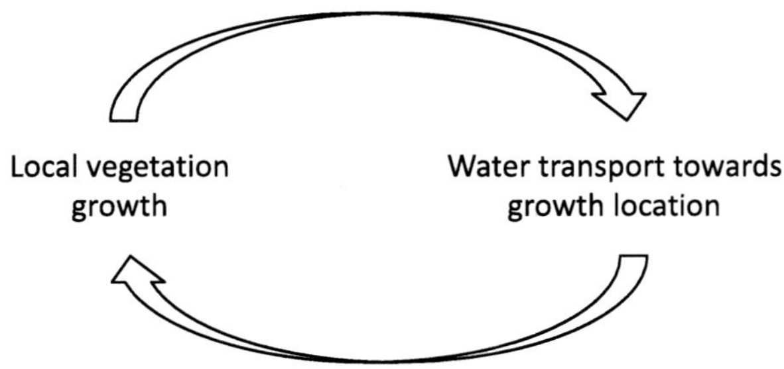

FIGURE 9.2: Schematic illustration of the general positive feedback that drives vegetation pattern formation in water limited systems. While accelerating vegetation growth in existing patches these precesses inhibit the growth in the patch surroundings, thereby favoring vegetation pattern formation.

in the slope or wind directions, which can lead to the formation of banded vegetation [260, 29].

We note that besides these pattern-forming feedbacks there can be additional positive feedbacks involving biomass and water that do not induce vegetation pattern formation. An example for such a feedback is reduced evaporation in a vegetation patch area by canopy and litter shading, which helps the vegetation growth and intensifies the shading effect. This positive feedback (hereafter the "shading feedback") does not lead to pattern formation because it does not involve water transport that induces long-range inhibition. However, it can induce bistability of uniform vegetation and bare soil, which allows for pattern formation or longed-lived transient patterns by a multiplicity of stable uniform states (see Section 7.2.1). We also note that pattern-forming feedbacks other than the biomass-water feedbacks discussed above have also been proposed, e.g., thermal convection in the context of grass patterns [304].

9.1.2.1 Infiltration feedback

Soil areas in water-limited systems that are devoid of vegetation (hereafter "bare soil" areas) are often covered by thin biogenic soil crusts, a few millimeters thick [330]. Depending on the precipitation regime, soil characteristics, and disturbances, these crusts may consist of one or more organisms, including cyanobacteria, green algae, fungi, lichens, and mosses. In arid soils that are more alkaline, cyanobacteria generally dominate [18]. Soil crusts reduce soil erosion by water and wind, provide a source of fixed carbon and nitrogen, and change the water distribution in the soil. It is the latter effect that is most significant to our discussion of vegetation pattern formation.

Soil crusts can induce overland water-flow by changing the rate of surface-water infiltration into bare soil. Crusts dominated by cyanobacteria, for example, can absorb water several times their dry weight in only a few

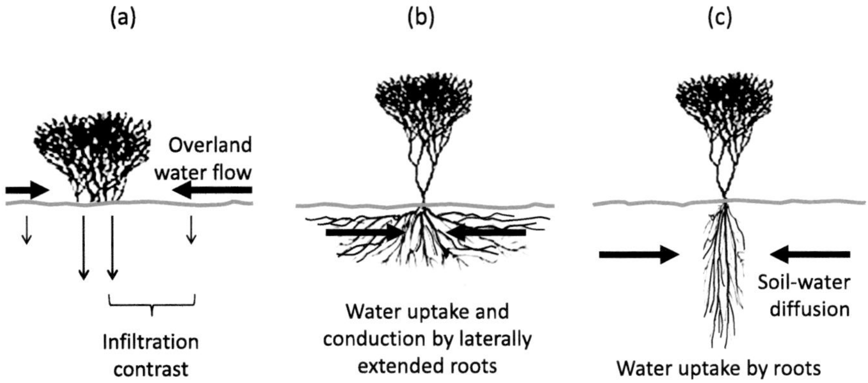

FIGURE 9.3: Schematic illustrations of three modes of water transport capable of inducing pattern-forming feedbacks in water-limited vegetation: (a) overland water flow induced by an infiltration contrast (infiltration feedback), (b) water uptake and conduction by laterally extended root systems (root-augmentation feedback), (c) fast soil-water diffusion relative to biomass spread (soil-water diffusion feedback).

seconds [38]. This results in crust swelling and soil-pore blocking and, consequently, in a significant reduction in water infiltration shortly after rain starts [323, 77]. Since cyanobacteria are photosynthetic organisms their growth is hindered by vegetation, which blocks sunlight. As a consequence, the infiltration rates in sparsely vegetated or bare-soil areas are lower than those in densely vegetated areas[1]. This infiltration contrast induces overland water flow toward densely vegetated areas (see Figure 9.3(a)), which accounts for the right arrow in Figure 9.2, i.e., enhancement of water transport by local vegetation growth, and closes the positive feedback loop. We refer to this pattern-forming feedback as the *infiltration feedback.*

The infiltration feedback is modeled by assuming a monotonically increasing dependence of the infiltration rate I on the above-ground vegetation biomass B, which asymptotes to a constant maximal value A [327, 128, 104],

$$I = A\frac{B + Qf}{B + Q}\,, \tag{9.1}$$

where Q and $f \in [0, 1]$ are additional constants. As Figure 9.4 shows, the dimensionless parameter f controls the infiltration contrast. When $f = 1$, $I = A$ is a constant independent of B. This limit represents no infiltration contrast between bare and vegetated soil areas and, therefore, no infiltration feedback. This limit may be approached in uncrusted sandy soils in which the

[1]Additional factors contributing to this outcome include soil mounds generated by dust deposition that help the interception of runoff at vegetated patches [281], higher soil porosity in vegetation patches and others [248, 294].

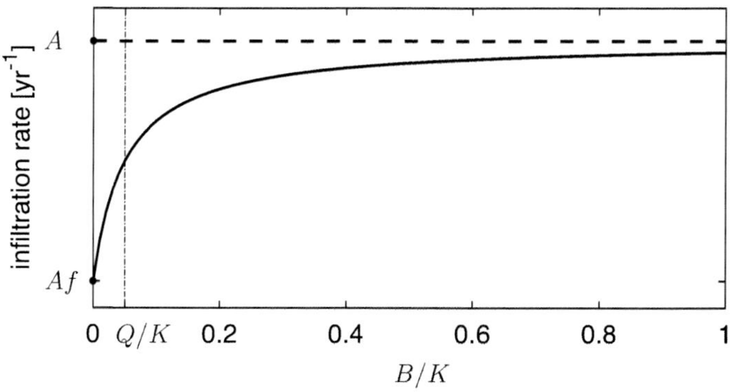

FIGURE 9.4: The infiltration rate $I = A(B + Qf)/(B + Q)$ as a function of biomass density B. When the biomass is diminishingly small ($B \ll Q$) the infiltration rate approaches the value of Af. When the biomass is large ($B \gg Q$) the infiltration rate approaches A. The infiltration contrast between bare and vegetated soil is quantified by the parameter f, where $0 < f < 1$; when $f = 1$ the contrast is zero and when $f = 0$ the contrast is maximal.

infiltration rate is high everywhere and no overland water flow takes place. The other extreme, $f \ll 1$, represents a high infiltration contrast, $I = Af \ll A$ in bare soil and $I \to A$ in densely vegetated soil, and, therefore, strong infiltration feedback. The parameter Q measures how fast the infiltration rate reaches its maximal value A.

9.1.2.2 Root-augmentation feedback

The biomass of a plant can be divided into two parts, above-ground or shoot biomass, and below-ground or root biomass. The two organ systems, shoot and root, are not independent; the relation between the two is expressed in terms of the root-to-shoot ratio, normally measured as the ratio of the root biomass to the shoot biomass. This ratio is monotonically increasing, but generally not constant; variations in root-shoot biomass allocation can occur among individuals of the same species due to plasticity [23], and even during the growth of a single individual [208] due to ontogenetic drift [87]. From a modeling point of view we may choose the above-ground biomass, B, to be the independent variable, and express the root system in terms of B. The lateral extension of the root zone as the shoot grows is another mechanism by which vegetation growth enhances water transport, for the extended root zone allows water uptake and conduction from a larger domain (see Figure 9.3(b)). This process, and the consequent accelerated vegetation growth, define the *root-augmentation feedback*.

The root-augmentation feedback is modeled by introducing a kernel function,

$$G(\mathbf{x}, \mathbf{x}') = G\left(\frac{|\mathbf{x} - \mathbf{x}'|}{S[B(\mathbf{x})]}\right), \tag{9.2}$$

that describes the spatial distributions of the roots in the lateral directions x and y. Here $\mathbf{x} = (x, y)$ represents the plant (shoot) location and $\mathbf{x}'$ a distant point. The kernel function tails off to zero as $|\mathbf{x} - \mathbf{x}'|$ increases beyond a characteristic length, S, which represents the lateral extension of the root zone. The root-augmentation feedback is captured by letting S increase monotonically with the above-ground biomass $B(\mathbf{x})$. The root-to-shoot ratio can then be associated, up to an allometric factor, with the derivative dS/dB. A specific form for G and the manners by which it couples to the biomass and soil-water variables to generate the pattern-forming feedback will be described in Section 9.2.1.

9.1.2.3 Soil-water diffusion feedback

There is yet another possible mechanism of water transport toward a patch of growing vegetation–water depletion by local uptake, followed by soil-water diffusion from the patch surroundings (see Figure 9.3(c)). The associated *soil-water diffusion feedback* also relies on the root-to-shoot property of plants, as the root-augmentation feedback does, except that the role of the roots is the creation of soil-water gradients due to local water uptake. This pattern-forming feedback may apply to plants with vertical roots and strong water uptake, and soil types that allow for fast lateral water diffusion relative to the rate of vegetation spread.

Viewing the vegetation spread as a biomass diffusion process, the soil-water diffusion feedback can be associated with a Turing mechanism where pattern formation results from the time-scale disparity of different diffusive processes. We have already mentioned this mechanism in the context of the FHN model (see Section 5.2.2), which represents an activator-inhibitor system. In that context a Turing instability of the zero state occurs as the inhibitor diffusion becomes fast enough relative to the activator diffusion. Under this condition the diffusion of the inhibitor outside the activator's growth area (where the inhibitor is produced) inhibits the activator's growth in the surrounding area and induces pattern formation. Vegetation systems can also be viewed as activator-inhibitor systems. The analogy is established by associating the vegetation biomass with the activator and the *lack* of water with the inhibitor.

9.2 A mathematical model for water-limited vegetation

We present here a PDE model that was introduced by Gilad et al. [104, 105] to describe vegetation patchiness formed by a single plant species in a water-limited system with arbitrary smooth topography. An extension of the model to a community of plant species will be presented in Section 12.1.1. In

Section 9.2.1 we present the model equations and explain the various terms they contain, focusing in particular on how they capture the three feedbacks described in the previous section. In Section 9.2.2 we introduce non-dimensional variables, space-time coordinates and parameters, and express the model equations in a non-dimensional form. There are two circumstances in which the model equations can be simplified considerably. The first pertains to situations where no significant overland water flow takes place, and the second to plant species with laterally confined root zones. The simplifications in the model equations that can be made in these circumstances are described in Section 9.2.3. Various other PDE models have been proposed to describe vegetation patchiness in water limited systems. We conclude in Section 9.2.4 by comparing the Gilad et al. model with some of the other models.

9.2.1 Model equations

The Gilad et al. model contains three state variables: a biomass variable, $B(\mathbf{X},T)$, representing the above-ground biomass per unit ground area, a soil-water variable, $W(\mathbf{X},T)$, describing the soil-water content available to the plants per unit ground area, and a surface water variable, $H(\mathbf{X},T)$, representing the height of a thin water layer above ground level. Here $\mathbf{X}=(X,Y)$ represents the spatial coordinates and T the time coordinate. The model equations are:

$$\partial_T B = G_B B\,(1-B/K) - MB + D_B\nabla^2 B\,, \tag{9.3a}$$

$$\partial_T W = IH - LW - G_W W + D_W\nabla^2 W\,, \tag{9.3b}$$

$$\partial_T H = P - IH - \nabla\cdot\mathbf{J}\,, \tag{9.3c}$$

where

$$\mathbf{J} = -2D_H H\nabla(H+Z)\,, \tag{9.4}$$

$\nabla = \hat{\mathbf{x}}\partial_X + \hat{\mathbf{y}}\partial_Y$, and G_B, I, L and G_W are either functions or functionals of the state variables to be described below. Temporal biomass changes are affected by water dependent plant growth $G_B B$, by mortality $-MB$, and by short-distance seed dispersal or clonal vegetation expansion from nearby plants $D_B\nabla^2 B$. The late biomass growth phase is also affected by species-specific constraints that impose a maximal standing biomass per unit ground area K. These can be genetic constraints, e.g., the stem strength of a woody plant, the maximal biomass an annual life form can attain during its short life cycle, etc. Temporal changes of the soil-water content are affected by the infiltration of surface water into the soil IH, by water loss due to biomass dependent evaporation or deep infiltration beyond the reach of the roots $-LW$, water uptake by plants' roots $-G_W W$, and by soil-water diffusion $D_W\nabla^2 W$. Finally, surface-water changes are affected by precipitation P, infiltration into the soil $-IH$, and overland water flow $-\nabla\cdot\mathbf{J}$, where the flux $\mathbf{J}$ depends on the ground surface height $\tilde{Z}(X,Y)$, which we describe in terms of the quantity $Z=\rho_w\tilde{Z}$, where ρ_w is the density of water. The quantity Z describes the

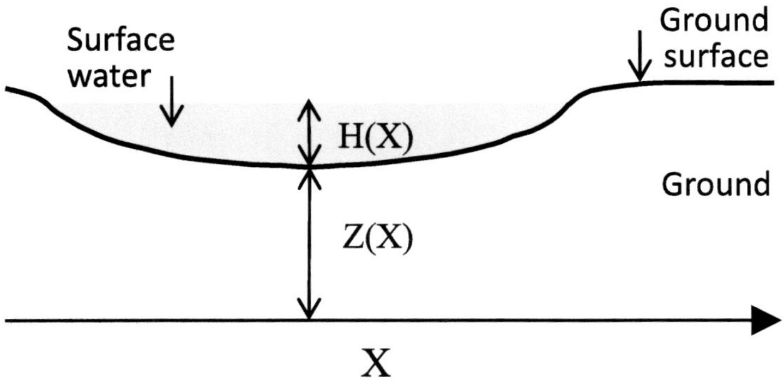

FIGURE 9.5: A schematic illustration of the topography function $Z(\mathbf{X})$ and of the surface water variable $H(\mathbf{X}, T)$. Surface-water flow is induced by gradients of the sum $H + Z$, provided H is nonzero.

landscape's topography as Figure 9.5 illustrates. The equation for the surface water is a particular case of the diffusive-wave approximation of the shallow water equations [266]. Note that for flat terrains for which Z is constant the transport term $-\nabla \cdot \mathbf{J}$ reduces to $D_H \nabla^2(H^2)$.

The explicit forms of I, L, G_B and G_W are chosen so as to capture the various feedbacks discussed in the previous section. We have already introduced the form of the infiltration rate in (9.1). The term $-IH$ in (9.3c), which creates surface water gradients in the case of strong infiltration contrasts ($f \ll 1$), together with the overland water flow term $-\nabla \cdot \mathbf{J}$, and the water dependence of the biomass growth rate G_B (see below), account for the infiltration feedback. The biomass dependence of the evaporation rate L is modeled as

$$L = \frac{N}{1 + RB/K}, \tag{9.5}$$

where R is a positive constant that quantifies the strength of the shading feedback; the larger R the smaller the evaporation rate and the higher the water retention.

The biomass growth rate G_B and the water uptake rate G_W are given by

$$\begin{aligned} G_B(\mathbf{X}, T) &= \Lambda \int_\Omega G(\mathbf{X}, \mathbf{X}', T) W(\mathbf{X}', T) d\mathbf{X}', \\ G_W(\mathbf{X}, T) &= \Gamma \int_\Omega G(\mathbf{X}', \mathbf{X}, T) B(\mathbf{X}', T) d\mathbf{X}', \end{aligned} \tag{9.6}$$

where Λ is the biomass growth rate per unit soil-water content, Γ is the soil-water uptake rate per unit above-ground biomass, $G(\mathbf{X}, \mathbf{X}')$ is the root kernel (9.2), which is determined by the root architecture of the particular plant species considered, and Ω is the lateral root zone. For the sake of concreteness

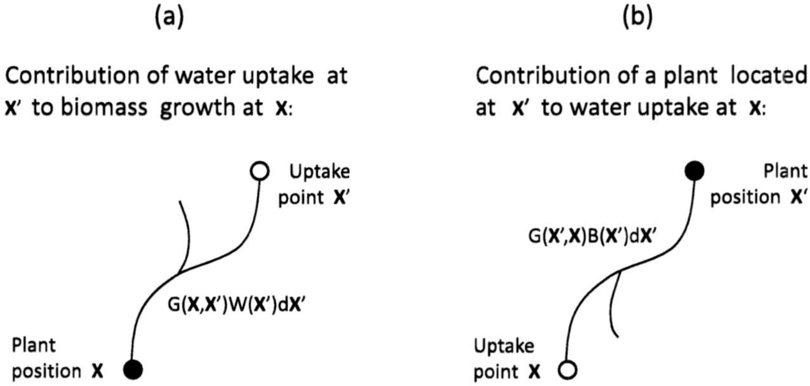

FIGURE 9.6: Schematic illustrations of nonlocal contributions to the biomass-growth rate G_B from water uptake at distant points (a), and to the water-uptake rate G_W from plants located at distant points (b). The rates G_B and G_W are obtained by summing up the contributions from all points $\mathbf{X}'$.

we will assume the following Gaussian form for the root kernel:

$$G(\mathbf{X},\mathbf{X}',T) = \frac{1}{\pi S_0^2} \exp\left[-\frac{|\mathbf{X}-\mathbf{X}'|^2}{S[B(\mathbf{X},T)]^2}\right], \tag{9.7}$$

where $S_0 = S(0)$ represents the lateral root-zone size of a seedling. The nonlocal forms of G_B and G_W reflect the water uptake by laterally extended root systems as Figure 9.6 illustrates. According to the form of G_B, the biomass growth rate depends not only on the amount of soil water at the plant location $\mathbf{X}$, but also on the amount of soil water at any point $\mathbf{X}'$ at the reach (Ω) of the plant's roots (Figure 9.6(a)). Similarly, contributing to the water uptake rate G_W at a point $\mathbf{X}$ are plants located at distant points $\mathbf{X}'$ whose roots extend to $\mathbf{X}$ (Figure 9.6(b))[2].

We recall that a monotonic biomass dependence of the root-zone size, $S(B)$, is crucial for capturing the root-augmentation feedback. This dependence is species dependent, varies during the plant growth phase, and is also affected by plant plasticity in varying environmental conditions. At the moment we simply present it in the form

$$S = S_0 s(B), \tag{9.8}$$

where s is a dimensionless function of the biomass that satisfies $s(0) = 1$ so that $S(0) = S_0$. To quantify the strength of the root-augmentation feedback

[2]The integration in the expressions for G_B and G_B are over the root zone Ω, but in practice we integrate over the entire domain.

we define the following parameter:

$$E = \frac{ds}{dB}|_{B=0} , \tag{9.9}$$

which provides a measure for the root-to-shoot ratio at the early growth phase of the plant.

The soil-water diffusion feedback is captured by choosing both E and the ratio D_W/D_B to be large enough. Large E is required to create strong local depletion of soil-water at the growing vegetation patch and steep soil-water gradients. Large D_W/D_B is required to guarantee soil-water diffusion toward the vegetation patch rather than patch expansion, which would act to smooth out the soil-water gradients.

For simplicity, we have not presented the model equations (9.3) in their most general form. The linear soil-water (W) dependence of the biomass-growth and water-uptake terms is a simplification of a more realistic dependence of the form $W/(W_0+W)$, where W_0 is a half saturation constant [254]. It reduces to a linear dependence when $W \ll W_0$. Biomass loss has been associated with mortality, but other drivers of biomass loss exist, which require the introduction of non-constant biomass-decay rates, e.g., grazing stress which can be modeled by a biomass-dependent decay rate $M = M(B)$. The biomass "diffusion" term represents short-distance seed dispersal or clonal expansion, but can be generalized to describe long-distance seed dispersal by introducing a nonlocal term involving an integral over a kernel function [305, 223]. The nonlocal term reduces to a diffusion term for sufficiently localized kernels. Finally, we will generalize (9.3) to describe a community of species that compete not only for water but also for light (see Section 12.1.1).

The dimensions of all quantities appearing in the model equations are given in Table 9.1. Note that although the dimensions of H and Z are $\mathcal{M}\mathcal{L}^{-2}$ and the dimension of P is $\mathcal{M}\mathcal{L}^{-2}\mathcal{T}^{-1}$, we can regard H and Z as having the dimensions of length, $\mathcal{L}$, and P as having the dimension of length over time $\mathcal{L}\mathcal{T}^{-1}$ by referring to the same quantities divided by the density of water, $\rho_w = 1g/cm^3$. With this convention $1kg/m^2$ is equivalent to $1mm$.

TABLE 9.1: Dimensions of all quantities appearing in the model equations in terms of the fundamental dimensions $\mathcal{L}, \mathcal{M}, \mathcal{T}$ (see Section 5.1.1).

Quantity	Dimension	Quantity	Dimension
B, W, H, Z, K, Q	$\mathcal{L}^{-2}\mathcal{M}$	D_B, D_W	$\mathcal{L}^2\mathcal{T}^{-1}$
M, N, A	$\mathcal{T}^{-1}$	D_H	$\mathcal{L}^4\mathcal{M}^{-1}\mathcal{T}^{-1}$
Λ, Γ	$\mathcal{L}^2\mathcal{M}^{-1}\mathcal{T}^{-1}$	E	$\mathcal{L}^2\mathcal{M}^{-1}$
P	$\mathcal{L}^{-2}\mathcal{M}\mathcal{T}^{-1}$	T	$\mathcal{T}$
X, Y, S_0	$\mathcal{L}$	R, f	1

Implicit in the model equations (9.3) are two additional assumptions. The first is that the landscape has a fixed topography. Thus, while past erosion-deposition processes, that shaped the topography, may have significant effects on water flow and vegetation growth, the latter processes are assumed to have no significant feedback on the topography [264]. The second assumption is that while rainfall may strongly affect vegetation growth, the total vegetation biomass is too small to feed back to the atmosphere and affect the rainfall [65]. When these conditions are not met additional state variables must be taken into account. In the terminology of Section 4.2.1 the state variables B, W and H represent the system and the fixed precipitation P and the topography Z, among other parameters, represent the environment.

9.2.2 Non-dimensional model equations

The model described above for water-limited vegetation belongs to the $\mathcal{LMT}$ class (see Section 5.1.1). The first step in bringing equations (9.3) to a non-dimensional form is to identify a set $\mathbf{A}$ of three parameters which are dimensionally independent. A possible choice is $\mathbf{A} = (K, M, S_0)$ as the dimensions $[K] = \mathcal{ML}^{-2}$, $[M] = \mathcal{T}^{-1}$ and $[S_0] = \mathcal{L}$ are obviously independent. The next step is to use (5.12) to construct non-dimensional quantities from the complementary set $\mathbf{B}$ of dimensionally dependent quantities, which includes the state variables, the space and time coordinates, and parameters not included in $\mathbf{A}$. Finally, expressing any dimensional quantity B_j in terms of its non-dimensional counterpart Π_j and inserting these expressions into the model equations results in a non-dimensional form.

There are other possible choices for the set $\mathbf{A}$. For example, instead of K we could use Λ, because M/Λ has the same dimension, $\mathcal{ML}^{-2}$. Another option is to mix the two subsets by scaling the biomass variable with K and the water variables with M/Λ. The benefit of this mixed scaling is the elimination of four parameters, K, M, Λ, S_0, rather than three. The non-dimensional model equations presented below are a result of mixed scaling of this kind, but with a small additional twist—replacing M with N in the scaling of the water variables. With this twist the instability threshold of the bare-soil state has the nice property of assuming the numerical value of unity, independent of any other model parameter, as we will see in Section 10.1.1.

Expressed in terms of the non-dimensional quantities defined in Table 9.2, which include in particular the non-dimensional state variables (b, w, h) and the space and time coordinates ($\mathbf{x}$,t), the non-dimensional model equations are [105]:

$$\partial_t b = G_b b(1-b) - b + \delta_b \nabla^2 b\,, \tag{9.10a}$$

$$\partial_t w = \mathcal{I}h - lw - G_w w + \delta_w \nabla^2 w\,, \tag{9.10b}$$

$$\partial_t h = p - \mathcal{I}h - \nabla\cdot\mathbf{j}\,, \tag{9.10c}$$

with

$$\mathbf{j} = -2\delta_h h \nabla(h+\zeta)\,. \tag{9.11}$$

TABLE 9.2: Relations between non-dimensional quantities and their dimensional counterparts.

Quantity	Scaling	Quantity	Scaling
b	B/K	p	$\Lambda P/MN$
w	$\Lambda W/N$	δ_b	D_B/MS_0^2
h	$\Lambda H/N$	δ_w	D_W/MS_0^2
q	Q/K	δ_h	$D_H N/M\Lambda S_0^2$
ν	N/M	ζ	$\Lambda Z/N$
α	A/M	ρ	R
η	EK	t	MT
γ	$\Gamma K/M$	$\mathbf{x}$	$\mathbf{X}/S_0$

Here, the non-dimensional evaporation rate l, infiltration rate $\mathcal{I}$, growth rate G_b and soil-water uptake rate G_w are given by

$$l = \frac{\nu}{1+\rho b}\,, \qquad \mathcal{I} = \alpha\frac{b(\mathbf{x},t)+qf}{b(\mathbf{x},t)+q}\,, \tag{9.12}$$

$$\begin{aligned} G_b(\mathbf{x},t) &= \nu\int_\Omega g(\mathbf{x},\mathbf{x}',t)w(\mathbf{x}',t)d\mathbf{x}'\,, \\ g(\mathbf{x},\mathbf{x}',t) &= \frac{1}{\pi}\exp\left[-\frac{|\mathbf{x}-\mathbf{x}'|^2}{\tilde{s}[b(\mathbf{x},t)]^2}\right]\,, \end{aligned} \tag{9.13}$$

and

$$G_w(\mathbf{x},t) = \gamma\int_\Omega g(\mathbf{x}',\mathbf{x},t)b(\mathbf{x}',t)d\mathbf{x}'\,, \tag{9.14}$$

where $\tilde{s}(b) = s(B)$ and the dimensionless counterpart of E is

$$\eta = \frac{d\tilde{s}}{db}|_{b=0} = \frac{ds}{dB}|_{B=0}K = EK\,. \tag{9.15}$$

The non-dimensional form of the precipitation parameter

$$p = \frac{\Lambda P}{MN}\,, \tag{9.16}$$

shows the equivalence of decreasing the precipitation rate, P, to increasing the mortality rate, M, or the evaporation rate, N.

9.2.3 Simplified versions of the model

Two aspects of the model equations (9.3) make them harder to analyze in comparison to earlier models [155, 325, 254, 171]. The first is the non-local integral terms (9.6) and the second is the nonlinear overland flow term involving the flux $\mathbf{J}$ in (9.3). Both types of terms slow down numerical solutions of (9.3) (or (9.10)) and require sophisticated methods to speed them up. For example, the difficulty with the integral terms is the biomass dependence of the root kernels, which rules out the use of the convolution theorem in solving the equations by a spectral method [103]. Another aspect that slows down numerical computations is the separation of time scales, with surface water being the fastest variable and biomass the slowest. Below we describe simplifications of the model equations that circumvent these difficulties and discuss the conditions under which they can be justified.

The first possible simplification involves the elimination of the infiltration feedback by setting $f = 1$. The infiltration rate becomes a constant, $I = A$, independent of the biomass B, and the equation for the surface water variable H decouples from those for B and W. This equation has a single stationary uniform solution, $H_0 = P/I$, which is always linearly stable. Since H is the fastest variable we can assume that on the much slower time scales over which B and W significantly change it has already equilibrated at H_0. Inserting the solution $H = H_0$ into the equation for W, equations (9.3) reduce to the two-variable model

$$\begin{aligned} \partial_T B &= G_B B\,(1 - B/K) - MB + D_B \nabla^2 B\,, & (9.17a)\\ \partial_T W &= P - LW - G_W W + D_W \nabla^2 W\,, & (9.17b) \end{aligned}$$

where G_B and G_B are still given by (9.6). This simplified model still captures two pattern-forming feedbacks, the root-augmentation and the soil-water diffusion feedbacks. It can be used in landscapes with uncrusted sandy soil, where the infiltration rate in bare areas is as high as in vegetated areas.

A different simplification can be made by assuming that $S_0 \ll 1$, i.e., that the root kernel $G(\mathbf{X}, \mathbf{X}', T)$ given in (9.7) is narrow in the lateral directions. If both B and W are approximately constant across the narrow root zone, we can substitute $W(\mathbf{X}', T) \approx W(\mathbf{X}, T)$ and $B(\mathbf{X}', T) \approx B(\mathbf{X}, T)$ inside the integrals (9.6) and integrate the remaining Gaussian function over $\mathbf{X}'$:

$$\int_\Omega G(\mathbf{X}, \mathbf{X}', T) d\mathbf{X}' = \frac{1}{\pi S_0^2} \left(\int_{-\infty}^{\infty} \exp\left(-\frac{X'^2}{S^2} \right) dX' \right)^2 = s(B)^2\,. \qquad (9.18)$$

Inserting this result in (9.6) we obtain the following local forms for the biomass-growth and water-uptake rates:

$$G_B = \Lambda W s(B)^p \qquad G_W = \Gamma B s(B)^p\,, \qquad (9.19)$$

with $p = 2$. We note that the exponent p depends on the dimensionality of the system, $p = 1$ for 1d systems and $p = 2$ for 2d systems. The simplified model then consists of equations (9.3) with G_B and G_W given by (9.19). In this model only the infiltration and soil-water diffusion feedbacks are captured. This simplification can be used for species with highly confined root zones in the lateral directions.

We can further combine the two simplifications made above to obtain a two-variable model with no integral terms, that is equations (9.17) with G_B and G_W given by (9.19). Out of the three pattern-forming feedbacks this simplification captures only the soil-water diffusion feedback.

9.2.4 Comparisons with other models

The simplified version (9.17) of the full model (9.3) with G_B and G_W given by (9.19) is a two-variable model that is similar in structure to a modified version [318] of a model introduced by Klausmeier [155]:

$$\partial_T B = G_B B - MB + D_B \nabla^2 B \,, \tag{9.20a}$$

$$\partial_T W = P - LW - G_W W + D_W \nabla^2 W \,, \tag{9.20b}$$

where

$$G_B = \Lambda W B \qquad G_W = \Gamma B^2 \,. \tag{9.21}$$

The original Klausmeier model does not include the water-diffusion term. Instead, it includes an advection term which accounts for water flow on a slope. On a flat terrain, the water-diffusion term is necessary for capturing a non-uniform stationary instability that leads to stationary vegetation patterns. The simplified version (9.17) formally reduces to (9.20) for $K/B \ll 1$ and $s = \sqrt{B}$. Note that with this form of $s(B)$ the parameter E defined in (9.9) diverges to infinity.

The three-variable version of the full model (9.3) with G_B and G_W given by (9.19) is comparable to the model introduced by Rietkerk et al. [254], which we write in the form

$$\partial_T B = G_B B - MB + D_B \nabla^2 B \,, \tag{9.22a}$$

$$\partial_T W = IH - LW - G_W W + D_W \nabla^2 W \,, \tag{9.22b}$$

$$\partial_T H = P - IH + D_H \nabla^2 H \,, \tag{9.22c}$$

where

$$G_B = \frac{\Lambda W}{W + W_0} \,, \qquad G_W = \frac{\Gamma B}{W + W_0} \,, \tag{9.23}$$

and I is given by (9.1). There is no limit at which the two models strictly coincide because of the different modeling of overland flow. In (9.3) the modeling is based on shallow-water theory while in (9.22) a linear diffusion term has been assumed, which is not derivable from shallow-water theory [266]. Apart from this difference the simplified model version (9.3) with (9.19) and equations (9.22) coincide when $W \ll W_0$, $s = 1$ (i.e., independent of B), and $B \ll K$.

All models considered above take into account the two lateral space dimensions (x, y) but do not include the vertical dimension that represents soil depth. A first step in that direction has recently been made by extending the Gilad et al. model [104, 105] to include two soil layers [14].

Chapter summary

In order to apply the methods of pattern-formation theory to spatially extended ecosystems we need mathematical models of these systems. We focus on drylands, where water stress results in vegetation pattern formation, and resort to continuum PDE models, which are amenable to the mathematical methods of pattern formation theory. Such models do not follow the dynamics of individual plants but rather the dynamics of the overall above-ground biomass per unit area that these plant individuals form. In the context of dryland vegetation, biomass variables of this kind can justifiably be regarded as continuous variables even for small plant populations.

To account for vegetation pattern formation, proposed PDE models should capture pattern forming feedbacks. Several feedbacks have been identified in the literature, and three of them are specifically addressed here. They are all positive feedbacks between local vegetation growth and water transport towards the growing vegetation, and differ only in the mechanism of water transport: overland water flow (infiltration feedback), water conduction by laterally extended root systems (root-augmentation feedback), and soil-water diffusion (soil-water diffusion feedback). These feedbacks are pattern forming because the transport of water toward the growing vegetation inhibits the growth in the vicinity of the growth location.

Various PDE models have been proposed with an increasing degree of complexity; a single-variable model for the dynamics of vegetation biomass, two-variable models that include in addition water dynamics, and three-variable models that distinguish between the dynamics of above-ground water and below-ground water. The three-variable models are advantageous for several reasons; they can capture additional pattern-forming feedbacks, introduce better defined and measurable parameters, provide a better account of the roles various physical factors play in pattern-formation processes, and exclude pattern-formation behaviors that are not realizable in particular contexts. On the other hand, they are less amenable to mathematical analysis.

To benefit from both aspects we consider a three-variable model that captures all three pattern-forming feedbacks mentioned above, but also identify specific contexts that allow simplifications of the model by eliminating one or two feedbacks. The model describes the dynamics of the above-ground biomass, below-ground (soil) water and above-ground (surface) water. We present the model equations both in dimensional and non-dimensional forms, describe the terms that are responsible for the three pattern-forming feedbacks, and discuss three simplifications thereof.

The first simplification corresponds to the elimination of the infiltration feedback by assuming high infiltration rates throughout the whole system. In this case, which can be realized in sandy soils, overland water flow becomes negligible and the model can be reduced to two variables—above-ground biomass and below ground water. The second simplification corresponds to the elimination of the root-augmentation feedback by assuming plant species with laterally confined root zones. In this case, all three state variables remain, but the terms involving integrals over the root kernel can be evaluated to yield simple algebraic forms. The last simplification is a combination of the earlier two, which leads to a two-variable model containing no integral terms in which the only remaining pattern-formation mechanism is the soil-water diffusion feedback. The simplified models bear strong similarity to other proposed models, which we highlight at the end of the chapter.

Chapter 10

Vegetation pattern formation

10.1 Uniform and periodic vegetation states along environmental gradients

Equipped with the tools of pattern formation theory and the model equations (9.10)–(9.14) for water limited vegetation, we can now study vegetation patchiness as a pattern-formation phenomenon. Linear stability analysis can tell us, for example, whether vegetation patchiness arises from a non-uniform instability of uniform vegetation as the precipitation rate decreases below a critical value. Numerical studies can confirm whether the pattern that appears beyond the instability point is a hexagonal pattern and whether the instability is subcritical, as the analysis of the SH equation in Section 7.1 predicts. These aspects of uniform and patterned vegetation states and the complete sequence of such states along the rainfall gradient are studied in Section 10.1.1 for flat terrains and in Section 10.1.2 for sloped terrains. In Section 10.1.3 we consider bistability ranges, where two alternative stable states coexist, and study an intriguing dynamical aspect of such ranges that occur on slopes, namely, the possible induction of global state transitions between alternative stable vegetation patterns by local disturbances. The pattern-formation approach to dryland vegetation also suggests a novel operational view of the concept of aridity and of the definitions of aridity classes, which we discuss in Section 10.1.4.

10.1.1 Vegetation states in flat terrains

Flat terrains are characterized by constant ζ values for which the flow term in (9.10c) becomes

$$-\nabla \cdot \mathbf{j} = \delta_h \nabla^2 (h^2)\,. \tag{10.1}$$

For simplicity we choose throughout this chapter a linear biomass dependence of the lateral root-zone size

$$\tilde{s} = 1 + \eta b = s = 1 + EB\,. \tag{10.2}$$

The simplest solutions of (9.10) are uniform stationary states. Setting the time and space derivatives to zero we find two physical solutions of this kind. The first describes bare soil and is given by $b = 0, w = p/\nu$ and $h = p/\alpha f$. The second describes uniform vegetation. The existence and stability ranges of the two solutions are described by the bifurcation diagram shown in Figure 10.1. We refer the reader to Ref. [105] for a detailed description of the linear stability analysis of these solutions that leads to Figure 10.1. The bare soil state is linearly stable for $p < p_c = 1$ and goes through a uniform stationary instability at $p = 1$. The bifurcation is subcritical (supercritical) when the quantity $2\eta\nu/[\nu(1-\rho)+\gamma]$ is greater (smaller) than unity. This condition reflects the role of the shading feedback in inducing a subcritical bifurcation; high values of ρ, the parameter that quantifies the feedback strength, act to stabilize vegetation by decreasing soil-water evaporation. The uniform vegetation solution (denoted by $\mathcal{V}$ in Figure 10.1) that bifurcates from the bare soil solution at $p = 1$ exists for $p > 1$ when the bifurcation is supercritical, and for $p > p_1$ when it is subcritical. The threshold $p_1 = 1$ denotes a fold bifurcation at which a uniform solution that is unstable to uniform perturbations (dashed line) coincides with a uniform solution that is unstable to non-uniform perturbations (dotted line). The latter becomes stable only beyond another threshold, $p = p_2 > p_1$. As p is decreased below p_2 the uniform vegetation solution undergoes a non-uniform stationary instability, as the growth-rate curves in Figure 10.2 show, which leads to a stationary periodic pattern.

The periodic pattern that appears at $p = p_2$ is a hexagonal pattern of vegetation gaps. Further decrease of p leads to additional thresholds where stripe and spot patterns appear. Stable spot patterns persist down to a threshold $p_0 < p_1$. Below $p = p_0$ bare-soil is the only stable state. Altogether, there are five basic vegetation states along the rainfall gradient as Figure 10.3 shows: uniform vegetation, gaps, stripes, spots and bare soil. This sequence

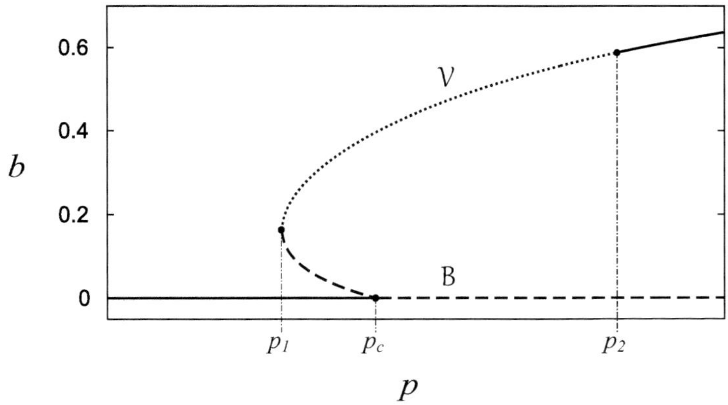

FIGURE 10.1: Bifurcation diagram for homogeneous stationary solutions of the model equations (9.10)–(9.14) showing the biomass b vs. the precipitation p for a flat terrain. The solution branches $\mathcal{B}$ and $\mathcal{V}$ denote, respectively, the bare-soil and uniform-vegetation solutions. Solid lines represent linearly stable solutions, dashed lines represent solutions which are unstable to uniform perturbations, and the dotted line represents solutions which are unstable to non-uniform perturbations. From [105].

of basic states has also been found in other vegetation models [325, 254, 170] and in the FHN model too [216, 156]. The analysis of stationary patterns in Section 7.1.1 provides a clue to the origin of this robust sequence of states. The analysis pointed toward the existence of a universal sequence of states, uniform-hexagons-stripes, in the vicinity of a non-uniform stationary instability of a uniform state (see Figure 7.4). Numerical continuation studies of (9.10) with G_B and G_W given by (9.19) in 1d indicate, as Figure 10.4 shows, that the periodic pattern branch that emanates from the uniform vegetation solution at $p = p_2$ terminates on the same solution branch at a lower precipitation threshold. Additional periodic solutions with smaller wave-numbers, emanating either from the uniform vegetation branch or from other periodic solutions at precipitation thresholds smaller than p_2, also terminate on the same uniform solution branch, but progressively closer to $p = 1$—the instability threshold of the bare-soil solution [66, 339, 318]. There are therefore two non-uniform stationary instabilities of the uniform-vegetation state at high ($p = p_2$) and low (near $p = 1$) precipitation values, and the solution sequence uniform-hexagons-stripes is expected to exist near both of them. Together they form the combined sequence [170, 111]: uniform vegetation, hexagonal gap pattern, stripe pattern, hexagonal spot pattern and low-biomass uniform vegetation, where the hexagonal spot and gap patterns correspond to the 0-hexagons and π-hexagons discussed in Section 7.1.1. Since the spot patterns generally persist at precipitation values significantly lower than $p = 1$, where the uniform-vegetation state no longer exists and the bare-soil state is stable,

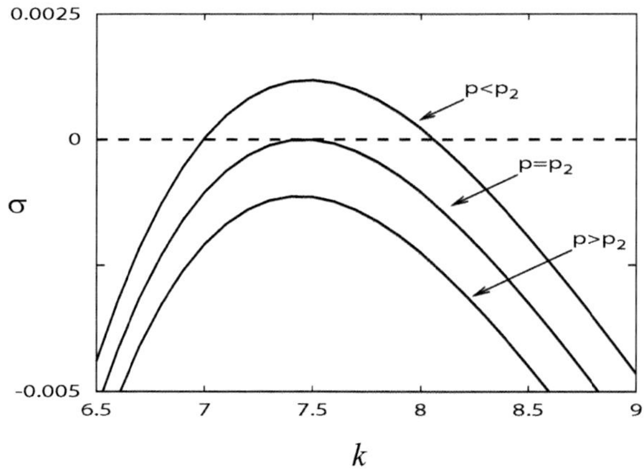

FIGURE 10.2: Growth-rate curves $\sigma = \sigma(k)$ for non-uniform spatial perturbations of the uniform vegetation solution of (9.10)–(9.14) (denoted by $\mathcal{V}$ in Figure 10.1), around the point $p = p_2$. When $p > p_2$ all wave-numbers k have negative growth rates and any perturbation decays. At $p = p_2$ a critical wave-number becomes marginal; perturbations having this wave-number neither grow nor decay. When $p < p_2$ there exists a band of wave-numbers with positive growth rates; the uniform solution is unstable and perturbations characterized by the critical wave-number grow faster than all others. From [105].

the sequence of basic vegetation states along the precipitation gradient generally consists of uniform vegetation, gaps, stripes, spots, and bare soil (rather than low-biomass uniform vegetation), as Figure 10.3 shows.

10.1.2 Vegetation states in hill slopes

Consider now a landscape with a constant slope in the x direction for which $\nabla\zeta = c\hat{\mathbf{x}}$, where c is the slope. The flow term in (9.10c) then becomes

$$-\nabla \cdot \mathbf{j} = \delta_h \nabla^2(h^2) + 2\delta_h c \partial_x h \,. \tag{10.3}$$

Studies of (9.10) with (10.3) lead to a bifurcation diagram for uniform solutions similar to that shown in Figure 10.1 for flat terrains [334, 105]. The periodic pattern states are also similar except for three significant differences: (i) the stripe (or band) patterns orient themselves in a direction perpendicular to the slope, (ii) they migrate uphill, and (iii) they occupy a wider precipitation range. The orientation perpendicular to the slope direction is a means by which the vegetation maximizes the amount of water it receives by runoff interception; patterning along the slope would result in water leakage out of the system and vegetation decay. The orientation dynamics are illustrated in

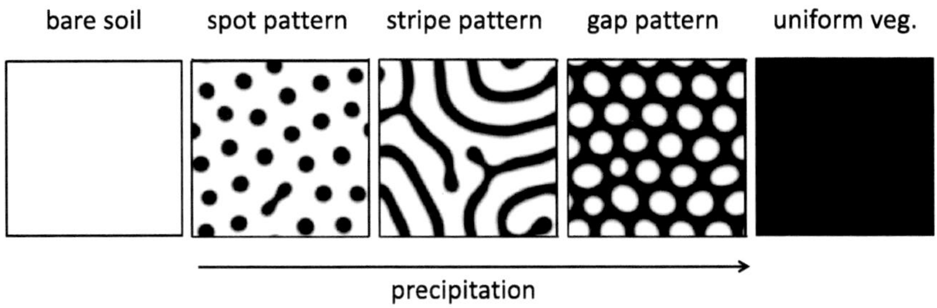

FIGURE 10.3: The five basic vegetation states along the rainfall gradient; uniform vegetation, gap pattern, stripe pattern, spot pattern and bare soil. The pattern states shown are numerical solutions of (9.10)–(9.14).

Figure 10.5 with a numerical solution of (9.10). The migration uphill is also easy to understand; while the top part of a vegetation band benefits from the interception of runoff and can expand uphill, the bottom part loses runoff and retracts. The migration uphill of vegetation bands has been confirmed in field observations [317, 64]. In variable environments the uphill dynamics may involve the expansion and contraction of vegetation bands, as the observations in the northeastern Chihuahuan desert shown in Figure 10.6 indicate. In wetter periods (panels of March 1991) colonization uphill leads to band expansion, while in dryer periods (panels of January 2003) dieback uphill leads to band contraction. The average migration speed over a period of 22 years in this site was found to be 20cm per year. The widening of the precipitation range of banded vegetation as the slope increases can be attributed to the more dominant role of the pattern-formation mechanism associated with downhill water advection, as the slope increases (see Section 9.1.2).

10.1.3 Bistability of stable vegetation states and state transitions

Another significant result predicted by (9.10) and earlier models [325, 170, 254] is that any consecutive pair of the basic vegetation states along the rainfall gradient has a precipitation range of bistability. That is, there are bistability ranges of bare soil and spot patterns, of spot and stripe patterns, stripes and gap patterns, and finally a bistability range of gap patterns and uniform vegetation. These bistability ranges are consequences of the positive feedbacks between biomass and water. For example, the increased infiltration of surface water into vegetated soil helps sustaining vegetation patches under conditions where bare soil is a stable state and seed germination cannot occur. Since the alternative stable states in each bistability range differ in their bioproductivity, transitions between these states that involve loss of

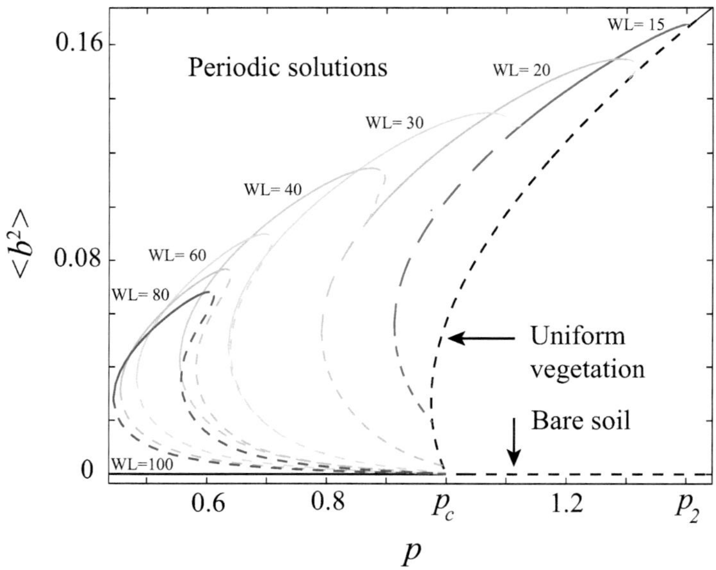

FIGURE 10.4: A bifurcation diagram for (9.10) with G_B and G_W given by (9.19) in 1d. The vertical axis represents the spatial average of b^2 while the horizontal axis represents the precipitation rate. The diagram shows the existence and stability information for uniform vegetation and bare soil solutions, and for periodic solutions that differ in their wavelengths (WL) as indicated. The periodic solution with the shortest wavelength emanates from the uniform-vegetation solution branch at $p = p_2$ in a non-uniform stationary instability, and terminates on the same solution branch at a lower precipitation value. Additional periodic solutions emanate either from the uniform vegetation solution or from other periodic solutions, and also terminate on the uniform-vegetation solution branch at lower p values that progressively increase toward the instability point $p_c = 1$ of the bare soil solution. From [339].

bioproductivity result in desertification (see Section 3.2.1). We postpone the discussion of this important topic to Chapter 11 and focus here on an intriguing model prediction associated with the bistablity of patterned states in sloped terrains.

Consider a bistability range of spot and stripe patterns on a hill slope, and imagine an initial stripe pattern oriented perpendicular to the slope with a local biomass-removal disturbance uphill, as the leftmost panel in Figure 10.7 shows. The subsequent dynamics, obtained by numerical integration of (9.10) with (10.3), are shown by the snapshots in the panels to the right. The initial biomass removal allows for more runoff to accumulate in the band segment just below it. As a result this segment grows taller, draws more water from

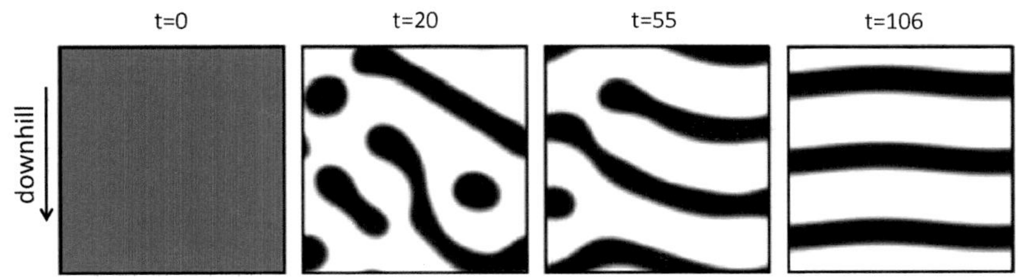

FIGURE 10.5: Development of vegetation bands migrating uphill from an unstable uniform vegetation state, obtained by numerical integration of (9.10). The panels show snapshots of this process at progressive times (t in years). From [105].

its surroundings and outcompetes the vegetation in the nearby segments of the same band. Runoff through the segments of decayed vegetation leads to a chain of similar processes downhill that culminates in a global transition from a stripe pattern to a spot pattern. In terms of pattern formation theory, the initial disturbance induced a local transition to the alternative stable state—the spot pattern, and created a front that propagates downhill. The mechanism of front propagation is tightly related to the overland water flow, suggesting that the slope plays a crucial role in this process. Indeed in a flat terrain a similar local disturbance of a stable stripe pattern has no global effect as Figure 10.8 demonstrates. This is consistent with a numerical finding of front pinning in a bistability range of spots and stripes [214]. Runoff downhill plays an important role also in transitions between stripe patterns of different periodicity [334].

The existence of five basic vegetation states and bistability ranges between any consecutive pair of states are concrete theoretical predictions that have not been tested yet in controlled laboratory experiments, although they appear consistent with field observations [63]. Good candidates for such experiments are *Paspalum Veginatum* and *Poa bulbosa*—perennial grasses that form patterns with a characteristic length of about 10cm, suggesting that grass trays as small as $1 \times 1\text{m}^2$ can already represent pretty large systems appropriate for pattern formation studies. Figure 10.9 shows pictures of *P. vaginatum* patterns taken in a residential neighborhood. The local appearance of all three pattern states, spots, stripes and gaps (Figure 10.9(b-d)) in the same lot suggests not only the possible observation of all five basic vegetation states with this plant species, but also a multiplicity of different stable states[1].

[1]It is interesting to point out that pattern formation in *P. Vaginatum* has also been attributed to a thermal-convection instability (see Section 2.1.2) induced by a thermal gradient between warm ground and cool air [304].

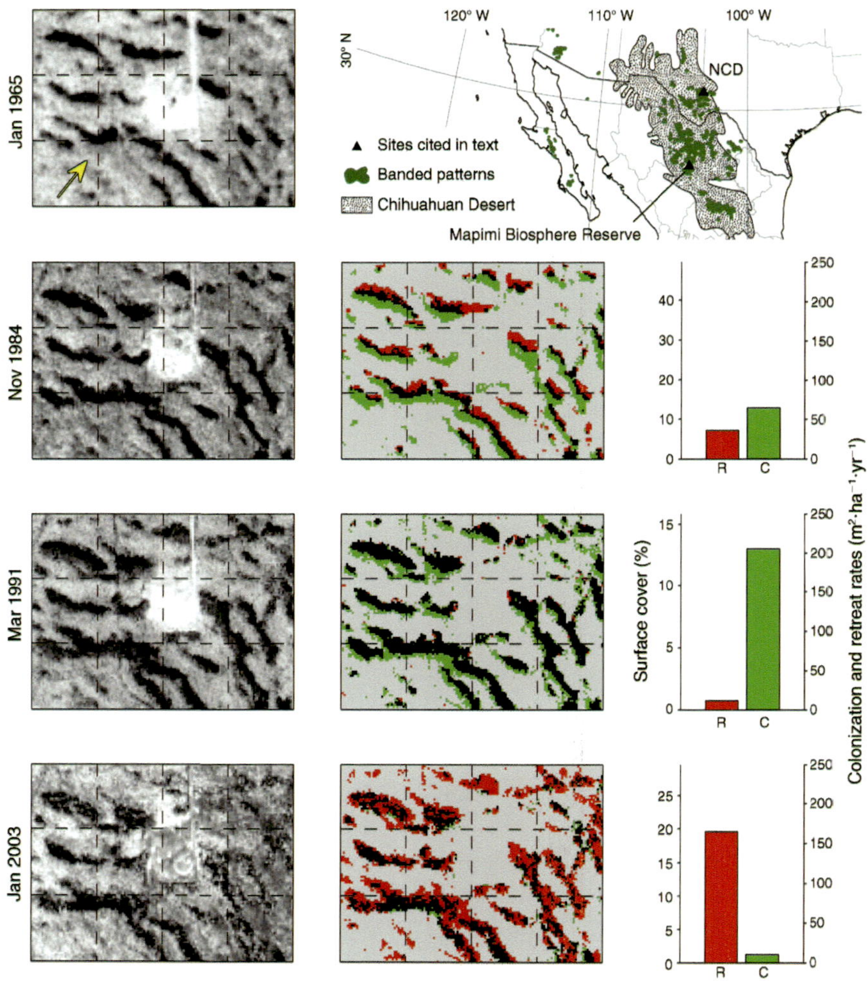

FIGURE 10.6: Uphill dynamics of vegetation bands in the northeastern Chihuahuan desert. The first column to the left shows aerial views of the same location on four different dates. These images were used to compute the transition maps shown in the second column: red, retreat; black, persistent vegetation; green, colonization; gray, persistent bare soil. A dashed reference grid with 100m intervals is included. The yellow arrow indicates the water flow direction. The third column summarizes respective colonization (C) and retreat (R) rates and surface percent cover for each time interval. For more details see [64].

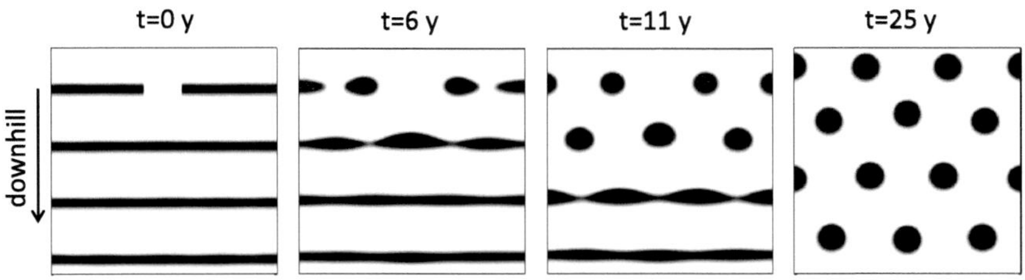

FIGURE 10.7: Local disturbance leading to a global state transition. A local clear-cut along the uppermost band of a linearly stable band pattern on a slope (leftmost panel) induces a chain process that culminates in a stable spot pattern (rightmost panel). The driving forces of the process are runoff and intraspecific competition. From [105].

FIGURE 10.8: A local biomass removal in a bistability range of spot and stripe patterns in a flat terrain has no global effect, as the initial and asymptotic states shown in the left and right panels, respectively, indicate. From [213].

10.1.4 Classification of aridity

The results described above suggest a novel operational view of the concept of aridity. The term refers to a permanent water deficit that bears on the bioproductivity of a given region; the more arid, the less productive the region is. Aridity has traditionally been quantified in terms of average precipitation rates. A more advanced estimator of aridity is the *aridity index* (AI) defined as the ratio, P/PET, between the mean annual precipitation (P) and the mean annual potential evapotranspiration (PET). The PET is defined as the amount of water that would be lost from water-saturated soil by plant transpiration and direct evaporation from the ground [188, 191]. Whatever choice of aridity estimator is made, the degree of aridity is defined by four successive intervals along the estimator axis representing hyper-arid, arid, semi-arid and dry sub-humid climates. For example, in terms of the aridity index the

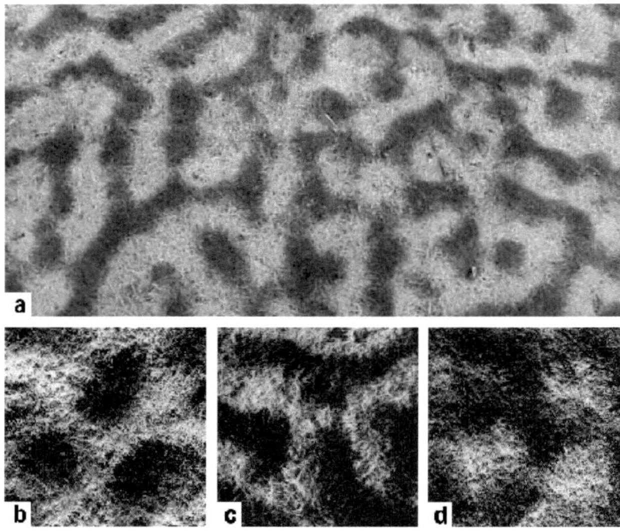

FIGURE 10.9: Patterns of *P. vaginatum* observed in the Northern Negev (200 mm mean annual rainfall): a labyrinth-like stripe pattern (a) and closeups showing spots (b), stripes (c) and gaps (d). The typical distance between spots and stripes is about 0.1 m. From [214].

four aridity classes are: $AI < 0.03$ - hyper-arid, $0.03 < AI < 0.20$ - arid, $0.20 < AI < 0.50$ - semi-arid, $0.50 < AI < 0.65$ - dry sub-humid [316].

The difficulty with this approach in classifying the aridity of drylands is twofold; it attempts to characterize a system state by a single water-related index and it associates aridity classes with thresholds that are, to some extent, arbitrary. Focusing on a single index such as AI does not take into account other factors that affect bioproductivity, such as biomass growth and mortality rates, or water redistribution by a non-uniform topography. It also does not take into account the possible coexistence of alternative stable states of different bioproductivity, e.g., bare soil and a periodic spot pattern. The results described in the previous sections motivate a different approach to classifying aridity—an approach that is based on the *inherent* states of the system. A possible classification along this line is illustrated in Figure 10.10 and goes as follows [325]:

Hyper-arid: A class characterized by a single unproductive stable state—the bare-soil state ($0 < p < p_0$).

Arid: A class characterized by coexistence of stable bare-soil state and a pattern state ($p_0 < p < p_c$).

Semiarid region: A class where the only stable states are vegetation patterns ($p_c < p < p_2$).

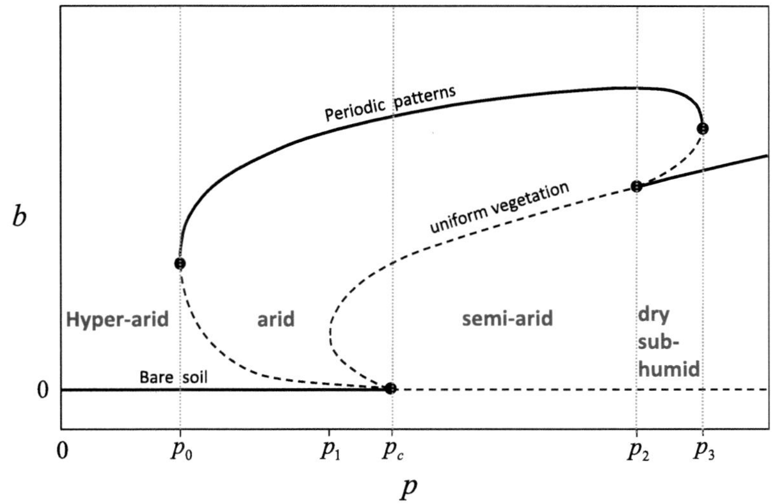

FIGURE 10.10: A classification of aridity based on the inherent vegetation states of the system. Shown is a schematic bifurcation diagram of basic vegetation states in which four precipitation ranges are delineated in color. The precipitation ranges define the four aridity classes: hyper-arid ($0 < p < p_0$), arid ($p_0 < p < p_c$), semi-arid ($p_c < p < p_2$) and dry sub-humid ($p_2 < p < p_3$). For simplicity only one branch of a periodic-pattern solution is shown.

Dry sub-humid: A class where vegetation patterns stably coexist with uniform vegetation ($p_2 < p < p_3$).

By focusing on the inherent states of the system, the aforementioned difficulties with the current approach are circumvented; by definition the aridity classes are related to the system states, which integrate the effects of various biotic and abiotic factors, and the thresholds that distinguish one class from another are no longer arbitrary. The information about coexistence of stable states has operative aspects in regard to vulnerability to desertification or the recovery potential of degraded regions. A region with patchy vegetation that is classified as arid is vulnerable to desertification and requires special attention, while a similar landscape that is classified as semi-arid is not vulnerable to desertification. Likewise, a bare-soil region classified as arid is recoverable, while a bare-soil region classified as hyper-arid is not recoverable. The implementation of the proposed classification of aridity requires long term observations of the diversity, coexistence and dynamics of vegetation patterns, from which the inherent system state may be inferred. In addition, small-scale experimentation may be needed in identifying alternative stable states.

10.2 Non-periodic patterns

The basic vegetation states described in the previous section are either uniform or periodic, while vegetation in nature is often non-periodic. To what extent can PDE models account for irregular patterns? In the following two sections we use the model equations (9.10) to study two mechanisms by which irregular patterns can appear in homogeneous systems. The first mechanism (Section 10.2.1) is associated with bistability ranges of different basic vegetation states, where spatial mixtures of the two alternative states can either lead to long-lived transient patterns, or to stationary non-periodic states. The second mechanism (Section 10.2.2) applies to conditions under which the competition for water becomes global and scale-free patterns develop. Dryland landscapes are never strictly homogeneous; quite often they consist of rock-soil mosaics where rock patches act as sources of runoff water to adjacent soil patches. In Section 10.2.3 we use equations (9.10) to study the interplay between self-organization and such rock-soil relationships in shaping vegetation patchiness.

10.2.1 Localized and hybrid states in bistability ranges

10.2.1.1 Transient patterns

The three pattern states, spots, stripes and gaps, represent periodic patterns, or nearly periodic patterns if the periodicity is disrupted by defects [55]. The bistability of any of these patterned states with other basic vegetation states allows for long-lived non-periodic transient patterns. Figure 10.11 shows numerical simulations of (9.10) in the bistability range of bare soil and hexagonal spot patterns that reveal long-lived transients in the course of a global transition from bare soil to a spot pattern. Figure 10.11(a–d) shows the dynamics that result from an initial condition consisting of two adjacent spots in an otherwise bare soil; the two spots split into four and then into eight. This spot-replication process continues (not shown in the figure) until a hexagonal spot pattern is obtained. Figure 10.11(e–h) shows the dynamics that result from an initial condition consisting of a single isolated spot for the same model parameters. The spot expands symmetrically in the radial direction and form a ring with a characteristic width that continues expanding. Here too the asymptotic state is a hexagonal spot pattern; the ring eventually disintegrates into self-replicating spots as it approaches other spots or if it were not perfectly symmetric initially. Spot-replication processes have been observed in other pattern formation contexts [168] and have been analyzed mathematically [230].

Vegetation rings are common in nature. Figure 10.12 shows ring patterns formed by different plant species. Studies of (9.10) suggest two mechanisms for vegetation ring formation from an initially small patch [285, 286]. The first

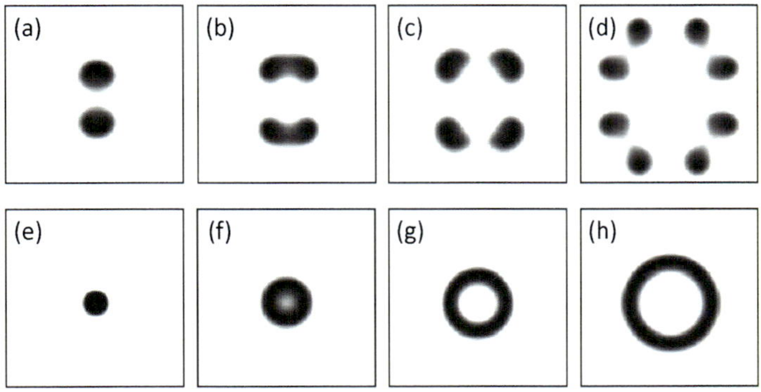

FIGURE 10.11: Snapshots of transient patterns in a bistability range of bare soil and a spot pattern, produced by numerical simulations of (9.10) in a flat terrain. (a–d) A pair of adjacent spots going through spot-replication processes (time proceeds from left to right). (e–h) A single spot developing into an expanding ring. Both simulations were conducted with the same model parameters; the only difference is the initial condition.

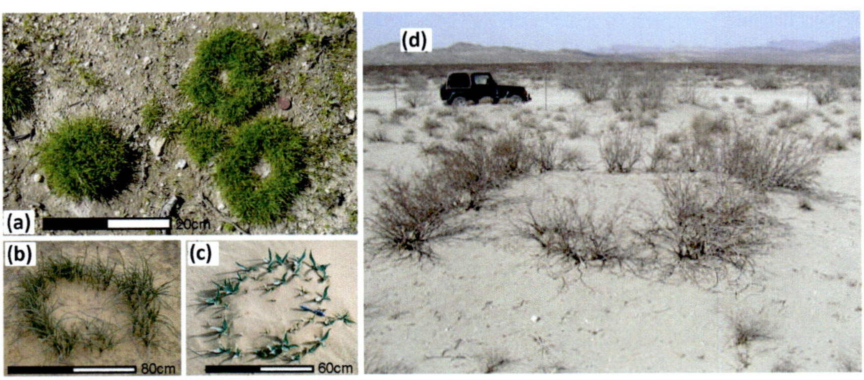

FIGURE 10.12: Ring patterns in nature. (a) Mixtures of rings and spots of *Poa bulbosa* L. observed in the Northern Negev, Israel (150 mm/yr); (b) An *Asphodelus ramosus* L. ring observed in the Negev desert, Israel (170 mm/yr); (c) A ring of *Urginea maritime (L.)* Baker observed in Wadi Rum, Jordan (50 mm/yr); (d) Ring of *Larrea Tridentata* (DC.) Coville in Lucerne Valley, California, USA (98 mm/yr). From [286].

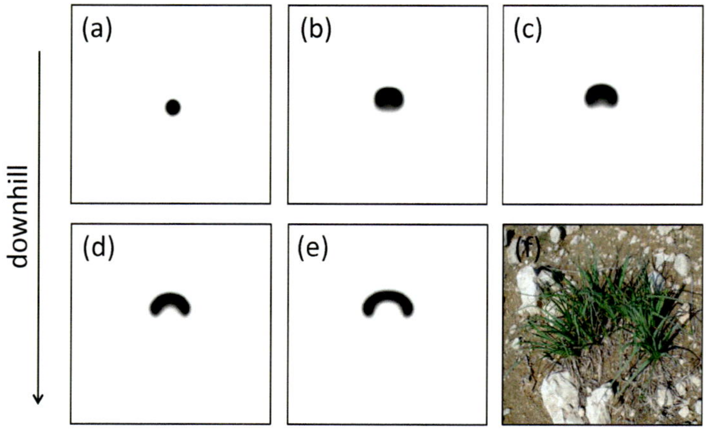

FIGURE 10.13: Crescent patch forms on a hill slope. (a–e) Snapshots of numerical simulations of (9.10) showing an initial spot forming in time a crescent-like patch that slowly migrates uphill (time increases from (a) to (e)). (f) A crescent-like patch of *Asphodelus ramosus* on a slope observed in the northern Negev (200 mm/yr). Typical patches are 20 cm long. From [215].

mechanism pertains to conditions of high infiltration contrast between vegetated and bare soil, under which overland water flow is intercepted mostly at the patch periphery. The decreasing amount of water that the patch core receives as the patch expands, leads to central dieback and ring formation. The second mechanism pertains to plants with large lateral root zones, and involves central dieback and ring formation due to increasing water uptake by the newly recruited individuals at the patch periphery whose roots extend to the patch core. In general the two mechanisms act in concert, but the relative importance of each mechanism depends on environmental conditions [286]. Mechanisms attributing ring patterns to negative plant-soil feedbacks in the patch core have also been proposed, e.g., a negative feedback between sediment deposition and vegetation growth [250], or a negative feedback between toxic compounds released by litter decomposition and vegetation growth [43]. Since basic biomass-water relations alone can account for ring formation, the importance of negative plant-soil feedbacks as primary drivers of ring formation in water-limited systems is not yet clear.

On a slope, transient crescent shapes migrating uphill can develop as the model simulations in Figure 10.13(a–e) show [215]. Ring shapes are unstable on slopes that are steep enough, as most of the runoff is intercepted in the upper ring part and does not reach the lower part. Crescent shapes are also commonly observed in nature. Figure 10.13(f) shows an example.

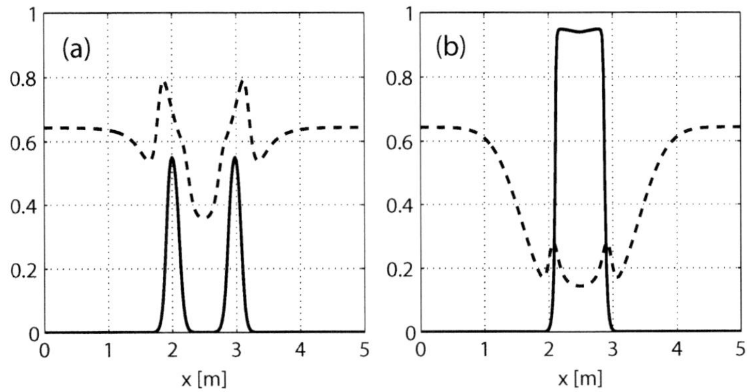

FIGURE 10.14: Transects of two-dimensional spatial distributions of biomass (solid lines) and soil water (dashed lines) for (a) an expanding ring solution ($\eta = 1.6$), and (b) a stationary spot solution ($\eta = 3.2$). From [215].

10.2.1.2 Stable stationary patterns

In Section 7.2.1.3 we studied pattern dynamics in a bistability range of a uniform zero state and a 1d periodic pattern, using the SH equation, and distinguished between two types of subranges, the snaking range, where fronts that separate the two states are pinned in place (Figures 7.12(b,e)), and the complementary subranges where fronts propagate. The front dynamics in the complementary subranges eventually lead to the zero state (Figures 7.12(c,f)) or to the patterned state (Figures 7.12(a,d)), depending on the value of the control parameter relative to the Maxwell point. Similar results hold in 2d within the bistability range of the zero state and the hexagonal-pattern state [179].

The vegetation model (9.10) shows richer behavior. While front pinning and homoclinic snaking is found in the bistability range of uniform vegetation and periodic patterns (hexagonal gap pattern in 2d) no pinning and snaking could be found in the bistability range of bare soil and periodic patterns (hexagonal spot pattern in 2d) [339]. We postpone the discussion of the former bistability range to Chapter 11 where we study gradual and incipient regime shifts, and address here the latter bistability range.

Although pinned fronts could not be found in numerical studies of (9.10) in the bistability range of bare soil and spot patterns, localized solutions consisting of a single stationary spot are easily found. One way to identify such solutions is to consider higher water uptake rates by increasing the parameter η. Figure 10.14 shows transects of the biomass and soil-water variables across an expanding ring solution and across a stationary spot solution obtained at larger η value. The strong soil-water uptake when η is large depletes the soil-water content in the patch neighborhood down to a level that does not allow patch expansion and results in a stationary spot structure.

Localized single-spot solutions, such as that shown in 10.14(b), can provide building blocks for a wide variety of non-periodic spot patterns. As Figure 10.14(b) shows, the impact of a vegetation spot on its environment, through soil-water depletion, is limited to a confined area around the spot. Beyond this impact zone spots do not interact with one another in any noticeable way and can form any pattern. Figure 10.15(top) shows a numerical simulation of (9.10) within the bistability range of bare soil and spot patterns, with an initial condition consisting of a sextet of spots whose impact zones overlap to some extent and a farther isolated spot. In the course of time the vegetation spots first contract because of the intra-spot competition for water. The sextet of spots contract more than the isolated spot because of additional inter-spot competition. After a relatively short phase of contraction the spots slowly move outward to more moist areas. Once their modified impact zones hardly overlap the spots become practically stationary[2], forming a configuration that is very close to the initial one. This pattern plasticity, i.e., the ability to yield to the initial pattern, does not exist outside the bistability range, where periodic spot patterns are the only stable states, as Figure 10.15 (bottom) shows; the same initial condition and any other initial condition eventually lead to a periodic spot pattern.

The wide variety of patterns that can be formed using the localized single-spot structures as building blocks is reminiscent of the multitude of stable hybrid states associated with homoclinic snaking in the SH equation, but there appears to be an important difference. The hybrid-state solutions in the snaking range consist of patterned domains separated by zero-state domains. Within any patterned domain the spots are coupled entities and the solutions are stationary because the fronts that separate the zero and patterned domains are pinned. This is unlike the non-periodic numerical solutions of the vegetation model (9.10), which are stationary because they consist of isolated non-interacting spots. Mathematically, the difference between the SH equation and the vegetation-model equations is that in the SH equation the periodic-pattern solution bifurcates from the zero solution, whereas in (9.10) it bifurcates from the (unstable) uniform vegetation solution, which is not one of the two alternative stable state.

Stationary non-periodic patterns are not limited to the bistability ranges of uniform and patterned states; a wide variety of such patterns also exist in the bistability ranges of two patterned states. The non-periodic patterns are spatial mixtures of the two patterned states: spots mixed with stripe segments in the bistability range of spot and stripe patterns, and stripe segments mixed with gaps in the bistability range of stripe and gap patterns. The mathematical theory of such patterns is not developed but numerical observations of pinned fronts, separating the two patterned states, suggest some forms of homoclinic snaking in these bistability ranges too [214]. Figure 10.16 shows examples

[2]In theory, the spots may still propagate outwards, but this propagation is negligibly small.

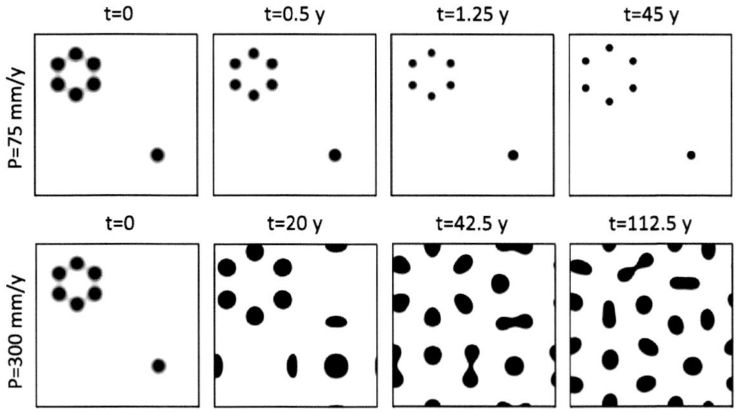

FIGURE 10.15: Pattern plasticity in a bistability range of uniform and patterned states. Shown are numerical simulations of (9.10) at low precipitation ($P = 75$mm/yr), within the bistability range of bare soil and spot patterns, and at high precipitation ($P = 300$mm/yr), where bare soil is an unstable state, using the same initial condition. In the bistability range (top) the system shows pattern plasticity; the asymptotic pattern is very similar to the initial one. In the monostability range of spots (bottom) the asymptotic state is a spot pattern irrespective of the initial condition. From [213].

of non-periodic solutions of (9.10) appearing in the four bistability ranges: bare-soil and spots, spots and stripes, stripes and gaps, and gaps and uniform vegetation. Figure 10.17 shows examples of similar patterns in nature.

10.2.2 Scale-free vegetation patchiness

10.2.2.1 Global competition and wide patch-size distributions

So far we have discussed three basic periodic patterns (spots, stripes, gaps) and a variety of additional non-periodic patterns in bistability ranges. Despite the wide variety of these patterns they all share one property—a characteristic length scale, whether it is the width of a stripe, the diameter of a spot or the diameter of a gap. Recent field studies in the Kalahari desert [267] and in Mediterranean ecosystems [149], however, have reported the observations of patterns with wide patch-size distributions, lacking any characteristic length. How can we reconcile the existence of these "scale-free" patterns with the pattern-forming feedbacks that dictate a length scale? The key to the answer of this question is *global competition* for the limiting water resource [267, 194, 324] as we explain below.

Studies of the model equations (9.10) [324] suggest that global competition and scale-free patterns can develop when the spatial distribution of the

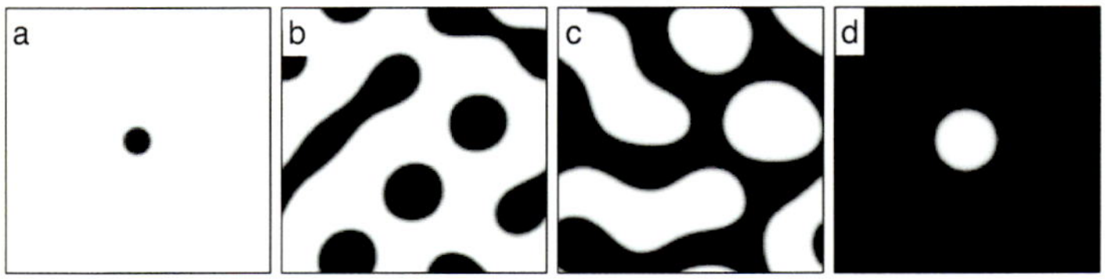

FIGURE 10.16: Bistability ranges of basic states give rise to stationary non-periodic patterns: bare soil and spots (a), spots and stripes (b), stripes and gaps (c), gaps and uniform vegetation (d). From [156].

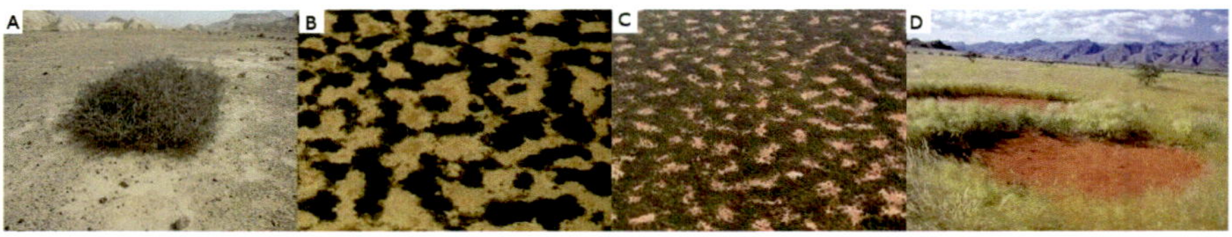

FIGURE 10.17: Non-periodic patterns in nature: an isolated shrub patch in the northern Negev, Israel (A), mixture of spots and stripes of woody vegetation in Niger (B), mixture of stripes and gaps of woody vegetation in Niger (C), and isolated gaps in the pro-Namib zone of the west coast of southern Africa (D). From [212].

water resource is fast relative to processes that exploit it. Two different realizations of this principle have been identified, fast surface-water flow relative to the infiltration of surface water into the soil, and fast soil-water diffusion relative to water uptake by plant roots. Consider first a system with a sharp infiltration contrast for which the infiltration feedback is the only driver of vegetation patchiness. When the condition of fast surface-water flow relative to infiltration is not satisfied, a small initial patch either grows to a vegetation spot of a characteristic size, or forms an expanding vegetation ring of a characteristic width, as we have already discussed in Section 10.2.1.1. The ring forms because most of the surface water flowing toward the growing patch infiltrates at the patch periphery and does not reach the patch center. Imagine now that the condition of fast surface-water flow relative to infiltration is satisfied. Then, by the time any significant infiltration takes place at a growing vegetation patch, surface water has already reached the patch center. As a result, central dieback is avoided and big patches can develop. Furthermore, at any vegetation patch where infiltration begins to take place, fast long-range overland flow immediately compensates for the local surface-water depletion, thereby inducing global competition. As a result, available surface water in the vicinity of a small patch can be exploited by farther bigger patches, preventing the growth of the small patch. The combination of these processes

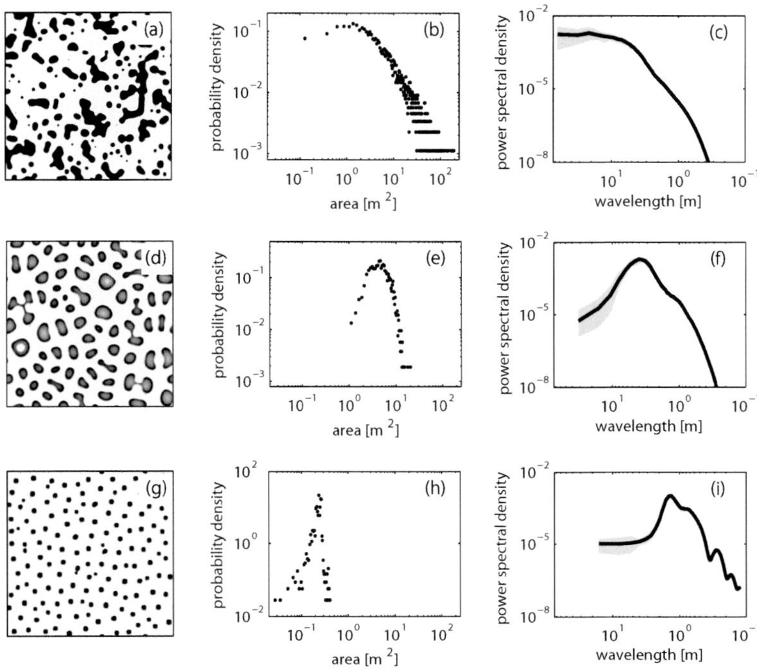

FIGURE 10.18: A typical scale-free pattern, obtained for global competition induced by fast surface-water flow relative to infiltration (a–c), and the appearance of characteristic patch sizes upon decreasing the competition range, either by making the surface-water flow slower (d–f), or by increasing the lateral root-zone size (g–i). Panels b, e, h show patch-size distributions. Panels c, f, i show average power spectra. For additional information see Ref. [324].

acts to maintain both large and small patches and therefore leads to wide patch-size distributions.

Figure 10.18(a–c) shows the results of simulating (9.10) under conditions of fast surface-water flow relative to infiltration and no root-augmentation feedback ($\eta = 0$), using small amplitude random initial conditions. The biomass pattern consists of amorphous patches (Figure 10.18(a)) whose size distribution spans a range of two orders of magnitude (Figure 10.18(b)) and lacks any characteristic length (Figure 10.18(c)). Slowing down the surface-water flow by decreasing δ_h or switching on the root-augmentation feedback by increasing η, reduces the competition range, narrows down the patch size distribution and singles out a characteristic length scale, as Figures 10.18(d–f) and 10.18(g–i) indicate, respectively.

The condition for global competition induced by fast overland flow relative to infiltration can be quantified in terms of model parameters using dimensional analysis (see Section 5.1). To this end we write the condition as

$$\chi_F \equiv \frac{\tau_I}{\tau_F} \gg 1\,, \tag{10.4}$$

where τ_I and τ_F are the time scales associated with infiltration and overland flow, respectively, and χ_F is a non-dimensional parameter that quantifies the degree of lacking a characteristic length. The infiltration time τ_I is simply the inverse of the infiltration rate and can be estimated as $\tau_I = A^{-1}$. To estimate τ_F we associate it with the time required for the flow to pass a distance S_0. This time is given by S_0/C_F, where C_F is the overland flow velocity. We now use dimensional analysis to express C_F in terms of dimensionally independent model parameters. The dimensional parameters that are relevant to overland water flow are those appearing in the equation for H (9.3c), that is, P, D_H and the parameters A and Q that appear in the expression (9.1) for the infiltration rate. Together with C_F they form the set $\mathbf{X} = (D_H, P, A, Q, C_F)$ of relevant dimensional quantities (see Section 5.1.1). The subset $\mathbf{A}$ of dimensionally independent parameters can be chosen to be $\mathbf{A} = (D_H, P, A)$. The complementary subset is then $\mathbf{B} = (Q, C_F)$. The dimensionless quantities associated with the subset $\mathbf{B}$ are

$$\Pi_1 = \frac{Q}{A^{-1}P}\,, \qquad \Pi_2 = \frac{C_F}{{D_H}^{1/2}P^{1/2}}\,, \tag{10.5}$$

and according to the Π-theorem these dimensionless quantities should be related, $\Pi_2 = \psi(\Pi_1)$. We thus obtain

$$C_F = \sqrt{D_H P}\cdot\psi\left(\frac{QA}{P}\right). \tag{10.6}$$

The function ψ is unknown, but its argument is typically very small, $QA/P \ll 1$ [105]. Assuming it is approximately constant (see Section 5.1.4) we obtain the scaling law $C_F = \sqrt{D_H P}$ (up to a constant factor). Inserting $\tau_I = A^{-1}$ and $\tau_F = S_0/C_F \sim S_0/\sqrt{D_H P}$ into (10.4) we find

$$\chi_F = \frac{\tau_I}{\tau_F} = \frac{\sqrt{D_H P}}{AS_0}\,. \tag{10.7}$$

It is interesting to note that χ_F also provides a measure for the ratio of the largest patch size, denoted S_P, to the smallest possible patch size, S_0. To show that let us estimate S_P by associating it with the largest wet patch, assuming that the growth is not seed-limited. The largest wet patch is given by the distance overland water can flow before infiltrating into the soil: $S_P = C_F \cdot \tau_I$. Using the expressions for C_F and τ_I we find

$$S_P = A^{-1}\sqrt{D_H P}\,, \tag{10.8}$$

a form that has been confirmed numerically using the model equations. Comparing (10.8) with (10.7) we find

$$\chi_F = \frac{S_P}{S_0}\,. \tag{10.9}$$

The condition (10.4) of fast overland flow relative to infiltration therefore implies $S_P \gg S_0$ or wide patch-size distributions. This condition holds for the simulations shown in Figure 10.18(a–c) (in which D_H has been assumed to be infinitely large [324]), but does not hold for the simulations shown in Figure 10.18(d–f) for which $\chi_F \approx 2$ [324].

Similar considerations can be applied to the second realization of global competition—fast soil-water diffusion relative to uptake. In this case the maximal patch size can be estimated as $S_P = C_D \cdot \tau_U$, where C_D is the hydraulic conductivity and τ_U is the uptake time. Dimensional analysis leads in this case to the scaling law $S_P = \sqrt{D_W E/\Gamma}$ and to the following condition for global competition and wide patch-size distributions [324]:

$$\chi_D = \frac{S_P}{S_0} = \frac{1}{S_0}\sqrt{\frac{D_W E}{\Gamma}} \gg 1\,. \tag{10.10}$$

We note that although global competition can induce scale-free patterns, it does not necessarily lead to such patterns; the asymptotic patterns strongly depend on the initial biomass distributions, and with appropriate choices regular patterns with narrow patch-size distributions can be induced.

10.2.2.2 Natural realizations of scale-free patterns

The scale-free patterns observed in the Kalahari desert [267] cannot be accounted for by the mechanism of fast overland flow relative to infiltration, mainly because the soil is sandy and the infiltration time is short. The mechanism of fast soil-water diffusion relative to uptake is harder to rule out because of the relatively high hydraulic conductivities in sandy soils, but further empirical data are needed before a conclusion can be made. A more convincing example is shown in Figure 10.19. The dark patterns represent biomass patchiness of the perennial grass *P. Bulbosa* at increasing rainfall values [284]. This clonal grass grows in dense clumps of genetically identical individuals with laterally confined root zones, suggesting a small η value. A density of about five individuals per cm^2, suggests S_0 values of the order of 1mm. The maximal patch size can be estimated using the relation $S_P = C_F/A$. Exact figures are not available for that particular case study, but flow velocities as low as $C_F \sim$ 1m/h and infiltration rates as high as 10mm/h, equivalent to $A \sim 1\text{h}^{-1}$, already provide a realistic order of magnitude for the largest patch size, $S_P \sim$ 1m, and a large χ_F value of order 10^3. These estimates suggest that fast overland flow relative to infiltration may possibly explain the scale-free patterns of *P. bulbosa*. Conclusive results, however, may only be obtained by conducting controlled laboratory experiments, using trays that can be tilted to control runoff velocity and soils with different infiltration properties.

The *P. bulbosa* patterns shown in Figure 10.19 are in sharp contrast to the spot and ring shapes of the *P. bulbosa* patches shown in Figure 10.12(a). The patches in Figure 10.19 are amorphous and lack any characteristic

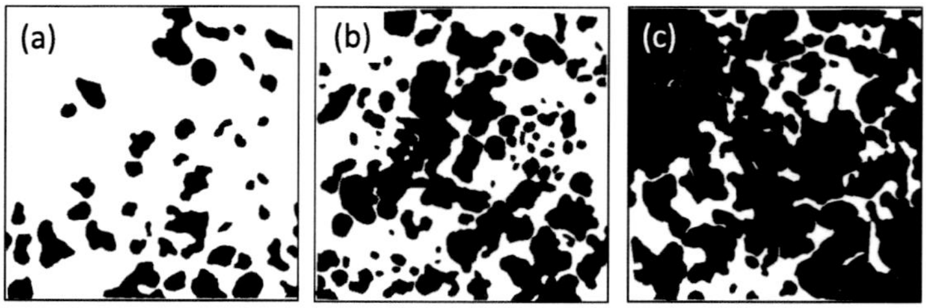

FIGURE 10.19: Field observations of *P. bulbosa* patchiness in $1 \times 1m^2$ fairly uniform plots in three sites along the rainfall gradient in Israel, Lehavim (200mm) (a), Adulam (400mm) (b) and Carmel Mt. (700mm) (c). From [284].

length, while the patches in Figure 10.12(a) are circular and appear to have characteristic lengths. A possible explanation for these differences is that the condition $\chi_F \gg 1$ of fast overland flow relative to infiltration is not satisfied at the particular location of Figure 10.12(a), but there is another possible explanation. Numerical studies of (9.10) along a precipitation gradient indicate, as demonstrated in Figure 10.20(a), that even when the condition $\chi_F \gg 1$ is satisfied, round-shape patches sharing a characteristic size can develop at sufficiently low precipitation. The narrowing down of the patch-size distribution can be attributed to curvature effects. Recall the linear approximation (7.58) of the velocity-curvature relation, which for $D > 0$ implies the existence of a critical radius $R_c = \kappa^{-1} = D/c_0$, below which a circular patch contracts ($c_n < 0$) and above which it expands ($c_n > 0$). A calculation that is based on the biomass equation only, treating the soil-water content W as a constant parameter, shows that the critical radius, R_c, is a decreasing function of W [284]. That is, R_c increases as precipitation, and consequently W, decreases. The larger critical radius at lower precipitation eliminates more vegetation patches, as more of them contract and disappear, and thus narrows down the patch-size distribution. As Figure 10.20(e) shows, a similar process occurs at sufficiently high precipitation with vegetation patches replaced by vegetation gaps. The narrowing down of the patch-size distribution in stressed systems has been found in other models too and has been suggested as a possible early indicator for imminent desertification [150] (see Chapter 11). Note that although the patterns in the intermediate precipitation range (Figure 10.20)(b–d) consist of amorphous patches spanning wide size ranges, they still capture the spot-, labyrinth- and gap-like appearances of nearly periodic patterns along the precipitation axis.

10.2.2.3 Patch coarsening

We have identified conditions that give rise to scale-free vegetation patterns but we have not studied yet their asymptotic behavior at long times. In Section 7.1.3 we found that the appearance of scale free patterns in the FHN

FIGURE 10.20: Numerical simulations of (9.10) at increasing precipitation values (from left to right) under conditions of fast overland flow relative to infiltration. Scale free patterns consisting of amorphous patches spanning a large range of sizes exist only at intermediate precipitation rates. At sufficiently low or high precipitation rates round biomass patches or bare-soil patches spanning much narrower ranges of sizes are found. From [284].

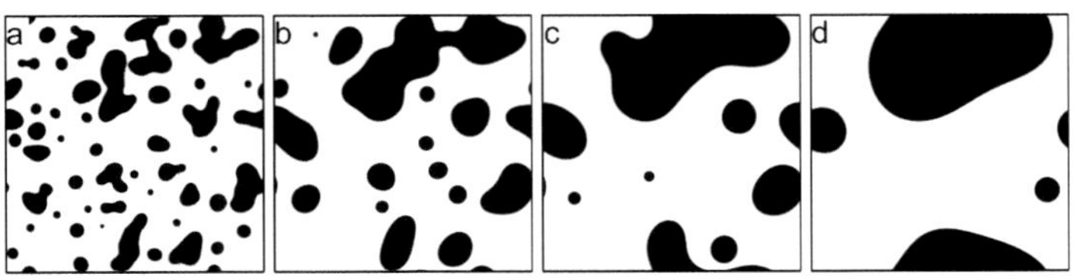

FIGURE 10.21: Patch coarsening of an initial scale-free vegetation pattern, obtained by numerical solution of (9.10). The snapshots were taken after 87 (a), 500 (b), 1500 (c) and 4,580 years (d). From [156].

model is followed by a phase coarsening process in which small spots shrink and disappear while big spots grow and merge together (7.5). Long-time simulations of (9.10), under conditions of fast surface-water flow relative to infiltration, reveal a similar phase- or patch-coarsening process as Figures 10.21 and 10.22 show. It is evident from these figures that while the overall fraction of area covered by vegetation (see inset in Figure 10.22) and the total biomass asymptote quickly to equilibrium constant values, determined by the actual precipitation rate, the pattern itself continues evolving, with growth and merger of patches, eventually approaching an asymptotic state containing a single circular patch. During this patch-coarsening process the average patch size grows linearly in time or, equivalently, the average patch radius grows like $\bar{R} \sim t^{1/2}$, as Figure 10.22 shows. This is the same power law that was found for the FHN model and is known as interface-controlled growth in two-phase mixtures [210]. The time scales involved in this process, however, can be very long—up to the order of 10^3 years for shrubs [156].

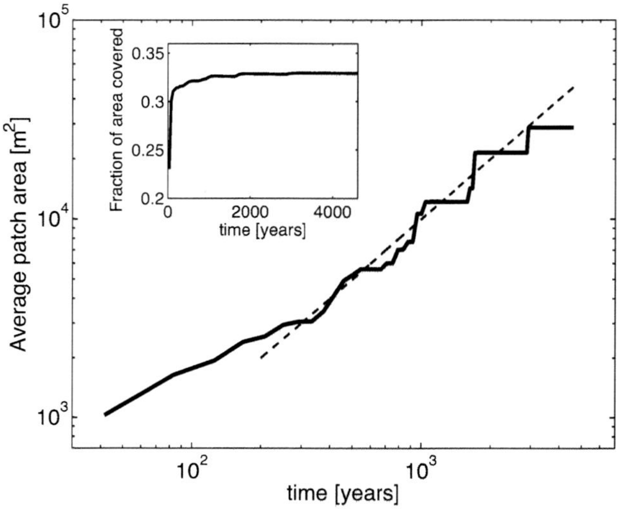

FIGURE 10.22: The average patch area as a function of time for the data of Figure 10.21. The dashed reference line indicates a linear growth with time. The inset shows the fraction of area covered by vegetation as a function of time. From [156].

10.2.3 Vegetation patchiness in heterogeneous environments

The discussion of vegetation patchiness has dealt so far with homogeneous environments where patchiness at the landscape level is a self-organization phenomenon associated with patch-level pattern-forming feedbacks. Highly heterogeneous environments, on the other hand, may interfere with self-organization and even dictate the form of vegetation patchiness. A typical example of the latter case is rocksoil mosaics in water-limited landscapes, where bare bedrock patches create run-off of water and nutrients that flow to adjacent soil patches. The resource-enriched soil patches provide favorable conditions for vegetation growth, and can lead to template-dictated vegetation patchiness [333, 219, 283]. In general, real landscapes comprise both pattern-forming feedbacks and source-sink relationships between abiotic and biotic patches, intermingling the two mechanisms of patchiness formation. The manner by which the two mechanisms act in concert is an important open problem. Some progress has recently been made by putting forward a conceptual model and testing it empirically and numerically using the vegetation-model equations (9.10) [283]. We briefly review here the main results of this study.

The development of vegetation patchiness in a heterogeneous landscape can be affected by four main factors: (1) the spatial extents of the soil patches in which the plant species grows, (2) the physiological-genetical size of the plant species, (3) the precipitation regime and (4) the source-sink resource

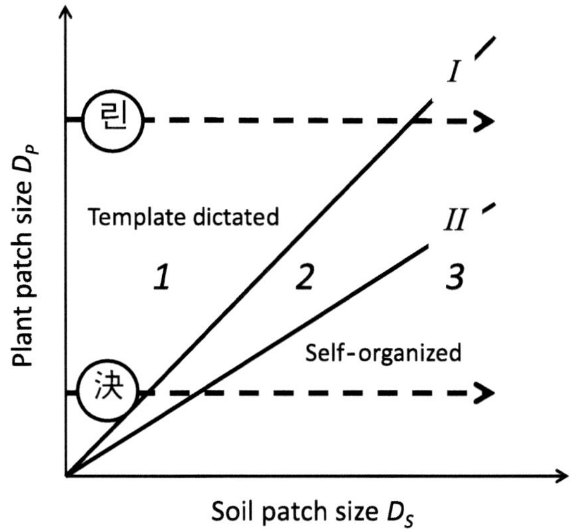

FIGURE 10.23: A conceptual model of vegetation patchiness in heterogeneous landscapes. Different domains in the D_s-D_p plane correspond to different mechanisms of vegetation patchiness: domain 1–patchiness dictated by the physical template, domain 2–patchiness affected by both the template and self-organization and domain 3–self-organized patchiness. The diagonal line *I* denotes the decline from full vegetation cover in domain 1 to a partial cover in domain 2. The diagonal line *II* denotes the convergence to the typical cover of a self-organized vegetation pattern in domain 3. Horizontal lines represent plants with large (a) and small (b) typical patch sizes.

relationships between different patch types within the physical template. The last three factors determine the typical "plant-patch size," defined as the spot diameter in a spot pattern, the stripe width in a stripe pattern and the distance between nearby gaps in a gap pattern. The ratio of this size to the size of the soil patch in which the vegetation grows determines the prevailing mechanism of vegetation patchiness, as Figure 10.23 illustrates.

Assuming first a fixed water input to a soil patch, we ask how vegetation coverage is affected by the ratio of plant-patch to soil-patch sizes (10.23). When the typical plant-patch size D_p is larger or about the size D_s of a soil patch, the soil patch will be fully covered (for simplicity, we consider D_s to be the diameter of a circular soil patch). If that is the case for all soil patches in the landscape, the spatial organization will be dictated by the physical template. In the plane spanned by D_s and D_p, this behavior is limited to the area for which $D_p \geq D_s$ (domain 1 in Figure 10.23). When the typical plant-patch size is much smaller than the size of a soil patch, $D_p \ll D_s$, the plants will self-organize to form a vegetation pattern. At the landscape level, the vegetation will still follow the physical template by occupying the network of soil patches, but within soil patches, the vegetation pattern will show self-organization into a patterned state. This behavior is limited to domain 3 in Figure 10.23. An intermediate regime of the plant patch size relative to the soil-patch size exists in which soil patches are partially covered by vegetation, but are too small for a vegetation pattern to develop (domain 2 in Figure 10.23).

The convergence at the origin of the transition lines that separate the three domains (I and II in Figure 10.23) reflects the increased probability to find self-organized patterns as the plant's patch size gets smaller. Thus, a plant species with a sufficiently small patch size (b in Figure 10.23) will develop self-organized patterns within most of the soil patches in the landscape, while a species with a large patch size (a in Figure 10.23) will uniformly cover most of the soil patches. In other words, while the vegetation patchiness associated with species b is mostly due to self-organization, the patchiness associated with species a is mostly dictated by the physical template.

The typical patch size of a plant species increases with water availability. Increased precipitation will therefore result in more fully covered soil patches and fewer soil patches showing self-organized patchiness. At sufficiently high precipitation, we may expect the patchiness to completely follow the pattern dictated by the physical template. Water availability in the space available for growth, however, is also determined by the source-sink relationships within the physical template, e.g., between rock and soil patches. Specifically, the redistribution of incoming rainfall in a rock-soil landscape will result in a differential increase of the water content in soil patches, with small patches having soilwater content significantly higher than large patches. This differential water availability increases landscape variability. Consider the extreme case of a precipitation regime that is too low for any form of vegetation to grow in a homogeneous landscape totally covered by soil. In a patchwork of soil and rock, large soil patches may remain bare because of insufficient run-off water input. In smaller soil patches, rainfall plus run-off may supply sufficient water to support vegetation patterns. Very small soil patches may even be fully covered. This will be the case when the typical plant-patch size exceeds the soil patch size because of the high water content in small soil patches. Thus, dry heterogeneous landscapes characterized by wide distributions of soil-patch sizes can be highly variable, showing all five basic vegetation states, and all corresponding hybrid states, in the same landscape.

As the precipitation rate increases, this variability decreases; the typical plant patch size increases, more soil patches become fully covered and larger, partially covered patches exhibit patterns typical of more favorable conditions, e.g., gaps and stripes rather than spots. Landscape variability can also change with the type of plant species considered. While a species with a small typical patch size will exhibit a highly variable landscape with soil patches exhibiting all or most types of vegetation states, a larger species, whose typical patch size exceeds the sizes of a wider range of soil patches, will exhibit lower landscape variability. Moreover, large soil patches, partially covered by the larger species, may tend to exhibit patterns characteristic of arid landscapes, such as spots rather than gaps, because of the lower soil-water content in large soil patches.

The conceptual model presented in Figure 10.23 has been tested using the model equations (9.10) for a heterogeneous environment under conditions of fast overland flow relative to infiltration, in order to relate the results to empirical studies of *P. bulbosa* [283]. These conditions give rise to scale-free patterns for which the concept of a typical plant-patch size is not well defined. This concept, however, is still applicable at low and high water availability as Figures 10.20(a,e) indicate, and, as we will shortly see, the predictions of the conceptual model extend to this case too. To describe a rock-soil system we consider squared domains, each containing a circular subdomain at its center within which the infiltration rate is non-zero and biomass dependent. In the surrounding area, the infiltration rate is set to zero to model a rock patch. Two main predictions of the conceptual model have been tested: (1) the possible increase of patchiness variability by source-sink relations in a landscape of variable soil-patch sizes and (2) the decreased variability and the appearance of arid-like patchiness for plants of larger patch sizes. These tests have been conducted by simulating (9.10) for various sizes of soil patches and plant patches as shown in Figure 10.24.

Figure 10.24(A) shows the effect of decreasing soil patch size on the patchiness of a species with a small plant-patch size. A large soil patch surrounded by a small rock patch (panel Ae) receives little run-off and remains relatively dry. If the precipitation rate is not too small for any vegetation form to grow, such a soil patch supports vegetation spot patterns typical of homogeneous arid landscapes (compare with Figure 10.20(a,b)). A smaller soil patch receives more run-off and shows labyrinth-like patterns characteristic of homogeneous semi-arid landscapes (panels Ac,d, and compare with Figure 10.20(c)), while a yet smaller soil patch is wet enough to support a gap pattern often observed in drysubhumid landscapes (panel Ab and compare with Figure 10.20(d,e))). A sufficiently small soil patch is not water limited and shows uniform vegetation coverage (panel Aa). These results are in line with the first prediction of the conceptual model: increased patchiness variability by source-sink relations in a landscape of variable soil-patch sizes.

Figures 10.24(BD) show the effects of increasing plant-patch size. The variability of vegetation patchiness decreases as more soil patches become fully covered by vegetation (panels Ba,b,Ca,b and Da,b) or completely bare (panels Ce, and De). In addition, soil patches that do support patchy vegetation of species with large plant-patch sizes show vegetation patterns characteristic of dry landscapes such as spots (panel Dd). These results are in line with the second prediction of the conceptual model: decreased variability and the appearance of arid-like patchiness for plants of larger patch sizes.

The suite of patterns generated by the model simulations (Figure 10.24) support the conceptual model as presented in Figure 10.23. Looking at the locations of these patterned states in the phase plane spanned by the soil-patch and plant-patch sizes, one can identify the transitions between the three areas

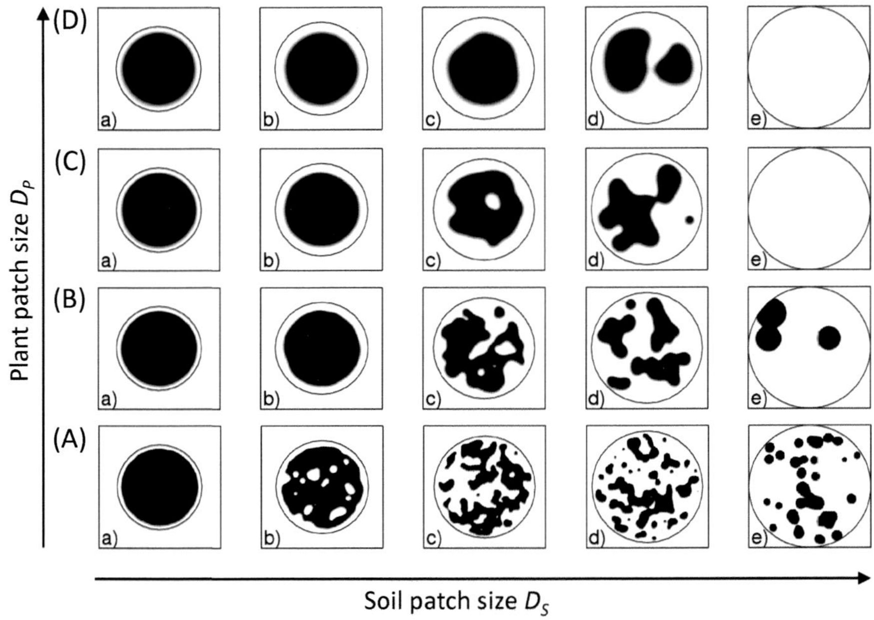

FIGURE 10.24: Vegetation patchiness in rocksoil domains of increasing soil-patch size (a–e, successively increasing in area by a factor 1.2) and increasing plant-patch size (A–D), obtained as solutions of (9.10). The circles represent soil patches that act as sinks for run-off produced by the surrounding rock areas. Small plant-patch sizes lead to highly variable self-organized patchiness (A). As plant-patch size increases, the patchiness becomes more template dictated (B, C, and D). From [283].

depicted in Figure 10.23 (lines I and II): self-organized patchiness (right bottom corner of the figure), patchiness controlled by both self-organization and physical template and patchiness solely controlled by the physical template (upper left corner).

Chapter summary

Vegetation pattern formation is a means by which dryland ecosystems cope with water scarcity. Being a self-organization phenomenon driven by positive feedbacks, it lends itself to the methods of pattern formation theory, using vegetation models of the kind introduced in the previous chapter. Analytical and numerical model studies of plain terrains predict five basic vegetation states along the rainfall gradient: uniform vegetation, hexagonal gap patterns, stripe patterns, hexagonal spot patterns and bare soil. On gentle slopes a similar sequence of states is found with a few differences: (i) the stripe patterns orient themselves in a direction perpendicular to the slope, (ii) they migrate uphill, and (iii) they occupy a wider precipitation range.

These inherent states of dryland landscapes suggest a new classification of aridity, which provides operational meaning to classified regions. For example, spotted vegetation in arid regions is vulnerable to desertification because of the existence of an alternative stable bare-soil state, which calls for landscape watch. Bare soil in hyper-arid regions is not restorable because of the lack of an alternative productive state, and so on.

Another model prediction is the existence of a bistability precipitation range between any consecutive pair of basic states. Landscapes in such bistability ranges are susceptible to state transitions that are inducible by rainfall fluctuations or disturbances. Such transitions will be discussed in Chapter 11 in the context of desertification. Here, a different aspect of state transitions that is relevant to sloped terrains only is highlighted; a local disturbance uphill can induce a gradual state transition from banded vegetation to spotted vegetation by front propagation downhill. In plain terrains such fronts are pinned and the disturbance effect remains local.

Bistability of different vegetation states can also increase landscape diversity. Spatial mixtures of the two alternative states can lead to long-lived non-periodic transient states, or to stationary non-periodic states. An example of the former is an expanding vegetation ring in a bistability range of bare soil and spot pattern, which develops from an initially small patch by central dieback. When the precipitation rate is sufficiently small, an isolated patch may stop expanding before developing into a ring, and form a stable localized spot structure of fixed size. Such spot structures constitute building blocks for a wide variety of non-periodic patterns, which are stationary in practice; once spots are sufficiently far apart their interactions become negligible and they behave as if they were isolated. Hybrid patterns, obtained when the fronts that separate the alternative stable states are pinned, is another mechanism by which stationary non-periodic patterns can appear. Front pinning is numerically observed in bistability ranges of different pattern states and in the bistability range of uniform vegetation and gap patterns, as we will see in Chapter 11.

The landscape diversity predicted by model studies is consistent with field observations, although an unequivocal empirical support of model predictions should be sought in controlled laboratory experiments. Field observations of dryland landscapes reveal another form of non-periodic patterns, which we refer to as scale-free patterns. These patterns lack any characteristic length-scale, such as the diameter of a spot or the width of a stripe, and are characterized by wide patch-size distributions. Model studies indicate that the development of such irregular patterns in uniform systems can occur under conditions of global competition for water, and that global competition develops when water transport is very fast relative to water exploitation. One realization of this condition is fast overland water flow relative to water infiltration. Model studies further predict a slow patch coarsening dynamics in which small patches disappear in time at the expense of growing big patches. This process culminates in a single big patch covering the same area as the total area covered by earlier scale-free patterns in the coarsening dynamics. Patch coarsening may not be noticeable in practice, however, because of the long time scales involved.

The discussion so far refers to homogeneous systems where patchiness is induced by self-organization. Dryland landscapes, however, are never strictly homogeneous; quite often they consist of rock-soil mosaics where rock patches act as sources of runoff water to adjacent soil patches. The chapter concludes with a study of the interplay between self-organization and such rock-soil relationships in shaping vegetation patchiness. Two aspects of this interplay are highlighted: (1) the relation between the size of a soil patch, and the size of a plant patch and (2) the dependence of the plant-patch size on genetic species constraints and on resource availability. The two aspects are coupled through source-sink resource relations within the rocksoil template; the smaller the soil patch, the more resources it receives and the larger the plant patches it can grow. When the plant-patch size exceeds the size of the biggest soil patch in the landscape, the vegetation patchiness is template dictated. When the smallest soil patch is large in comparison with the plant-patch size, the patchiness is self-organized. It is further found that source-sink resource relations increase patchiness variability by allowing all forms of vegetation patterns to coexist in the same landscape, and that species of larger patch sizes are associated with decreased variability and the appearance of arid-like patchiness.

Chapter 11

Regime shifts and desertification

11.1 The common view

The responses of ecosystems to small environmental changes are generally divided into two categories, small and reversible, or large, abrupt, and irreversible. Various examples of the latter response, commonly termed as a "critical transition" or "regime shift," have been reported, including sudden loss of transparency and vegetation in shallow lakes subject to human-induced eutrophication, coral reefs overgrown by fleshy macroalgae, and desertification induced by climate changes or human disturbances [270]. The abrupt response is often detrimental to the ecosystem, involving loss of biodiversity and bioproductivity [181, 71]. Using a simple dynamical system we first explain the conceptual view of regime shifts as transitions between two alternative stable states (Section 11.1.1), and then describe a few indicators that have been proposed as early warning signals for impending regime shifts (Section 11.1.2).

11.1.1 The concept of regime shifts

Large sudden responses to small environmental changes have been attributed to the coexistence of two alternative stable states in some range of environmental conditions [129, 207, 270]. This bistability range is bounded by instability points where one alternative stable state either loses stability or ceases to exist, as the bifurcation diagram in Figure 3.1 illustrates (see also Figures 2.8, 2.9(f) and 2.11). A small environmental change that moves the system across an instability point can induce an abrupt transition, or regime shift, to the other alternative stable state as the solid arrows in these figures illustrate. Another scenario of regime shift consists of disturbances that move the system to the basin of attraction of the alternative stable state as the dashed arrow in Figure 3.1 illustrates.

Bifurcation diagrams such as those depicted in Figure 2.8, 2.9(f) and 2.11 are derivable from simple dynamical systems that have Lyapunov functions. As explained in Section 2.2.5, the Lyapunov function is a mathematical entity that provides information about the steady states of a dynamical system and their stability properties, including the basins of attraction of its stable states. Lyapunov functions are generally not known for realistic ecosystem models, but the concept of a Lyapunov function is instrumental for understanding the gross aspects of regime shifts.

Consider a simple dynamical system, such as (2.10). This system can be cast in the form

$$\frac{du}{dt} = -\frac{dL(u;b)}{du}\,, \tag{11.1}$$

where

$$L(u;b) = -\frac{\lambda}{2}u^2 + \frac{1}{4}u^4 - bu + L_0\,, \tag{11.2}$$

is the Lyapunov function, which satisfies $\dot{L} \leq 0$, and L_0 is an arbitrary constant. Figure 11.1 shows the surface $L = L(u;b)$ in the 3-dimensional space spanned by L, u and b. Cuts of this surface at constant b values describe "stability landscapes" [270] that contain information about the stable states of the system—the minima, and their resilience—the basins of attraction. The stability landscapes are also useful for inferring about statistical properties, such as temporal auto-correlations and skewness of time-series distributions, which we will discuss in the next section in the context of warning signals for impending shifts.

The manners by which stability landscapes change with environmental conditions provide information about regime shifts. Within the range $b_1 < b < b_2$ (see Figure 11.1) the landscape is characterized by two minima, which correspond to the two alternative stable states, $u_\pm(b)$ in the bifurcation diagram shown in Figure 2.9(f). As b is decreased below this range or increased above it one of the minima merges with the maximum in between and disappears. As a result, a regime shift takes place. This is because the instantaneous value of u is no longer at a minimum of L and u changes in time according to (11.1). Since L cannot increase in time ($\dot{L} \leq 0$) the system must approach the other minimum, that is, it must go through a regime shift. An intuitive way to understand this scenario of regime shift is provided by the analogy to an object that moves under the influence of a gravitational potential L that has two wells (see Section 2.2.5); when the well in which the object resides disappears, by merging with the potential barrier that separates the two wells, the object must move by gravity to the other well. The stability landscape $L = L(u;b)$ also helps to understand regime shifts induced by disturbances. The role of

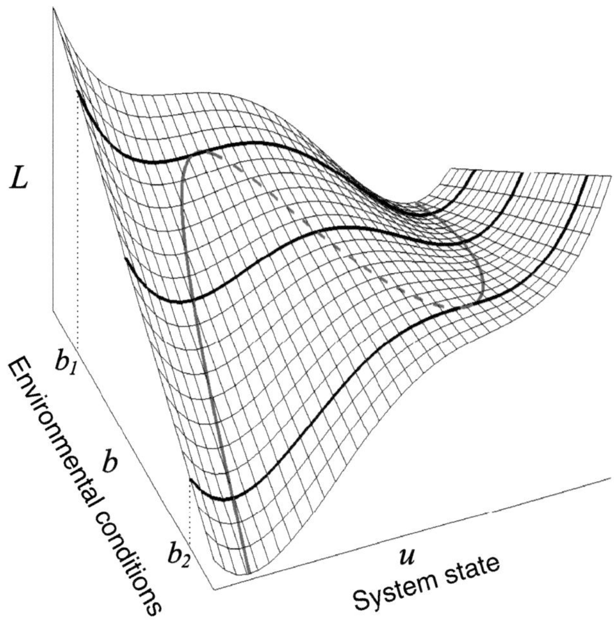

FIGURE 11.1: The concept of regime shifts illustrated with a Lyapunov function. Shown is the surface $L = L(u; b)$ where L is the Lyapunov function (11.2) and u and b represent, respectively, the system state and the environmental conditions. The cuts at constant b values represent stability landscapes. Also shown are the loci of the extremum points of L, where the solid lines represent the minima of L and the dashed line its maximum. The two minima in the range $b_1 < b < b_2$ represent alternative stable states. Driving the system outside this range results in the merging of two extremum points (in a fold bifurcation) and the disappearance of one alternative stable state. The non-vanishing slope of L leads to a shift to the remaining stable state.

such a disturbance is to move the system across the potential barrier. As the well that the system resides in becomes shallower and closer to the potential barrier, the feasibility of a disturbance to overcome the barrier increases, making the shift more likely.

The relation between the stability landscapes and the corresponding bifurcation diagram is established by looking at the loci of extremum points of L in the (u, b) plane as Figure 11.1 shows. The minima of L correspond to the solid lines (stable states) of the bifurcation diagram, the maximum of L corresponds to the dashed line (unstable state), and the disappearance of a well, which occurs by the merging of a minimum and a maximum of L, corresponds to a fold bifurcation.

11.1.2 Warning signals for impending regime shifts

The heightened concern about potential ecosystem deterioration from global climate change has motivated vigorous research efforts aimed at devising early indicators of impending degradation processes [269]. The basic idea behind most of these efforts is to identify properties of a dynamical system that change significantly as the system approaches an instability point, and to use them to construct measurable early indicators. We focus here on a few proposed indicators that are easily explainable using the simple view presented in the previous section.

Consider Figure 11.1 and the stability landscapes (cuts of $L = L(u; b)$ at constant b values) as b approaches the value b_2 from below. The shape of the well that disappears at $b = b_2$ goes through two major changes as the bifurcation point is approached; it becomes shallower and more asymmetric. The flatter basin of attraction associated with a shallower well results in a weaker restoring force, $-dL/du$[1]. This, in turn, implies larger deviations from the steady state following a disturbance of a given strength and slower convergence back to the steady state. A time-series analysis of a state variable of a system that is continuously subjected to random disturbances should therefore show increased variance and autocorrelation as the system approaches a fold bifurcation. The variance should increase because the disturbance is faced with a weaker restoring force, and the autocorrelation should increase because of the longer time the system spends in perturbed states and the resulting increase in the system's memory. Accordingly, both variance and autocorrelation of time series have been proposed as early indicators of impending shifts. These indicators have been successfully tested in several model studies and also in analyzing real data [269].

The asymmetry that develops in the restoring force, $-dL/du$, as the fold bifurcation is approached, results in an asymmetric distribution of the time the system spends away from the stable steady state. Consider, as an example, the steady state $u_+(b)$ (right well in Figure 11.1) with b being close to the fold bifurcation at b_2. Since the restoring force at u values lower than $u_+(b)$ is weaker than that at values higher than $u_+(b)$, the time the system spends at $u < u_+(b)$ is longer than the time it spends at $u > u_+(b)$, following the same disturbance strength. This asymmetry can be measured by the skewness of the time series probability distribution, which is another quantity that has been proposed as an early warning signal for impending shifts [115].

Indicators that exploit spatial information have also been studied using spatially coupled models of bistable systems, where the spatial coupling is either diffusive or in the form of a dispersal kernel. Proposed indicators include the increase of spatial correlation [59], spatial variance [116, 70], and spatial skewness [116], as fold bifurcations are approached.

---

[1]In the analogy to an object that moves in a gravitational field, L is the gravitational potential and the restoring force, $-dL/du$, is the actual force of gravity.

The common view of regime shifts described above is limited in two respects. It first assumes that the system is variational and thus excludes the rich dynamics that non-variational systems can show, such as sustained oscillations, traveling waves and various forms of chaotic dynamics (see Sections 7.1.2, 7.2.2 and 7.3.2). Secondly, it ignores pattern formation aspects that both variational and nonvariational systems share, such as non-uniform stationary instabilities and front dynamics (see Sections 7.1.1 and 7.2.1). Indeed model tests of temporal and spatial indicators have been successful for relatively simple models, for which the alternative stable states are spatially uniform and the dynamical behaviors are typical of variational systems, and failures of proposed indicators have been reported for non-variational systems showing oscillations [126] and for systems showing spatial self-organization [60]. For recent reviews of early warning indicators the reader is referred to Refs. [272] and [26].

11.2 Spatial aspects of regime shifts

An extension of the common view of regime shifts to spatially extended ecosystems should take into account two important aspects. The first is strong but local disturbances that form confined spatial domains of the alternative stable state, even far from the instability points. As discussed in Section 11.2.1 below, such disturbances can lead to front dynamics and gradual regime shifts. The second aspect is spatial patterning of one of the alternative stable states (or both) and the possible existence of a multitude of stable hybrid states. The implications of this aspect to regime shifts is discussed in Section 11.2.2.

11.2.1 Gradual shifts involving front dynamics

We first consider bistable ecosystems for which the two alternative stable states are spatially uniform. Bistability of this kind can be realized in plankton systems that show two stable biomass distributions at different depths [262]. The vertical (depth) direction is not uniform because of nutrient and light gradients, but the lateral directions can be assumed to be uniform. Another example is coastal vegetation on a salinity gradient that consists of salinity-tolerant and salinity-intolerant life forms [140]. Here too the spatial uniformity refers to the coastal direction, i.e., perpendicular to the inland salinity-gradient direction.

Applying the common view of disturbance-induced regime shifts to spatially extended ecosystems implies the consideration of global disturbances that shift the system as a whole. Such disturbances are possible, but are less likely for large systems, especially far from the tipping point to the other stable state, where the disturbances should be strong. More likely are disturbances that are confined to small areas, but are still strong enough to induce local transitions to the alternative stable state, e.g., surging sea waves covering coastal vegetation. The response of a bistable ecosystem occupying one of the two uniform states to a strong local disturbance can generally be divided into two phases; a fast local transition to the alternative stable state followed by a slow phase of either contraction or expansion of the the alternative-state domain that has formed. The latter dynamics depend on the fronts that bound the alternative-state domain.

Figure 7.8 shows three different forms of front dynamics following the establishment of the initial alternative-state domain, obtained by solving numerically equation (7.30). Apart of a particular value, λ_m, of the control parameter λ, the so called Maxwell point, the fronts that separate the two alternative states propagate. The front propagation is such that on one side of the Maxwell point one state expands into the second, while on the other side of the Maxwell point the propagation direction is reversed - the second state expands into the first. We note that the concept of a Lyapunov function, which now becomes a functional $L = L[u]$ (see (7.32)), is still useful. We no longer can draw a stability landscape, because the spatial extent of the system implies infinite degrees of freedom, but, as explained in Section 7.2.1, the constraint $\dot{L} \leq 0$ provides information on the direction of front propagation. The front propagates so as to gradually switch the whole system to the state that has lower L value, and it is stationary at the Maxwell point, λ_m, for which the two states have equal L values.

In general ecosystems are subjected to local disturbances at different locations which result in several confined domains of the alternative stable state as Figure 11.2(a) ($t_0 = 0$) illustrates. As the fronts that bound these domains propagate, some of them approach one another and begin to interact. When the interactions are attractive the fronts keep propagating until the alternative-state domains collapse and the fronts disappear. Such an event occurs in Figure 11.2(a) between t_1 and t_2 and between t_2 and t_3, and would have occur again after t_3 had the numerical simulation run long enough. Front collision events of this kind eventually induce a global transition to the alternative stable state albeit in a gradual manner.

The scenario of a global gradual shift depicted above occurs anywhere below the Maxwell point, λ_m, even far away from the fold bifurcation at λ_f, as Figure 11.2(c) shows. Since the early warning indicators discussed in Section 11.1.2 are all based on the close proximity to the fold bifurcation, they cannot probe on going gradual regime shifts in general. A similar picture holds for the reverse shift as Figures 11.2(b,c) show. When the front interactions are repulsive gradual shifts may terminate before the uniform alternative stable state is reached (see Section 7.2.1.2).

In non-variational systems strong local disturbances can lead to traveling wave phenomena and even to spatiotemporal chaos. A prototype model that shows such behaviors is the FHN model for bistable systems (see Figure 5.2). As discussed in Section 7.3.1, along with the two stable uniform states, traveling waves may exist too, provided the system is in the Bloch-fronts regime.

Moreover, close to the NIB front bifurcation spiral-vortex pairs may nucleate spontaneously and lead to a state of spatiotemporal chaos. Another interesting aspect is the effect of demographic noise, which may apply to populations of plants that are not adapted to water stress (see Section 9.1.1). Depending on the rate at which environmental conditions change, desertification can be continuous or abrupt [329].

11.2.2 Gradual and incipient shifts through hybrid states

When one of the two alternative stable states is spatially patterned additional dynamical behaviors are possible. In Section 7.2.1.3 we addressed this case using the SH equation (7.1) and described a subrange, $\lambda_1 < \lambda < \lambda_2$, within the bistability range, in which fronts that separate the uniform and pattern states do not propagate. This is unlike the case of two uniform stable states where the fronts are stationary only at a particular control-parameter value - the Maxwell point λ_m. Within this subrange a multitude of stable hybrid states exist in addition to the two basic stable states. They describe zero domains of increasing size in an otherwise periodic pattern (hole solutions), patterned domains of increasing size in an otherwise zero state (localized structures), and various combinations thereof. The bifurcation diagram in Figure 7.13 shows one branch of hole solutions that snakes down from the periodic-pattern solution towards the zero solution. We refer the reader to Section 7.2.1.3 for additional details and focus the attention here on the implications of this complex solution structure on regime shifts.

The bistability range $\lambda_f < \lambda < 0$ in the bifurcation diagram of Figure 7.13 can be divided into three subranges where different behaviors apply (see Figure 7.12): $\lambda_f < \lambda < \lambda_1$ - patterned (zero) domains contract (expand), $\lambda_1 < \lambda < \lambda_2$ - patterned and zero domains remain fixed in size, and $\lambda_2 < \lambda < 0$ - patterned (zero) domains expand (contract). Within the subrange $\lambda_f < \lambda < \lambda_1$ confined zero-state domains, form by strong local disturbances, expand and merge repeatedly until a global shift to the zero state is obtained as Figure 11.3(a-d) shows. Like in the case of two uniform alternative stable states, this is an example of a global gradual shift; it is a slow process, involving front propagation and front collisions, that eventually leads to a global shift to the alternative zero state. This gradual shift stands in sharp contrast to the abrupt shift shown in Figure 11.3(e-h) that occurs outside the bistability range, that is, for $\lambda < \lambda_f$.

Within the range $\lambda_1 < \lambda < \lambda_2$ fronts are pinned and alternative-state domains form a multitude of hybrid states. In this range global regime shifts are not expected to occur in steady environments. The system rather shows spatial plasticity; local disturbances are likely to shift the system to the closest hybrid pattern, which is a stable stationary state and therefore involves no further dynamics—the regime shift is incomplete or incipient. It is worth noting that incipient shifts can also occur as a result of global uniform environmental changes, such as a precipitation drop or a uniform disturbance, provided the

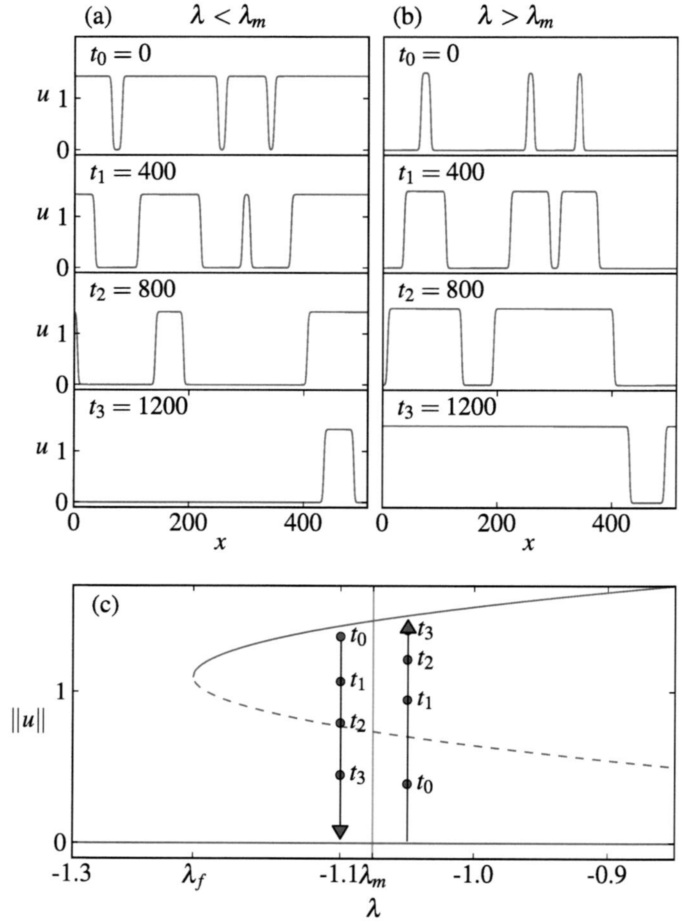

FIGURE 11.2: Gradual state transitions for bistability of two uniform states illustratyed with numerical solutions of equation (7.30). To the left of the Maxwell point, $\lambda < \lambda_m$, a disturbed up state gradually shifts towards the zero state, as the snapshots in panel (a) show. To the right of the Maxwell point, $\lambda > \lambda_m$, a near-zero state gradually shifts to the up state, as the snapshots in panel (b) show. The gradual shift can occur at λ values significantly higher than λ_f, the threshold of abrupt shift, as the downward arrow in panel (c) shows. Likewise, the gradual reverse shift can occur at λ values significantly lower than $\lambda = 0$, the threshold of abrupt reverse shift, as the upward arrow shows. The solid circles on the arrows denote the values of $\|u\|$ at the times the snapshots in panels (a) and (b) were taken. From [16].

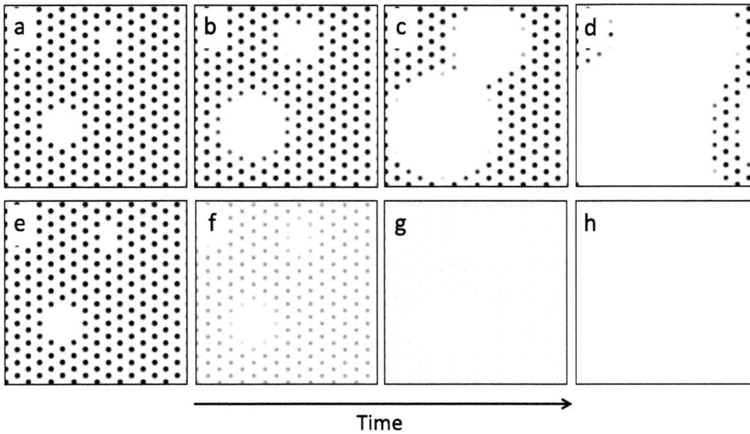

FIGURE 11.3: Illustration of gradual vs. abrupt global regime shifts using the SH equation (7.1). Panels (a–d) show a gradual transition from a disturbed pattern state (a) to a zero uniform state, within the bistability range, but outside the hybrid-state range, i.e., $\lambda_f < \lambda < \lambda_1$ in Figure 7.13. The gradual transition occurs by the local expansion and coalescence of the disturbed domains and is slow for large systems since it involves the local propagation of fronts. Panels (e–h) show an abrupt transition from the same initial state to the zero uniform state obtained by decreasing the control parameter below the bistability range, i.e., $\lambda < \lambda_f$ in Figure 7.13. This transition is fast because it proceeds globally. Note that the abrupt transition in panels (e–h) is so fast, relative to the gradual transition in panels (a–d), that no noticeable domain expansion occurs during the whole transition. Despite the different time scales of the two types of shifts both of them are global in the sense that they culminate in a zero state encompassing the whole system (panel (d) is still a transient). From [339].

initial pattern is not perfectly periodic, e.g., an hexagonal spot pattern containing defects. A transition of this kind, obtained with a vegetation model, is shown in Figure 3.4.

In reality, ecosystems are subjected both to disturbances and to environmental variability. The combination of the two can lead to gradual global transitions even within the hybrid-state range, $\lambda_1 < \lambda < \lambda_2$, as we now demonstrate. Consider the response of a hole solution, corresponding to an hybrid state up the snaking branch in Figure 7.13, to a time-periodic modulation of the control parameter, $\lambda = \lambda_0(1 + \epsilon \cos(\omega t))$ [16, 98]. In the context of dryland vegetation, for example, the hole solution may represent a vegetation pattern with locally removed biomass patches, while the modulated control parameter may represent a series of droughts. The two parameters λ_0 and ϵ are chosen such that $\lambda_0(1 - \epsilon) < \lambda_1$ and $\lambda_0(1 + \epsilon) < \lambda_2$, where $\lambda_1 < \lambda_0 < \lambda_2$. This choice amounts to a periodic escape of the control parameter out of the left side of the hybrid-state range.

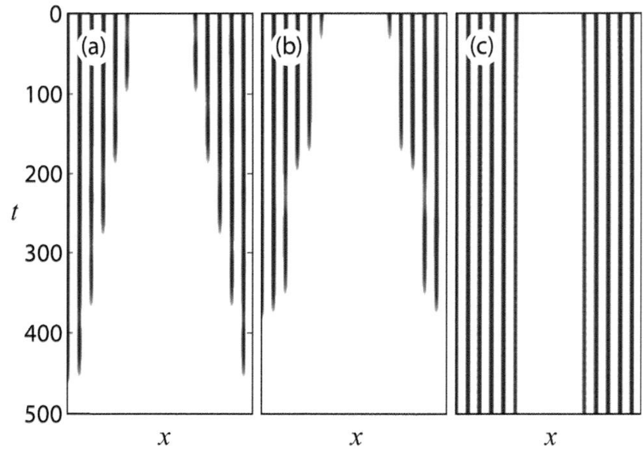

FIGURE 11.4: Gradual shifts within a hybrid-state range under time-periodic modulation of the control parameter. Shown are space-time plots obtained by solving numerically the SH equation (7.1). When the modulation time scale is comparable to the transition time between adjacent hole solutions, the transition involves hopping from one hole solution to the next at each modulation cycle (a). When the modulation time scale is sufficiently longer than the transition time, the hopping is to every second hole solution at each cycle (b). When the modulation time scale is sufficiently shorter than the transition time, no transition is taking place (c). From [16].

A possible response of the system is shown in Figure 11.4(a). The initial domain of the zero state expands into the adjacent domains of the pattern state by eliminating a pair of edge humps. This process occurs every cycle of the periodic modulation, during the time interval outside the hybrid-state range. The result is a step-like dynamics down the snaking branches with extended pauses in each branch that is visited. For this scenario to happen the time scale of the periodic modulation should be of the same order of magnitude as the state transition time scale (e.g., vegetation-patch mortality time). If the modulation is too slow the dynamics can skip one or more intermediate states in the descent towards the zero state (Figure 11.4(b)), and if it is too fast the zero domain remains confined and no global shift takes place (Figure 11.4(c))
.

11.3 Regime shifts in water limited landscapes

The two spatial aspects discussed in the previous section are particularly relevant to desertification processes in drylands. The alternative stable state to bare soil is generally a spot pattern rather than uniform vegetation, and the alternative state to uniform vegetation is generally a gap pattern rather than bare soil. Moreover, local disturbances leading to confined bare-soil domains in spotted vegetation or gaps in uniform vegetation are quite feasible. Examples of such disturbances are clear-cutting, fires, overgrazing, and infestation, sometimes in conjunction with rainfall variability [30]. In Section 11.3.1 we use a simplified version of the vegetation model (9.3) to study the applicability of gradual and incipient regime shifts to desertification and to the recovery from desertification in water-limited vegetation. We focus primarily on the bistability range of uniform vegetation and gap patterns where desertification amounts to the appearance of barren gaps in uniform vegetation. In Section 11.3.2 we address a related problem—vegetation restoration by water-harvesting methods, that is, vegetation recovery assisted by periodic landscape modulations that concentrate the limited water resource (see Section 3.3.3). We study vegetation restoration using a different simplification of the vegetation model (9.3), focusing on the bistability range of bare soil and spot patterns.

11.3.1 Desertification

Desertification is a regime shift involving loss of bioproductivity (see Sections 3.2.1 and 3.3.2). We focus on the bistability range of uniform vegetation and gap patterns, where abrupt desertification may occur as a result of a severe drought that takes the system outside the range of stable uniform vegetation and induces a transition to the less-productive gap-pattern state. Can moderate droughts, along with local disturbances, induce gradual or incipient regime shifts as discussed in Section 11.2.2? We address this question by considering a concrete example, the Namibian fairy-circle ecosystem [312, 144]. This ecosystem shows grasslands with circular gaps of barren soil—the "fairy circles" (see Figure 1.1c and Figure 10.17D), which often form nearly hexagonal gap patterns [101]. While there is still a debate about the formation mechanism of fairy circles, there is an increasing agreement that this is a pattern formation phenomenon associated with a non-uniform stationary instability of a uniform grassland that results in a periodic gap pattern [183, 54, 101, 89, 340].

To study the feasibility of gradual and incipient desertification in the Namibian fairy-circle ecosystem we use the vegetation model (9.3). Applying this model to the fairy-circle ecosystem allows for two significant simplifications. The first simplification is related to the reported observations of confined root zones [144] of the perennial grasses that form the vegetation matrix. This allows the elimination of the integrals over the laterally extended root zones. The second simplification pertains to the sandy soil in the Namibian fairy-circle ecosystem and the absence of significant overland flow. Assuming a negligible contrast between the infiltration rates of surface water into bare and vegetated soil, the equation for H can be eliminated (see Section 9.2.3 for detailed explanations of these types of simplifications).

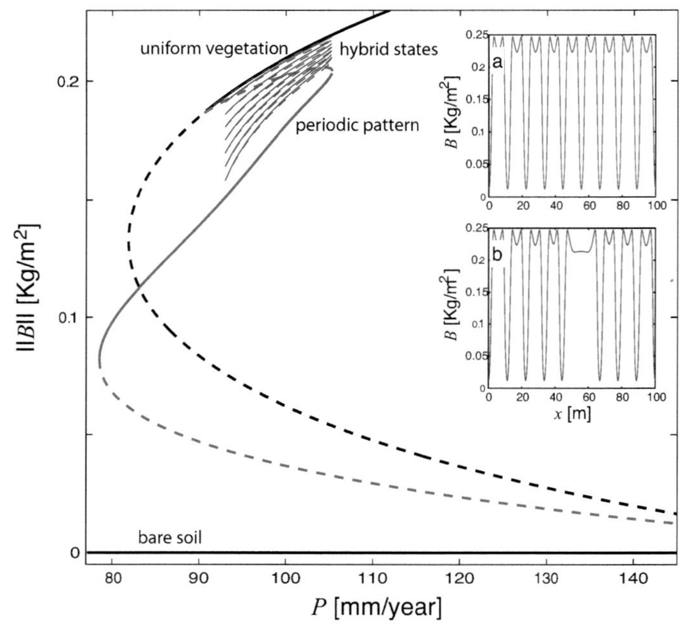

FIGURE 11.5: Bifurcation diagram for equations (11.3) in 1d, using parameters that describe the Namibian fairy-circle ecosystem. The vertical axis is the L^2 norm of the biomass variable. Solid (dashed) lines represent stable (unstable) solutions. The diagram shows a precipitation (P) range where both uniform vegetation and periodic vegetation pattern are stable. Within this range there exists a subrange of hybrid states, i.e., periodic gap solutions with missing gaps. The insets show spatial profiles of a 1d periodic gap pattern (a) and of a hybrid state consisting of a periodic gap pattern with a missing gap (b). Note that the bare-soil solution is stable throughout the whole precipitation range that is shown. Adapted from [340].

These simplifications lead to a 2-variable model described by (9.17), (9.19) and (10.2). Using the same scaling as in Table 9.2, accept for the spatial coordinates, which we now scale as $\mathbf{x} = \mathbf{X}\sqrt{M/D_B}$, and the soil-water diffusivity, $\delta_w = D_W/D_B$, the following dimensionless equations are obtained:

$$\partial_t b = \nu w b(1-b)(1+\eta b)^2 - b + \nabla^2 b \tag{11.3a}$$

$$\partial_t w = p - \nu(1-Rb)w - \gamma b w(1+\eta b)^2 + \delta_w \nabla^2 w\,, \tag{11.3b}$$

where we assumed a flat terrain ($Z = const.$) and $Rb \ll 1$.

A typical bifurcation diagram for 1d stationary solutions of (11.3), using parameter values that fit the Namibian fairy-circle ecosystem, is shown in Figure 11.5 [340]. It includes uniform solutions representing bare soil and uniform vegetation and periodic solutions which at the high precipitation range represent 1d gap patterns. Note the existence of a precipitation range where both uniform vegetation and periodic gap patterns are stable. Within this range there is a subrange with many more stable non-uniform solutions representing hybrid states. These are "hole solutions" consisting of gap patterns with increasing numbers of missing gaps as the solutions snake up towards the uniform vegetation state.

The bifurcation diagram shown in Figure 11.5 is similar to that of Figure 7.13 in that both contain a bistability range of uniform and patterned states and a subrange of hybrid states (snaking range). In Figure 7.13 the hybrid states describe hole solutions consisting of spot patterns with missing spots, whereas in Figure 11.5 the hybrid state describe hole solutions consisting of gap patterns with missing gaps. We may therefore expect the discussion in Section 11.2.2 of gradual and incipient shifts through hybrid states to apply to (11.3) and possibly to the Namibian fairy-circle ecosystem, which equations (11.3) represent. Indeed, numerical simulations of (11.3) with a time dependent precipitation, $p(t)$, that takes the system out of the hybrid-state subrange and back in periodically in time, show gradual regime shifts as the snapshots in Figure 11.6 indicate. The top row (a) shows the response of a uniform vegetation state, locally disturbed to form a few gaps, while the bottom row (b) shows the response of a nearly periodic gap pattern.

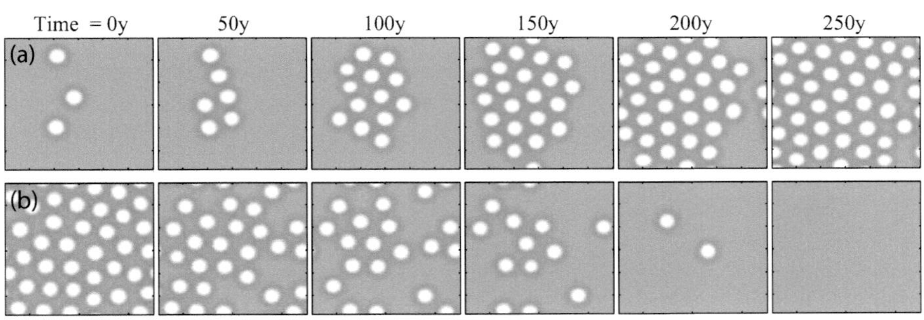

FIGURE 11.6: Gradual regime shifts. Shown are snapshots of simulations of (11.3) with time-periodic precipitation that takes the system outside the hybrid-state range in each oscillation cycle. In the top series (a) the initial state is a locally disturbed uniform vegetation and the escape from the hybrid-state range is to lower precipitation. The system responds by a succession of hybrid-state downshifts and the convergence to a periodic gap pattern. In the bottom series (b) the initial state is a nearly periodic gap pattern and the escape from the hybrid-state range is to higher precipitation. In this case the system responds by a succession of hybrid-state upshifts and the convergence to a uniform vegetation state. The domains shown are $60m^2$ and the times are in years. Adapted from [340].

We focused here on the bistability range of uniform vegetation and gap patterns because it appears most relevant to the Namibian fairy-circle ecosystem, a relatively uniform and undisturbed ecosystem that can serve as an excellent field laboratory for exploring various scenarios of gradual and incipient regime shifts, in addition to abrupt shifts. Indeed, the bifurcation diagram shown in Figure 11.5 hints towards additional possible responses to rainfall variability in this ecosystem. Unlike the bifurcation diagrams shown in Figures 10.1 and 10.4, the uniform bare-soil state remains stable throughout the entire existence range of periodic vegetation patterns and has a wide bistability range with the uniform vegetation state.

The bistability of uniform vegetation and bare soil suggests the feasibility of gradual regime shifts of the type shown in Figure 11.2. Such a shift can proceed provided $p < p_m$, where p_m is the Maxwell point (see Section 7.2.1.1), that is, when bare-soil domains expand into uniform-vegetation domains. These dynamics, however, may not necessarily culminate in a global transition to the uniform bare-soil state. As discussed in Section 7.2.1.2 and demonstrated in Figure 7.10, strong repulsive interactions between the fronts that bound contracting domains of the uniform-vegetation state can result in an asymptotic vegetation pattern. Such a scenario describes an incipient shift in which initial large domains of uniform vegetation contract but do not disappear; they rather lead to a stationary pattern consisting of small vegetation patches dispersed in bare soil. On the other side of the Maxwell point, $p > p_m$, uniform-vegetation domains expand into bare-soil domains. These dynamics may result in a global shift to uniform vegetation, or, if the front interactions are strongly repulsive, to an asymptotic pattern consisting of bare-soil gaps in an otherwise uniform vegetation. The formation of such patterns by front interactions has been proposed recently as a mechanism for the formation of fairy circles [89]. This is an alternative mechanism to the mechanism presented above, which views fairy circles as a gap pattern that appears in a non-uniform instability. The bifurcation diagram shown in Figure 11.5 suggests that both mechanisms may be realized in the same system at different precipitation ranges.

Whether gradual regime shifts are feasible in the other bistability range of uniform and patterned states, i.e., that of bare soil and spot patterns, is still an open question. Numerical studies of several vegetation models, including the simplified versions of (9.3) have failed to identify front pinning and homoclinic snaking [339]. This difference between the two bistability ranges may be related to the finding that the periodic solution branches are not connected to the bare-soil solution branch, but rather terminate on the uniform-vegetation solution branch very close to the bare-soil instability point (see Figure 10.4).

Moreover, front propagation in flat terrains has been found to be unidirectional throughout the whole bistability range; periodic patterns propagate into bare soil but bare soil has never been found to displace patterns. Front propagation in both directions (at different parameter values) can occur, however, on slopes [288]. We remind the reader that stationary isolated spot solutions do exist and can serve as building blocks for sparse patterns (see Section 10.2.1.2). As a consequence, the boundary between bare soil and sufficiently sparse spots can appear to be stationary not because of front pinning but rather because the spots hardly interact with each other.

Rather than homoclinic snaking, the typical behavior that is found within the coexistence range of periodic patterns and stable bare soil is that shown in Figure 10.4 [339, 318], i.e., a band of periodic solutions forming a Busse balloon (see Section 6.2.1) that starts at the non-uniform instability of uniform vegetation at high precipitation, and extends down to the precipitation threshold at which a single pulse solution disappears in a fold bifurcation [318]. The availability of stable periodic solutions within the Busse balloon, with wavenumbers decreasing to zero as the precipitation parameter decreases, bears on the dynamics of desertification [292]. A slow precipitation decrease accompanied by strong environmental noise (disturbances) at relatively high precipitation, lead to small wavenumber adaptations and slow desertification. The adaptation process is less gradual under conditions of either low precipitation, fast precipitation decrease or weak noise. Such conditions can lead to large wavenumber changes involving period or wavelength doubling, which result in faster desertification.

We restricted the discussion to homogeneous systems showing self-organized patchiness. Taking into account environmental heterogeneity may extend the range of vegetation patchiness to yet lower precipitation ranges, and make the dynamics of desertification more gradual [335].

11.3.2 Reversing desertification

The bistability range of periodic vegetation patterns and bare soil provides the context for most studies of vegetation restoration. Restoration studies focus primarily on factors that affect the threshold of the transition back to the productive vegetation state, and on means of inducing such a transition [297]. As discussed in Section 3.3.3, inducing a transition back to the productive state is often achieved by water harvesting methods [184]. These methods consist of spatially periodic ground modulations, e.g., in the form of parallel linear embankments, that intercept overland water and along which vegetation is planted. Such restoration practices fall in the class of spatial-resonance problems that involve the entrainment of symmetry-breaking patterns to external forcing patterns (see Section 8.1.2). Spatial resonances have received considerable attention recently in various physical contexts, including thermal convection [279, 96, 135], magnetohydrodynamics [97] and chemical reactions [68], but the concept has never been applied to the problem of reversing desertification.

Underlying the ground modulation approach to restoration is the presumption that the modulations determine the locations of vegetation growth and therefore dictate the pattern of restored vegetation. However, the studies

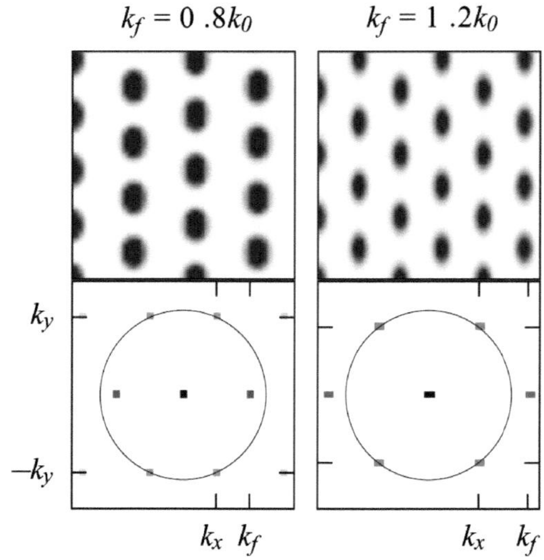

FIGURE 11.7: Resonant rhombic vegetation patterns and their power spectra obtained by numerical integration of (11.4). The patterns lock their wave-vector components in the forcing direction at exactly $k_x = k_f/2$ both for $k_f < k_0$ (left panels) and $k_f > k_0$ (right panels), where k_0 is the wavenumber that the unforced system forms. Adapted from [202].

described in Section 8.3.2 predict that periodic spatial forcing of a pattern-forming system can induce new patterns that differ both from the forcing pattern and the patterns that the unforced system forms. As Figure 8.19 shows these are resonant rhombic or oblique patterns with the property that their wave-vector components in the forcing direction, k_x, are locked to the forcing in a 2:1 resonance, that is $k_x = k_f/2$. To study such patterns in the context of vegetation restoration we use the vegetation model (9.3). For simplicity we eliminate the root-augmentation and soil-water diffusion feedbacks by setting $E = 0$. The periodic landscape modulations are introduced through the infiltration rate, I, rather than the topography function $Z = Z(x, y)$, which is taken to be constant. Modulating the infiltration rate amounts in practice to periodic soil-crust removal, which imitates the effect of vegetation in preventing crust growth. The resulting equations in dimensionless form (using D_B to rescale space) are

$$\partial_t b = \nu w b(1-b) - b + \nabla^2 b \tag{11.4a}$$

$$\partial_t w = Ih - \nu w - \gamma b w + \delta_w \nabla^2 w \tag{11.4b}$$

$$\partial_t h = p - Ih + 2\delta_h \nabla \cdot [h\nabla(h+\zeta)] \,, \tag{11.4c}$$

where

$$I(b) = a\frac{b+qf}{b+q}\,, \quad f = f_0\left[1 + \frac{\gamma_f}{2}(1+\cos k_f x)\right]\,, \tag{11.5}$$

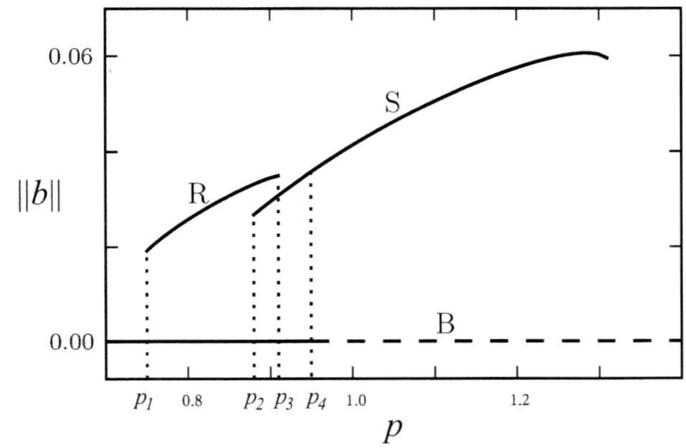

FIGURE 11.8: Bifurcation diagram obtained by numerical integration of (11.4). The vertical axis is the L^2-norm of the biomass field ($\|b\| \propto \int b^2 d\mathbf{x}$) while the horizontal axis represents the precipitation parameter. Shown are the bare soil solution (B), which is stable (unstable) for $p < p_4$ ($p > p_4$), the stable part of the solution that represents resonant rhombic pattern (R), $p_1 < p < p_3$, and the stable part of the solution that represents resonant stripe patterns (S), $p > p_2$. Note the existence of a tri-stability range, $p_2 < p < p_3$. From [206].

and all dimensionless quantities are given in Table 9.2, accept for the spatial coordinates, $\mathbf{x} = \mathbf{X}\sqrt{M/D_B}$, and the soil-water diffusivity, $\delta_w = D_W/D_B$. Figure 11.7 shows resonant rhombic vegetation patterns obtained for k_f smaller and bigger than k_0, the wavenumber that the unforced system forms, which confirms the prediction of the analysis of the forced SH equation; the pattern locks the wave-vector component in the forcing direction at $k_x = k_f/2$ and form two oblique modes $\mathbf{k}_\pm = -k_x\hat{\mathbf{x}} \pm k_y\hat{\mathbf{y}}$ that lie on the circle $k_x^2 + k_y^2 = k_0^2$ and satisfy the resonance condition $\mathbf{k}_+ + \mathbf{k}_- + k_f\hat{\mathbf{x}} = 0$ (see Section 8.3.2).

We now use equations (11.4) to study the vegetation states that prevail at relatively low precipitation values p and the resilience of these states to rainfall fluctuations [206]. Figure 11.8 shows a bifurcation diagram obtained by integrating (11.4) in time. The diagram shows that resonant rhombic patterns persist at significantly lower precipitation values, as compared with resonant 1:1 stripe patterns. It further shows that there are bistability ranges of rhombic patterns and bare soil and of stripe patterns and bare soil, and that there is a tri-stability range, $p_2 < p < p_3$, of rhombic patterns, stripe patterns and bare soil.

Which of the two types of patterns, rhombic or stripe, are preferable for restoration? To answer this question we studied the response of stripe patterns within the tri-stability range ($p_2 < p < p_3$) to precipitation downshifts, bearing in mind that the unproductive bare-soil state is stable and is

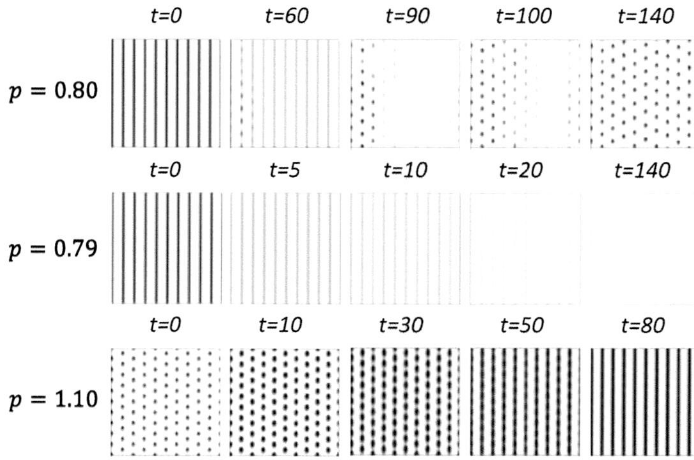

FIGURE 11.9: Responses of stripe patterns to precipitation shifts. The top two rows show snapshots of numerical simulations of (11.4) at two precipitation values in the range $p_1 < p < p_2$ starting with stripe patterns computed at $p > p_2$ (see Figure 11.8). A precipitation downshift to $p = 0.80$ (top row) involves a substantial biomass decrease before the pattern recuperates to a rectangular form, whereas a downshift to p=0.79 (middle row) results in a quick collapse to the bare soil state. A precipitation upshift to $p = 1.10$, starting with a rhombic pattern computed at $p < p_2$ results in a smooth transition to a stripe pattern. From [206].

therefore a possible attractor of the dynamics. As Figure 11.9 demonstrates, a moderate downshift to the bistability range of rhombic patterns and bare soil ($p_1 < p < p_2$) can lead to a significant degradation of the stripe pattern before the system recovers to a rhombic pattern (top row), while a stronger downshift can lead to a complete collapse of the stripe pattern to the bare-soil state (mid row).

To gain a deeper insight into the collapse of stripe patterns to bare soil we consider the coupled dynamics of the two oblique modes, $\mathbf{k}_{\mp}$, and the stripe mode, $\mathbf{k}_f$, and approximate a solution of (11.4) as

$$U(\mathbf{x},t) = U_0 + U_1\, B\, e^{ik_f x} + U_2\, a\, e^{i\mathbf{k}_-\cdot\mathbf{r}} + U_3\, b\, e^{i\mathbf{k}_+\cdot\mathbf{r}} + c.c.\,, \tag{11.6}$$

where $U = (b, w, h)$, U_0, U_1, U_2, U_3 are constant vectors in $\mathbb{C}^3$, $k_x = k_f/2$ and $k_y = \sqrt{k_0^2 - k_x^2}$. The dynamics of the amplitudes are governed by universal equations of the form (8.40), albeit with different coefficients. The parameter ϵ in these equations is a control parameter, analogous to the precipitation parameter p in the vegetation model, while the other parameters can be

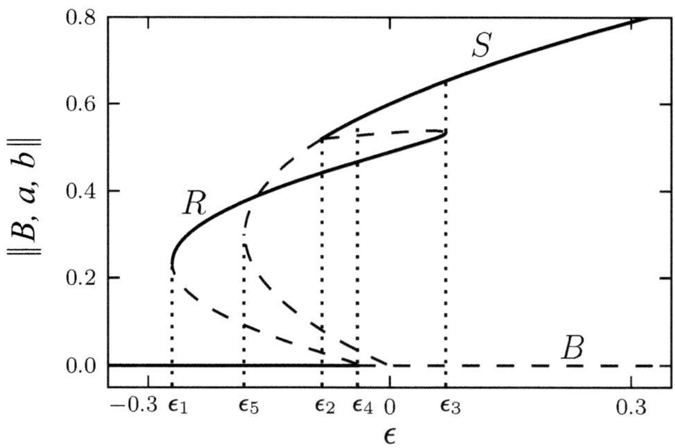

FIGURE 11.10: Bifurcation diagram derived from the amplitude equations (8.40). Shown are stationary solutions and their stability properties as obtained by linear stability analysis: stable (solid lines) and unstable (dashed lines) branches of the bare soil solution (B), rhombic solutions (R) and stripe solutions (S). The vertical axis represents the L^2 norm $||B,\ a,\ b|| = \sqrt{|B|^2 + |a|^2 + |b|^2}$. Note the existence of a tri-stability range, $\epsilon_2 < \epsilon < \epsilon_4$. From [206].

calibrated so that (8.40) reproduce a bifurcation diagram similar to the numerical bifurcation diagram shown in Figure 11.8.

The bifurcation diagram obtained from (8.40) is shown in Figure 11.10. It has stable solution branches similar to those appearing in Figure 11.8, and complements the latter with unstable solution branches. This additional information is crucial for understanding the vegetation collapse that follows precipitation downshifts (see Figure 11.9). Consider the tri-stability range, $\epsilon_2 < \epsilon < \epsilon_4$, of bare soil, rhombic patterns and stripe patterns. Starting with an initial stripe pattern within this range we ask what makes the system converge to bare soil rather than to rhombic patterns following a downshift to the bistability range, $\epsilon_1 < \epsilon < \epsilon_2$, of these two states? Figure 11.11 shows the stationary states (fixed points) and the vector field of (8.40) in the plane $(\rho_S,\ \rho_R)$, where $\rho_S = |B|$ and $\rho_R = |a| = |b|$, for three different ϵ values. The first value is in the range $\epsilon_5 < \epsilon < \epsilon_2$ (Figure 11.11(a)), where a pair of large-amplitude and small-amplitude stripe solutions exist but are unstable, and the other two values are in the range $\epsilon_1 < \epsilon < \epsilon_5$, where the stripe solutions do not exist (have disappeared in the fold bifurcation at $\epsilon = \epsilon_5$). Also shown in Figure 11.11 are phase trajectories of numerical solutions of (8.40), starting with a stripe solution that was computed at $\epsilon > \epsilon_2$ as an initial condition. These trajectories represent the responses of stripe patterns to ϵ downshifts of increasing strengths.

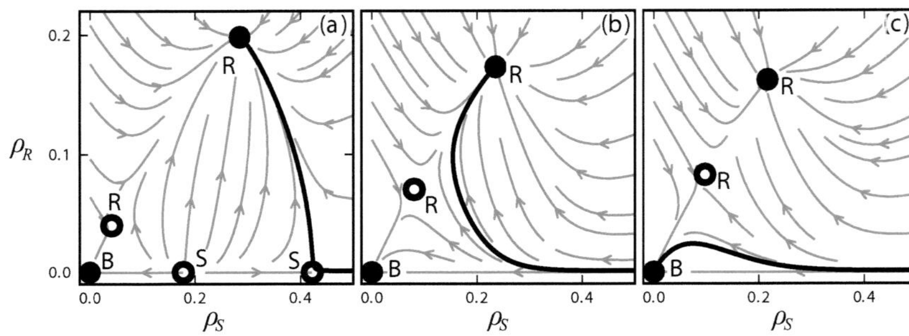

FIGURE 11.11: Projections of the phase-space dynamics of the amplitude equations (8.40) onto the phase plan spanned by $\rho_S = |B|$ and $\rho_R = |a| = |b|$ at (a) $\epsilon = -0.15$, which lies in the range $\epsilon_5 < \epsilon < \epsilon_2$ where a pair of unstable stripe solutions exist, (b) $\epsilon = -0.22$, which lies in the range $\epsilon_1 < \epsilon < \epsilon_5$ where the unstable stripe solutions no longer exist, and (c) $\epsilon = -0.24$, which lies in the same range as in (b). The solid (hollow) circles denote stable (unstable) stationary states. The labels B, S, and R denote the zero state (bare soil), stripe patterns and rhombic patterns, respectively. The arrows denote the projection of the vector field of (8.40). The responses of a stable resonant stripe pattern obtained at $\epsilon = \epsilon_2$ to ϵ downshifts is shown by the thick black phase portraits: (a) a small downshift to a range where unstable stripe solutions still exist, results in a smooth transition to a rhombic pattern. (b) a moderate downshift beyond the existence range of unstable stripe solutions involves temporary approach to the zero state followed by convergence to a rhombic pattern. (c) a yet stronger downshift places the initial stripe condition in the attraction basin of the zero state and leads to a complete collapse. From [206].

A downshift of ϵ to the range $\epsilon_5 < \epsilon < \epsilon_2$ results in a smooth transition to a rhombic pattern as Figure 11.11(a) shows. The unstable large-amplitude stripe solution plays a crucial role in this response; its unstable manifold, which represents the growth of the two oblique modes, acts as a separatrix that prevents convergence to the stable bare-soil state. Downshifts to the range $\epsilon_1 < \epsilon < \epsilon_5$ can be divided into two size groups. A moderate downshift results in an initial condition that still lies in the basin of attraction of the rhombic pattern; the system initially degrades towards the bare-soil state but then recovers, as Figure 11.11(b) shows. A stronger downshift places the initial condition in the basin of attraction of the bare-soil state and leads to a complete vegetation collapse, as Figure 11.11(c) shows.

This analysis shows that the common intuitive restoration approach in 1:1 stripe patterns suffers from poor resilience to precipitation downshifts (droughts). Restoration in a rhombic pattern, which facilitates the growth of the inherent oblique modes, circumvents this difficulty and does not suffer from precipitation upshifts; as Figure 11.9 indicates, the transition from rhombic patterns to stripes is smooth. Note that no compromise in the total bioproductivity is made in rhombic restoration, despite the lower (and more economic) vegetation coverage; as Figure 11.8 indicates, the L^2 norm of the biomass of a rhombic pattern is about the same as that of a stripe pattern, and the same result holds also for the total biomass.

Chapter summary

The response of an ecosystem to a small environmental change can be large and abrupt when it involves a transition to an alternative stable system state. Such transitions, or regime shifts, become feasible near instability points, where small environmental fluctuations can render the system state unstable or nonexistent and disturbances can kick the system out of the attraction basin of its stable state. Identifying and analyzing transitions of this kind is crucial to the understanding of detrimental ecosystem responses. Much effort has been devoted to devising effective indicators of impending regime shifts. Most proposed indicators rely on a generic property of a system approaching a bifurcation point—the vanishing of the largest eigenvalue (or its real part), which is accompanied by the slow and asymmetric decay of perturbations about the stable state.

Regime shifts are generally conceived as abrupt transitions between two alternative stable states occurring uniformly across the ecosystem. However, the response of an ecosystem to environmental changes is not necessarily uniform; spatially confined disturbances can induce local transitions to the alternative stable state, and varying external conditions can induce spatial instabilities that result in self-organized patterns.

The dynamics of a disturbed ecosystem with confined domains of the alternative stable state are determined by the properties of the fronts that bound these domains. When the two alternative states are spatially uniform, such fronts generally propagate. Varying the control parameter across the bistability range results in a change of the direction of front propagation at the so-called Maxwell point. Depending on which side of the Maxwell point the system is, confined domains of the alternative state either contract and disappear, or expand and gradually induce a global shift to the alternative stable state. A few exceptions to this behavior exist and are briefly discussed.

The behavior is more intricate when one of the alternative stable states is spatially patterned. In this case there might exist a subrange of the control parameter within the bistability range where fronts do not propagate but are rather pinned. Within this subrange a multitude of stable hybrid states exist, which can affect the dynamics of regime shifts. Local disturbances within the hybrid-state subrange result in the convergence to stable hybrid states with no further dynamics, unlike the case of two uniform alternative states. However, environmental variability that occasionally takes the system out of the hybrid-state subrange, where fronts are no longer pinned, can induce a gradual regime shift. Since gradual regime shifts can occur far from the instabilities points that define the bistability range, early warning signals based on the proximity to instability points are not applicable.

The two spatial aspects discussed above are particularly relevant to desertification processes in drylands. The alternative stable state to bare soil is generally a spot pattern rather than uniform vegetation, and the alternative state to uniform vegetation is generally a gap pattern rather than bare soil. Using a simplified version of the vegetation model, fitted to describe the Namibian fairy-circle ecosystem, a bistability range of uniform vegetation and periodic patterns, which includes a subrange of hybrid states, can be identified. Numerical simulations of this model in 2d with a time dependent precipitation that takes the system out of the hybrid-state subrange and back in periodically in time, show gradual regime shifts; from disturbed uniform vegetation to gap patterns, representing desertification, and from nearly periodic gap patterns to uniform vegetation, representing recovery from desertification. Interestingly, gradual regime shifts through hybrid states could not be found in the bistability range of bare soil and periodic patterns.

The recovery of degraded areas can be accelerated using water harvesting methods, which involve periodic ground modulations, that intercept overland water flow. These methods fall in the class of spatial resonance problems. Using the vegetation model with periodic infiltration modulation that mimic periodic crust removal, and earlier results obtained with a parametrically forced SH equation, the following conclusion is reached: plain realizations of water-harvesting methods, assuming a complete vegetation response to the imposed modulation pattern, suffer from poor resilience to rainfall variability. By contrast, less intuitive realizations, based on the inherent spatial modes of vegetation growth, can be highly resilient.

Chapter 12

Species coexistence and diversity in plant communities

12.1 Modeling plant communities

The discussion of vegetation patchiness in Chapter 10 was restricted to a single life form. In this chapter we extend the discussion to plant communities involving small or large numbers of life forms. Depending on the particular context, we will refer to a life form as representing a single species or a functional group, i.e., a group of species that share similar values of certain functional traits. We will first extend the model equations (9.3) to a community of life forms. In extending the model we will include inter-specific competition for light in addition to the competition for water, and selective grazing stress. These extensions are described in Section 12.1.1. Several simplifications of the model equations can be made to make their study easier. These simplifications are described in Section 12.1.2.

12.1.1 Model equations for a community of plant life forms

A straightforward extension of the model equations (9.3) to describe a community of life forms leads to the following equations:

$$\begin{aligned} \frac{\partial B_i}{\partial T} &= G_B^i B_i\left(1 - B_i/K_i\right) - M_i B_i + D_{B_i}\nabla^2 B_i \qquad i = 1,\ldots,n \\ \frac{\partial W}{\partial T} &= IH - LW - W\sum_i G_W^i + D_W \nabla^2 W \\ \frac{\partial H}{\partial T} &= P - IH - \nabla\cdot\mathbf{J}\,, \end{aligned} \tag{12.1}$$

where B_i stands for the above-ground biomass per unit ground area of the ith life form, the summation is over all life forms, and the overland water flux $\mathbf{J}$ is given by (9.4). The biomass dependence of the infiltration rate is a generalization of (9.1):

$$I = A\frac{\sum_i Y_i B_i + Qf}{\sum_i Y_i B_i + Q}\,, \tag{12.2}$$

where Y_i represents the contribution of the ith life form to the increased infiltration rate in a vegetation patch and $Y_1 = 1$. Likewise the biomass dependence of the evaporation rate L is a generalization of (9.5) and reads

$$L = \frac{N}{\left(1 + \sum_i R_i B_i/K_i\right)}\,. \tag{12.3}$$

Finally, the biomass-growth rate and the water-uptake rate given by (9.6) are generalized to

$$G_B^i(\mathbf{X},T) = \Lambda_i \int_\Omega G_i(\mathbf{X},\mathbf{X}',T)W(\mathbf{X}',T)d\mathbf{X}'\,, \tag{12.4}$$

$$G_W^i(\mathbf{X},T) = \Gamma_i \int_\Omega G_i(\mathbf{X}',\mathbf{X},T)B_i(\mathbf{X}',T)d\mathbf{X}'\,, \tag{12.5}$$

where

$$G_i(\mathbf{X},\mathbf{X}',T) = \frac{1}{\pi S_i^2}\exp\left[-\frac{|\mathbf{X}-\mathbf{X}'|^2}{[S_i(1+E_iB_i(\mathbf{X},T))]^2}\right]\,. \tag{12.6}$$

Equations (12.4) can be brought to a non-dimensional form using the same scaling relations as in Section 9.2.2 or straightforward extensions thereof [102].

Plant life forms in water-limited systems often compete for sunlight in addition to water, as taller plants reduce the availability of light to shorter plants by shading. The positive feedback between shoot growth and light availability results in an inter-specific competition for light that can lead to the dominance of the taller plant species [308]. To capture this competition we introduce the following form for the biomass growth rate parameter that appears in (12.4) [164]:

$$\Lambda_i(B) = \Lambda_{0i}\left(1 - \frac{\Sigma_{j\neq i} B_j}{\Sigma_j B_j + h}\right) . \tag{12.7}$$

Here, Λ_{0i} represents the growth rate of the ith life form in the absence of competitors and h is a positive constant serving as a reference value for the total biomass beyond which light becomes a limiting resource for small plants.

An important factor affecting community structure and diversity is grazing by herbivores, which acts selectively towards certain functional traits by removing above-ground biomass mostly from tall plants with palatable leaf tissue [139, 72, 238]. We therefore model grazing as a biomass dependent term in the total biomass loss rate,

$$M_i = M_{0i} + M_{1i} B_i , \tag{12.8}$$

where M_{0i} is the biomass-loss rate of the ith species due to mortality and $M_{1i}B_i$ is the loss rate due to grazing.

12.1.2 Model simplifications

In Section 9.2.3 we presented a few simplifications of the single-species model (9.3), and the same simplifications can be applied to the extended model (12.1). In particular, for uncrusted sandy soils and life forms with confined root zones the surface-water equation can be eliminated and the nonlocal terms that consist of integrals over laterally extended root zones can be replaced by local algebraic terms. The model equations (12.1) reduce then to the simpler form:

$$\frac{\partial B_i}{\partial T} = G_B^i B_i(1 - B_i/K_i) - M_i B_i + D_{B_i}\nabla^2 B_i , \tag{12.9a}$$

$$\frac{\partial W}{\partial T} = P - LW - W\sum_i G_W^i + D_W \nabla^2 W , \tag{12.9b}$$

where

$$G_B^i = \Lambda_i W(1 + E_i B_i)^p , \tag{12.10a}$$

$$G_W^i = \Gamma_i B_i (1 + E_i B_i)^p , \tag{12.10b}$$

with $p = 1$ for a 1d system and $p = 2$ for a 2d system (assuming a Gaussian root kernel).

Another simplification involves the consideration of functional groups rather than species. Plant species differ from one another in many respects. Many of these differences can be captured by model parameters, such as growth rates (Λ_i), uptake rates (Γ_i), mortality rates (M_i), root-to-shoot ratios (E_i), maximal standing biomass (K_i), seed dispersal (D_{B_i}) and others. Quite often, however, the questions to be addressed pertain to certain aspects of ecosystem function, e.g., the ability of an ecosystem to produce biomass, and not all species traits are equally relevant to the particular aspect of interest. For example, the functional traits that control resource capture are likely to be more significant for biomass production in a resource limited environment than many other species traits. In such cases species pools can be divided into functional groups that differ only in a few selected functional traits. The shift from species to functional groups entails a corresponding shift in community-level properties, such as diversity. Indeed, considerable effort has been devoted recently to studies of functional diversity rather than species diversity, as an entity that is more directly related to ecosystem function [200] (see also Section 3.2.2).

As a concrete example consider the trait parameters K and E in (9.3). The parameter K controls the plant's investment in above-ground biomass whereas the parameter E controls its investment in below-ground biomass. Both traits are functional in the sense that they affect the plant's performance under various environmental conditions. High K values provide a competitive advantage in capturing the light resource in wet environments, whereas high E values provide a competitive advantage in capturing the water resource in dry environments. We can thus characterize the species pool by functional groups that differ from one another by the values of the trait parameters K and E.

Another simplification yet can be made by assuming that in resource limited systems the plant's investment in above-ground biomass comes at the expense of investment in below-ground biomass and vice versa. This tradeoff allows the replacement of the two functional traits by a single dimensionless tradeoff parameter, $0 \leq \chi \leq 1$, defined implicitly through the relations

$$\begin{aligned} E(\chi) &= E_{min} + \chi^{\alpha}(E_{max} - E_{min}), \\ K(\chi) &= K_{min} + (1-\chi)^{\alpha}(K_{max} - K_{min}). \end{aligned} \qquad (12.11)$$

The parameter α describes different tradeoff curves in the K, E plane, as Figure 12.1 shows, and may represent different species pools.

The parameter χ can be used to define the functional groups that comprise the community as follows. Discretizing χ uniformly along the interval [0,1]: $\chi_i = i/n$, $i = 1, \ldots, n$, we define the ith functional group as the point χ_i and the small increment $\Delta\chi = 1/n$ that precedes it. In the model equations (12.9) the ith functional group is described by a biomass variable B_i that satisfies equation (12.9a) with $K_i = K(\chi_i)$ and $E_i = E(\chi_i)$, where all other parameters assume the same values for all functional groups. In general, there might be additional tradeoff parameters that characterize each functional group [295,

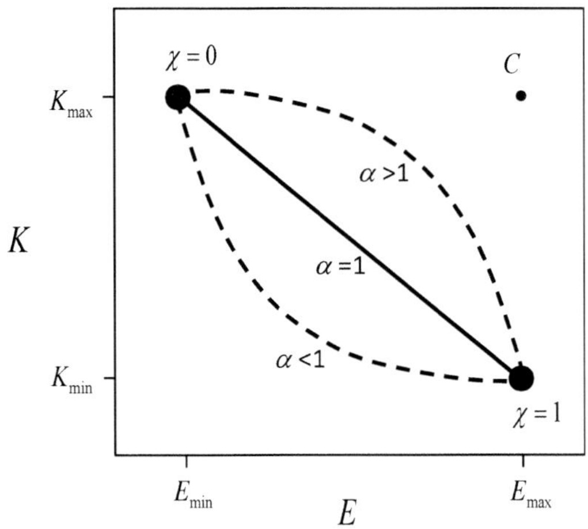

FIGURE 12.1: Root-shoot ($E - K$) tradeoff curves obtained from (12.11) for three different α values. The big solid circles represent the two extreme tradeoff cases, $\chi = 0$ and $\chi = 1$. The small solid circle C represents an ideal case (both shoots and roots are long) that serves as a reference point—species described by tradeoff points that are closer to C have larger fitness.

19]. For simplicity, we will consider in the following a single tradeoff, focusing on the $E - K$ tradeoff (12.11) as an example.

12.2 Species coexistence induced by ecosystem engineers

Associated with self-organized vegetation patchiness are non-uniform soil-water distributions. The redistribution of the water resource by a pattern-forming plant species can modify the growth conditions of other species, and, in certain circumstances, may facilitate their growth. Organisms that affect the growth of other organisms by modifying the abiotic environment are called *ecosystem engineers* [141, 142]. Numerous examples of organisms that act as ecosystem engineers exist [57]. In dryland plant communities, ecosystem engineering has been studied mostly in the context of shrubs and herbs that grow in the vicinity of and under the shrub canopy [247, 277].

In Section 1.2 we suggested that pattern formation processes may play important roles in mediating the relationships between the abiotic environment and biodiversity (see also Figure 1.2). Here we study a few examples of such processes, all associated with water redistribution by a pattern forming ecosystem engineer. Using the single-species model (9.10), we first explore the capacity of a woody life form to act as an ecosystem engineer (Section 12.2.1). We study the interplay between pattern-forming feedbacks that have counteracting effects on the soil-water distribution (Section 12.2.1.1), the resilience of ecosystem engineers (Section 12.2.1.2), and how ecosystem engineering changes along the rainfall gradient (Section 12.2.1.3). We then use the model equations (12.1) for two life forms to study species coexistence in woody-herbaceous systems. We explore the uniform and patterned vegetation states that can be realized in such systems (Section 12.2.2), and study conditions for species coexistence at the single-patch level (Section 12.2.3) and at the landscape level (Section 12.2.4).

12.2.1 Plants as ecosystem engineers

12.2.1.1 Counteracting feedbacks

The soil-water content in a vegetation patch, as compared with the content in bare soil, strongly depends on the relative strength of the infiltration and root-augmentation feedbacks, because of their opposite effects on the water distribution within the patch area. While the infiltration feedback acts to increase the soil-water content by overland water flow toward the patch, the root-augmentation feedback acts to deplete it by strong water uptake. Figure 12.2 shows biomass and soil-water distributions of an isolated vegetation patch for different values of the parameters that control the two feedbacks, f and η. The distributions were obtained by solving the model equations (9.10) for a single species that represents a small woody plant such as a shrub [102]. Strong infiltration feedback and weak root-augmentation feedback lead to high soil-water concentration in the vegetation-patch area (Figure 12.2(C)), whereas weak infiltration feedback and strong root-augmentation feedback lead to soil-water depletion in the patch area and its surroundings (Figure 12.2(D)). By concentrating the water resource beyond the bare-soil level, the ecosystem engineer forms an ameliorated micro-environment that may increase the fitness of other plant species and result in higher species richness.

When the engineer's patch area is sufficiently large, the strength of the infiltration feedback may change across the patch because most of the overland water flow is intercepted at the patch periphery and does not reach the patch core. As a result, soil-water concentration occurs mostly at the patch periphery, forming a heterogeneous micro-environment across the patch. Large patch areas can be obtained in several ways, including an increase in precipitation rate or in infiltration contrast. Figure 12.3 shows cross-sections of the soil-water distribution in two patches of different size obtained as solutions of (9.10) with different infiltration contrasts. The soil-water distribution in the smaller patch patch is single-humped while that of the larger patch is double-humped. The two types of water distributions represent

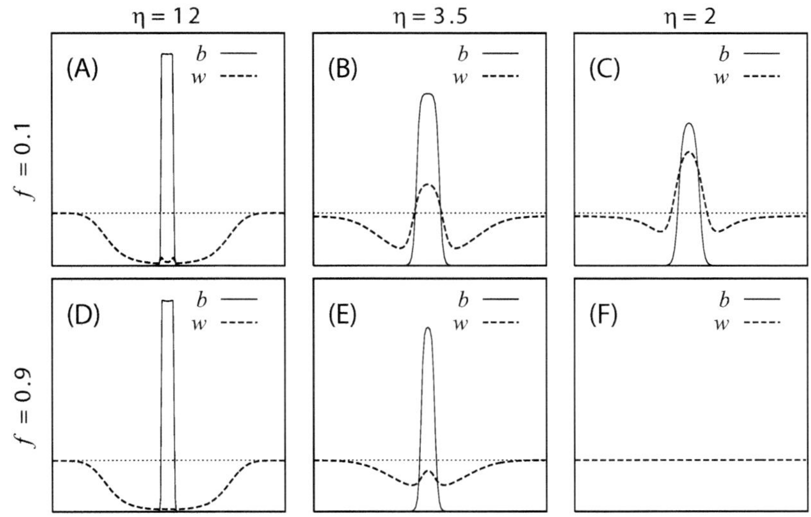

FIGURE 12.2: Spatial profiles of the above-ground biomass b and the soil-water content w as affected by the parameters that control the infiltration feedback, f, and the root-augmentation feedback, η. The profiles are cross-sections of two-dimensional simulations of (9.10). In all panels, the horizontal dotted lines denote the soil-water level at bare soil. Weak root-augmentation feedback lead to non-resilient strong ecosystem engineers (panels C,F). Moderate root-augmentation feedback can lead to resilient ecosystem engineers (panels B,F), while strong root-augmentation feedback results in soil-water depletion and no engineering, irrespective of the infiltration-feedback strength (panels A,D). Panels A) and D) span a horizontal range four times larger than the range in all other panels. From [105].

different micro-habitats for herbaceous species [278]. Double-humped distributions will favor the growth of herbaceous vegetation in the woody-patch periphery rather than within the patch. This may result in ring-shaped ecosystem engineering. Herbaceous vegetation responding to such ecosystem-engineering patterns have indeed been observed in field studies [174].

Counter acting feedbacks can be realized also for plant species with laterally confined root zones. In this case the root-augmentation feedback is no longer effective, but the soil-water diffusion feedback can substitute for it in counteracting the infiltration feedback [152].

12.2.1.2 Ecosystem engineering vs. resilience

The results shown in Figure 12.2 also indicate that conditions that maximize ecosystem engineering (small η and f values) can result in loss of resilience. As pointed out earlier, small values of f can represent the existence

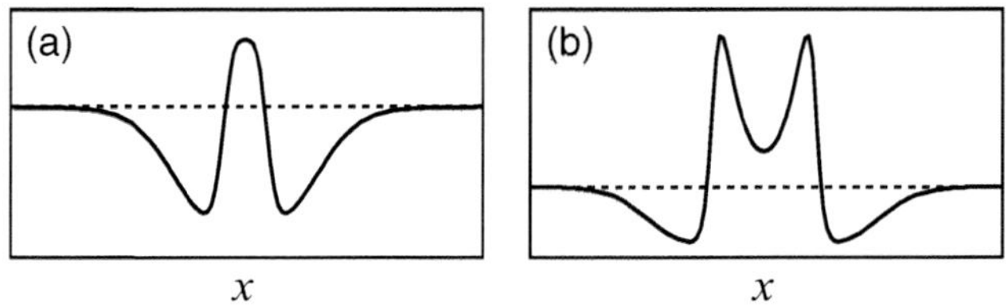

FIGURE 12.3: Cross-sections of two typical spatial distributions of soil water at the scale of a single patch, obtained by solving (9.10) with different infiltration contrasts: (a) $f = 0.1$, leading to a small biomass patch, (b) $f = 0.01$, leading to a large biomass patch. The dotted lines denote the soil-water content in bare soil. From [105].

of a biogenic soil crust that reduces the infiltration rate in bare soil, and large f values can simulate crust-removal disturbances. Imagine now conditions of strong engineering (Figure 12.2(C)) and a crust-removal disturbance that increases the value of f. As Figure 12.2(F) shows such a disturbance can result in the mortality of the ecosystem engineer. There are two reasons for the lack of resilience under these conditions: (i) surface water infiltrates equally well everywhere and, as a result, the plant patch does not benefit any longer from the interception of runoff, (ii) the engineer's roots are too short to collect water from the surrounding area. Resilient ecosystem engineers can be obtained with strong infiltration feedbacks and moderate root-augmentation feedbacks as Figure 12.2(B,E) shows. Removal of the crust (i.e., increasing f) destroys the micro-habitats that the engineer has created (the soil-water content becomes smaller than the content in bare soil), but the engineer persists. Once the crust recovers the ecosystem engineer resumes its capability to concentrate water and the micro-habitats recover as well. It is also of interest to comment that when the uptake feedback is too strong, the plant persists but it no longer functions as an ecosystem engineer even at high infiltration contrasts (Figure 12.2(A,D)).

12.2.1.3 Facilitation in stressed environments

The capability of shrubs to ameliorate their micro-environments and facilitate the growth of herbaceous vegetation, such as grasses and annuals, understorey often develops under conditions of water stress [302, 7]. Facilitation in stressed environments has been observed with other types of organisms and environmental stresses, such as alpine plants at high elevations, salt-tolerant plants in salt marshes, mussel beds and others [22, 37, 36, 53, 35], and has been hypothesized to be a general behavior [186]. The amelioration of the shrub's micro-environment as aridity increases has been attributed to many factors including redistribution of water and nutrients, hydraulic lifting, chemical interactions with the soil and changes in soil structure [278]. Although most of

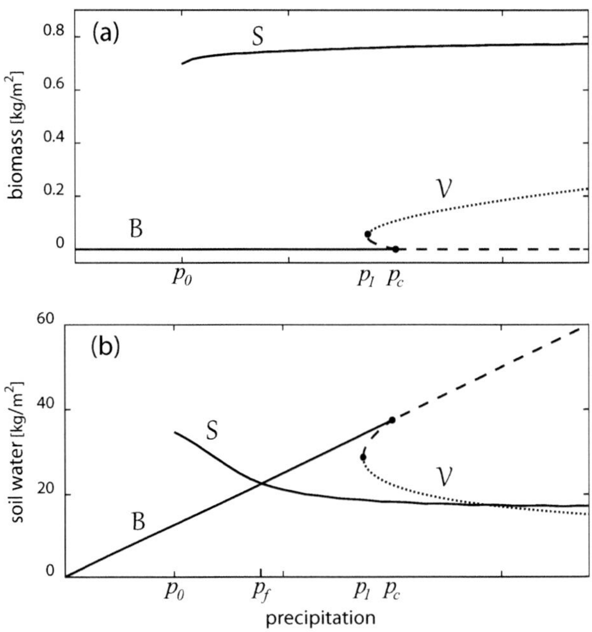

FIGURE 12.4: Bifurcation diagrams for the single-species model (9.10) showing (a) the biomass and (b) the soil-water content associated with bare-soil ($\mathcal{B}$), uniform vegetation ($\mathcal{V}$) and a periodic spot pattern ($\mathcal{S}$). The spot pattern branch, $\mathcal{S}$, denotes the amplitude of the biomass pattern in diagram (a) and the soil-water content at the center of a biomass spot in diagram (b). The two solution branches $\mathcal{B}$ and $\mathcal{S}$ in diagram (b) intersect at $p = p_f$. Below p_f the soil-water content at the center of a vegetation spot is higher than the content in bare soil, indicating ecosystem engineering. From [215].

these factors are not included in the model equations (9.10), the equations do capture the appearance of ecosystem engineering along a rainfall gradient, suggesting a more fundamental mechanism of facilitation in dryland woody-herbaceous vegetation.

Figure 12.4 shows a bifurcation diagram similar to that shown in Figure 10.1 for uniform solutions, except that it also includes the stable part of a periodic spot solution (the branch denoted by $\mathcal{S}$) and an analogous diagram with the biomass variable replaced by the soil-water variable (part b). In the latter diagram the line $\mathcal{B}$ shows the soil-water content in bare soil while the line $\mathcal{S}$ shows the maximal soil-water content under a biomass patch in a periodic spot pattern. The two lines intersect at $p = p_f$ implying a crossover from conditions of competition at high precipitation ($p > p_f$), where the soil-water content in the patch area is lower than in bare soil, to conditions of

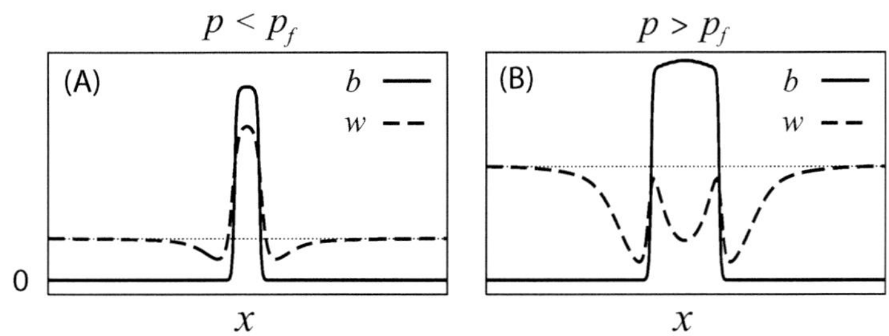

FIGURE 12.5: Solutions of (9.10) showing the spatial profiles of the biomass (solid lines) and soil-water (dashed lines) variables across a vegetation spot below (A) and above (B) the crossover-over point $p = p_f$ to facilitation. From [102].

facilitation at low precipitation ($p < p_f$), where the soil-water content in the patch area exceeds that of bare soil. Another view of this crossover is provided in Figure 12.5, which shows the spatial biomass and soil-water distributions across an isolated vegetation patch for p bigger and smaller than p_f, obtained as numerical solutions of (9.10).

The model offers the following explanation for this crossover. As the system becomes more arid, the biomass patch contracts (see Figure 12.5) and the water uptake from any unit area within the patch decreases significantly as there are fewer individuals in the patch. The infiltration rate at the reduced patch area, however, decreases only slightly because of its weak biomass dependence for $b \gg q$ (see Figure 9.4). This uneven change in the soil-water uptake and the surface-water infiltration is supported by the numerical calculations shown in Figure 12.6 of the water-uptake rate G_w and the infiltration rate I within a biomass patch as a function of the precipitation rate p. As a result, while at high precipitation the soil-water content within the patch is dominated by the strong uptake and results in water depletion relative to bare soil, at low precipitation it is dominated by the infiltration of surface water and results in soil-water concentration relative to bare soil.

Why does not the ecosystem engineer exhaust the soil water for its own growth? According to the model two factors act against that; the maximum standing biomass K which limits the local growth of the ecosystem engineer, and the depletion of soil water in the immediate vicinity of the engineer's patch which prevents its expansion. We remind the reader that K represents genetic plant-specific limitations (e.g., stem strength) and therefore while they restrict the growth of the ecosystem engineer they do not affect the growth of the additional life forms that benefit from the ecosystem engineering.

The transition from competition to facilitation as precipitation decreases (Figure 12.5) was obtained using (9.10) with a constant root-shoot parameter

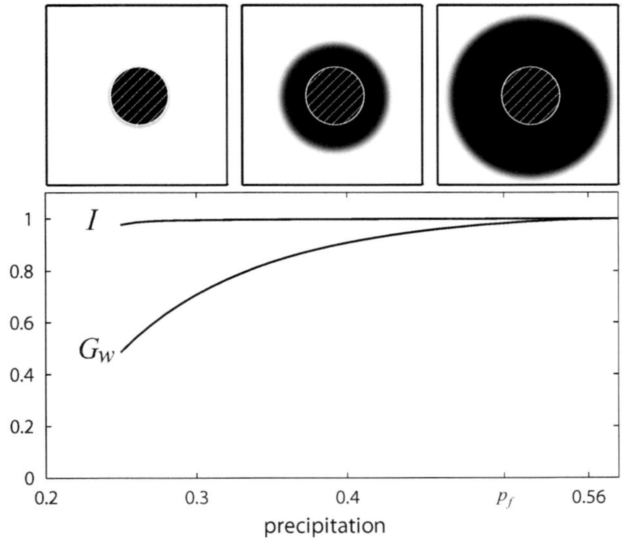

FIGURE 12.6: The water balance in an area of a fixed size within a biomass patch (hatched circles in upper panels) along the rainfall gradient. While the infiltration rate I, integrated over the fixed biomass area, hardly changes as precipitation decreases, the water-uptake rate G_w, integrated over the same area, significantly decreases due to the smaller vegetation coverage (fewer 'individuals') outside the area. As a consequence, the soil water content increases relative to the water content in bare soil as Figure 12.5 shows. The values of G_w and I were normalized with respect to their maximal values. From [102].

η that is independent of the precipitation p. Plant roots, however, can show a high degree of plasticity and can respond to water stress by increasing the root-to-shoot ratio. Using a decreasing dependence of η on p in (9.10), at the lowest precipitation range for which vegetation still survives, leads to a transition back to competition at sufficiently low p values [102]. Thus, both directions of plant-interaction change, competition to facilitation and facilitation to competition, may be realized along the same environmental gradient, as some field observations suggest [185, 192].

12.2.2 Uniform and patterned vegetation states

The micro-habitats that the woody engineer forms suggest the possible facilitation of herbaceous vegetation growth. To test the possible emergence of woody-herbaceous coexistence we study equations (12.1) for a woody life form characterized by a biomass variable B_1 (or its non-dimensional counter-part

$b_1 = B_1/K_1$ [102]) and an herbaceous life form characterized by a biomass variable B_2 ($b_2 = B_2/K_2$). We distinguish between the two life forms by choosing the maximum standing biomass of the woody species (K_1) to be an order of magnitude higher than that of the herbaceous species (K_2) while its growth and mortality rates are taken to be significantly slower. We confine ourselves to the case of strong infiltration feedback ($f \ll 1$) and moderate root-augmentation feedback of the woody species ($\eta_1 = E_1K_1 \sim \mathcal{O}(1)$) so that it can act as an ecosystem engineer (see Figure 12.2(B,C)). These conditions are often realized in drylands where biological soil crusts increase the infiltration contrast and the woody vegetation consists of shrubs [282]. With this parameter choice we find a strong asymmetric relations between the two life forms; herbaceous vegetation is strongly affected by the woody vegetation, but the woody vegetation is hardly affected by the herbaceous one. In the studies to be described below we do not take into account inter-specific competition for light, i.e., we set $h \to \infty$ in (12.7). The combined effect of inter-specific competition for water and light will be considered in Section 12.4.

Equations (12.1) for $n = 2$ have four stationary uniform solutions: $\mathcal{B}$ – bare soil ($b_1 = 0,\ b_2 = 0$), $\mathcal{V}_1$ – uniform woody vegetation ($b_1 \neq 0,\ b_2 = 0$), $\mathcal{V}_2$ – uniform herbaceous vegetation ($b_1 = 0,\ b_2 \neq 0$), and $\mathcal{M}$ – uniform mixed woody-herbaceous vegetation ($b_1 \neq 0,\ b_2 \neq 0$). A linear stability analyzes of these solutions lead to the bifurcation diagram shown in Figure 12.7 with $p = \Lambda_1 P/NM_1$, the dimensionless precipitation, as the bifurcation parameter. The bare-soil solution, $\mathcal{B}$, is given by

$$b_1 = 0\,, \quad b_2 = 0\,, \quad w = \Lambda_1 P/N^2\,, \quad h = \Lambda_1 P/NfA\ . \tag{12.12}$$

If $M_2\Lambda_2 < M_1\Lambda_1$ (the case we considered here) the bare-soil solution loses stability to the growth of herbaceous vegetation in a uniform stationary instability. The threshold of this instability is $p_{b_2} = M_2\Lambda_1/M_1\Lambda_2$. Figure 12.8(A) shows the growth-rate curve, $\sigma = \sigma(k)$, below, at and above p_{b_2}. The curve at p_{b_2} shows that the first wavenumber to grow is $k = 0$ indicating a uniform instability. Beyond this threshold a uniform stationary solution ($\mathcal{V}_2$), representing herbaceous vegetation, appears[1].

The uniform woody solution ($\mathcal{V}_1$) is stable only beyond a higher threshold $p = p_T$ as Figure 12.7 shows. Below this threshold the uniform woody vegetation loses stability in a non-uniform stationary instability as the growth-rate curve in Figure 12.8(B) indicates. This instability leads to a periodic woody-gap pattern whose solution branch is not shown in the bifurcation diagram (Figure 12.7). Shown instead is a non-uniform solution branch at a lower precipitation range ($\mathcal{S}$), evaluated by solving (12.1) (with $n = 2$) numerically. This solution branch describes a woody-spot pattern, which in a limited precipitation range includes the herbaceous life form too, as the bend in the solution branch in the b_2 direction hints. An example of such a coexistence

[1] If $M_2\Lambda_2 > M_1\Lambda_1$, the bare-soil solution loses stability to the growth of woody vegetation at $p = p_{b_1} = 1$.

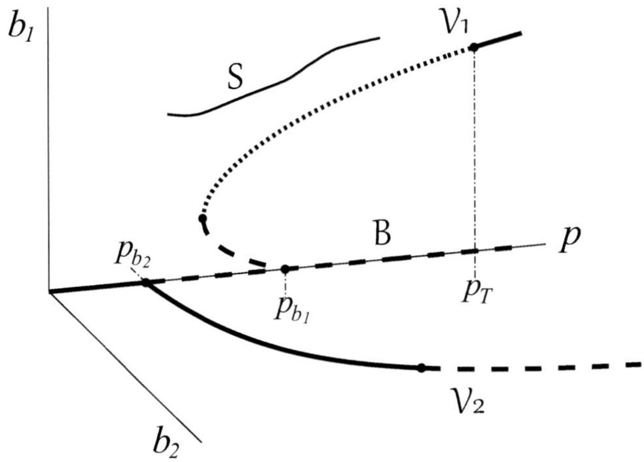

FIGURE 12.7: Bifurcation diagram showing solution branches of a woody-herbaceous system. The solution branches $\mathcal{B}$, $\mathcal{V}_1$ and $\mathcal{V}_2$ represent, respectively, uniform bare-soil, uniform woody vegetation and uniform herbaceous vegetation. The branch $\mathcal{S}$ represents the amplitudes of spots patterns. Solid lines represent stable solutions, and dashed and dotted lines represent solutions that are unstable to uniform and non-uniform perturbations, respectively. From [102].

pattern is shown in Figure 12.10(b). As will be discussed in Section 12.2.4, it demonstrates ecosystem engineering at the landscape level where patch density plays a role. The uniform mixed solution, $\mathcal{M}$, is found unstable for the parameter choice of this study and is not shown in the diagram.

12.2.3 Species coexistence: Single-patch scale

In Section 12.2.1.3 we studied the redistribution of the soil-water content that a woody ecosystem engineer induces along a rainfall gradient. We found that while at high precipitation ($p > p_f$) the woody engineer depletes the soil-water content in the area it occupies and in its immediate neighborhood, at low precipitation ($p < p_f$) it can increase the soil-water content beyond the level of bare soil (Figure 12.5). We now use (12.1) for a woody-herbaceous system to study the response of herbaceous vegetation to the soil-water distributions that the woody engineer forms (see Figure 12.5). We focus here on woody-herbaceous interspecific interactions at the level of a single patch, choosing precipitation values within the bistability range of uniform herbaceous

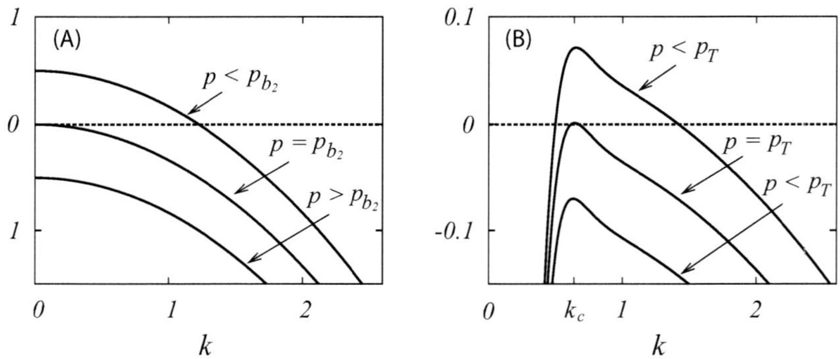

FIGURE 12.8: Growth-rate curves calculated from linear stability analysis (A) of the bare-soil solution $\mathcal{B}$ near the bifurcation point $p = p_{b_2}$, and (B) of the uniform woody-vegetation solution $\mathcal{V}_1$ near the bifurcation point $p = p_T$ (see Fig 12.7). As p is increased past $p = p_{b_2}$ a uniform mode ($k = 0$) begins to grow, implying an instability of the bare-soil solution to uniform perturbations (A). As p is decreased past $p = p_T$ a non-uniform mode ($k = k_c$) begins to grow monotonically in time, implying a non-uniform stationary instability of the uniform woody-vegetation solution (B).

vegetation and a periodic woody pattern. As Figure 12.9 shows, at high precipitation (but still in the bistability range) the herbaceous life form grows in the open areas only, i.e., away from the woody patch, while at low precipitation the same herbaceous life form grows only "understorey," i.e., within the woody patch area. Projecting to the case of a community of herbaceous life forms we may expect to find herbaceous vegetation occupying both open and

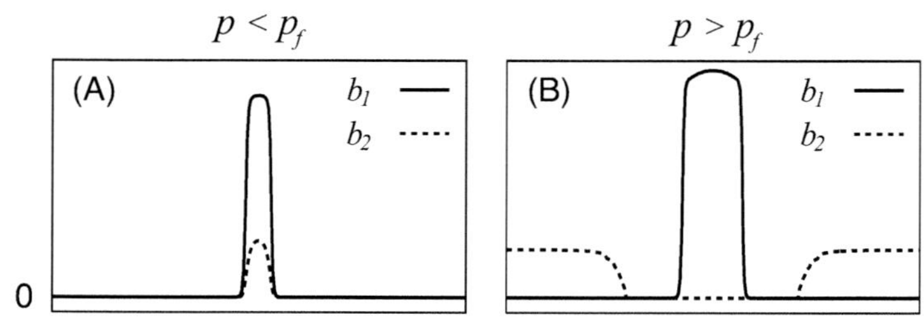

FIGURE 12.9: The response of herbaceous vegetation to a woody ecosystem engineer that redistributes the soil water as shown in Figure 12.5. Shown are solutions of (12.1) (with $n = 2$) for the woody biomass (solid lines) and the herbaceous biomass (dashed lines) below (A) and above (B) the crossover point $p = p_f$ to facilitation. From [102].

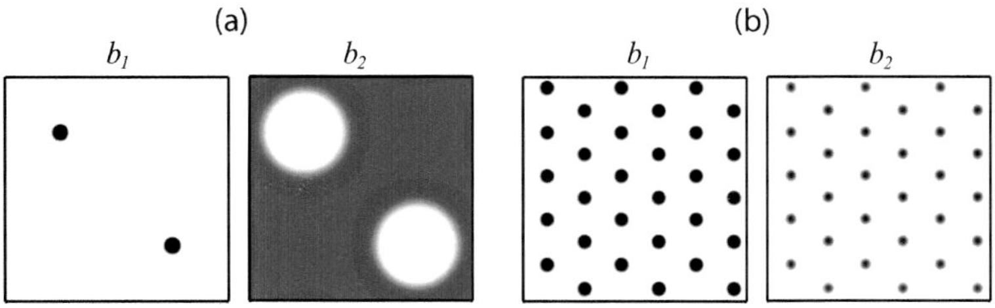

FIGURE 12.10: Woody herbaceous coexistence patterns. (a) A savanna-like hybrid state, in which the woody life form excludes the herbaceous life form. (b) A mixed periodic spot pattern, in which the woody life form facilitates the growth of the herbaceous life form. The patterns were obtained by solving numerically (12.1) in the bistability range of a woody spot pattern and uniform herbaceous vegetation (the overlap range of the solid branches $\mathcal{V}_2$ and $\mathcal{S}$ in Figure 12.7). From [102].

understorey areas, but with variable proportions along the rainfall gradient; higher abundance in open areas at high precipitation and higher abundance understorey at low precipitation. These expectations, solely based on the soil-water redistribution by the woody ecosystem engineer, are inline with field observations [133].

12.2.4 Species coexistence: Landscape scale

The discussion so far has related facilitation to increasing environmental stress. We now show that in situations where multiplicity of stable states can be realized, facilitation can result from transitions from one stable state to another for fixed environmental conditions. Figure 12.10 shows the effect of diluting a woody spot pattern (b_1) on the growth of herbaceous vegetation (b_2). Stable diluted patterns can be realized as stable hybrid states in bistability ranges of woody spot patterns and uniform herbaceous vegetation (see Section 7.2.1.3). Woody patches that are sufficiently sparse and effectively isolated can competitively exclude the herbaceous life form, as Figure 12.10(a) shows. However, when the patches are dense enough coexistence of the two life forms within the patches becomes possible (Figure 12.10(b)). This is because the competition for water reduces the woody patch size and consequently the soil-water uptake (see Figure 12.6). As a result, more soil-water is left for the herbaceous life form allowing its coexistence with the woody life form.

Another example of species coexistence induced by a transition from one stable state to another is shown in Figure 12.11. In Section 10.1.3 we discussed a transition from vegetation bands on a slope to a spot pattern induced by local biomass removal (see Figure 10.7). Here we consider a similar context—bistability of banded and spotted patterns on a slope and a local disturbance

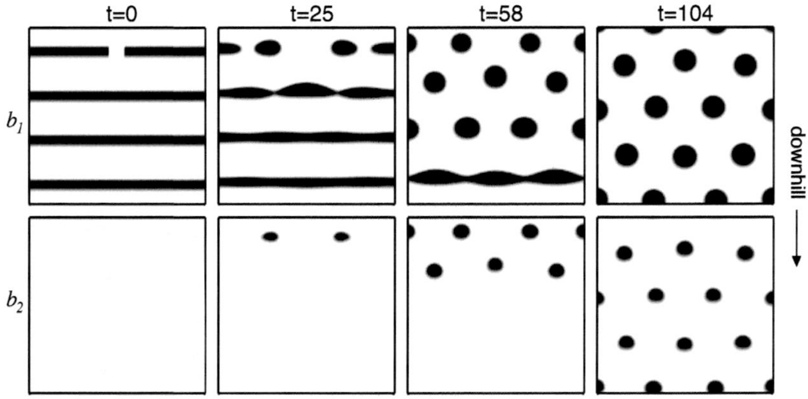

FIGURE 12.11: Species coexistence induced by a pattern transition. Shown are snapshots at different times (t is in years) of the woody biomass b_1 (top row) and the herbaceous biomass b_2 (bottom row) obtained by solving (12.1) numerically. The snapshots describe a transition from vegetation bands to vegetation spots on a slope induced by local removal of the woody biomass uphill. In the banded pattern the woody life form excludes the herbaceous life form, but in the spotted pattern they coexist due to enhanced runoff concentration. From [102].

uphill, but in woody-herbaceous systems, where the local disturbance involves biomass removal of the woody life form. The herbaceous life form is excluded as long as the woody life form self organizes in bands, but appears understorey once the woody bands break into spots. The reason for that becomes clear when considering the soil-water redistribution associated with this transition; the spot formation is accompanied by the interception of larger amounts of runoff because each spot experiences a bare area uphill twice as large as the bare area between successive bands [105]. The transition from banded to spotted vegetation involves, in effect, a facilitation front that propagates downhill; as bands gradually break into spots, patches with higher soil-water content form and facilitate the growth of the herbaceous life form. This is an example of a cross-scale effect where pattern transitions at the landscape scale change interspecific interactions at the single-patch scale (see Figure 1.3).

12.3 Savanna-like forms of species coexistence

The bifurcation diagram depicted in Figure 12.7 shows a bistability range of a uniform herbaceous state $\mathcal{V}_2$ (grassland) and a spot pattern $\mathcal{S}$ (shrubland). In Section 7.2.1.3 we discussed bistability ranges of uniform and patterned states in general and pointed out the possible existence of a multitude of stable hybrid states within homoclinic-snaking subranges. In the context of woody-herbaceous systems such states describe savanna landscapes, i.e., confined domains of woody spots in otherwise uniform herbaceous vegetation. In this section we consider generalized forms of such coexistence, i.e., confined domains of a pattern-forming life form in areas uniformly populated by another life form, and refer to them as "savanna-like coexistence," even when no association to woody and herbaceous vegetation can be made. Since the model equations (12.1) used to produced the bifurcation diagram in Figure 12.7 are not easily amenable to numerical continuation methods needed to trace snaking solution branches, we resort to the simplified model equations (12.9) for two life forms that make different tradeoffs in capturing sunlight and soil water. In Section 12.3.1 we study uniform and periodic solutions of (12.9) and identify a bistability range of uniform and patterned population states. In Section 12.3.2 we study the possible existence of homoclinic snaking in this range and the implications it bears for species coexistence.

12.3.1 Bistability of uniform and patterned population states

We consider the system (12.9) for two life forms in 1d, where G_B^i, G_W^i are given, respectively, by (12.10a), (12.10b) with $p = 1$, and $\Lambda_i(B)$ is given by (12.7). We focus on species that make different compromises in their investments in above and below-ground biomass, and characterize them by χ values close to zero and close to unity. Specifically, we assign a value $\chi_1 \ll 1$ to the species with biomass B_1, and a value $\chi_2 = 1$ to the species with biomass B_2. The χ_1 species represents a superior competitor for light (high K and low E) while the χ_2 species represents a superior competitor for water (high E and low K). The two species are kept identical in regard to all other trait parameters and, in practice, represent two distinct functional groups.

The system (12.9) (for two life forms) has three types of uniform stationary states $U = (B_1, B_2, W)$:

- The trivial bare soil state $U_0 = (0, 0, W_0)$
- Pure population states $U_1^* = (B_1^*, 0, W_1^*)$ and $U_2^* = (0, B_2^*, W_2^*)$,
- Coexistence states of the form $\bar{U} = (\bar{B}_1, \bar{B}_2, \bar{W})$.

A linear stability analysis of these solutions provides their existence and stability ranges [164]. Figure 12.12 displays a bifurcation diagram that shows the bare-soil and the two pure-population solution branches (the coexistence states are unstable and are not shown). At very low precipitation values P, the bare-soil state, U_0, is the only stable state. At sufficiently high P values there is a bistability range of the two uniform pure-population states, U_1^* and U_2^*. The latter loses stability at $P = P_2^M$ to a mixed-population state, leaving the pure-population state U_1^* of the superior light competitor, $\chi_1 \ll 1$, as

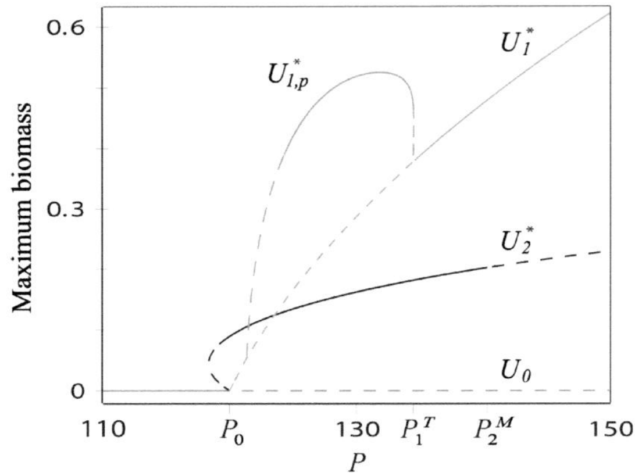

FIGURE 12.12: Bifurcation diagram showing biomass-precipitation solution branches for the bare-soil state, U_0, the two pure-population uniform states, U_1^* and U_2^*, and for the pure-population periodic-pattern state, $U_{1,p}^*$, for which the maximum biomass is plotted. Solid (dashed) lines denote stable (unstable) states. The two uniform states bifurcate from the bare-soil state at $P = P_0$. The periodic-pattern state bifurcates from the corresponding uniform state at $P = P_1^T$ and reconnects to it at a lower precipitation value. From [164].

the only stable state in the range $P > P_2^M$. The stability range of the uniform state U_1^* is bounded below by a non-uniform stationary instability at P_1^T that leads to a stationary periodic pattern. The uniform state U_2^* can also go through such an instability but at lower precipitation values. These are Turing instabilities that in the present context require strong water uptake and fast soil-water diffusion relative to the rate of seed dispersal or clonal growth (soil-water diffusion feedback, see Section 9.1.2.3).

To study the solutions that appear below the Turing instability at P_1^T we resort to numerical continuation and stability methods. Besides the spatially periodic solution, $U_{1,p}^* = (B_1^*(x), 0, W_1^*(x))$, that emanates from the uniform-solution branch, U_1^*, at P_1^T, there are additional periodic solutions, not shown in the bifurcation diagram, that emanate from the U_1^* branch at $P < P_1^T$. The periodic solution branches all reconnect to the U_1^* branch at lower P values. Note the wide bistability range of the uniform U_2^* state and the periodic $U_{1,p}^*$ pattern. We address this range in the next section.

12.3.2 Front pinning and species coexistence

As pointed earlier, bistability ranges of uniform and periodic-pattern states often give rise to stable hybrid states. These are localized solutions involving confined domains of one state in a system otherwise occupied by the alternative state. In the bistability range of the uniform U_2^* state and the periodic

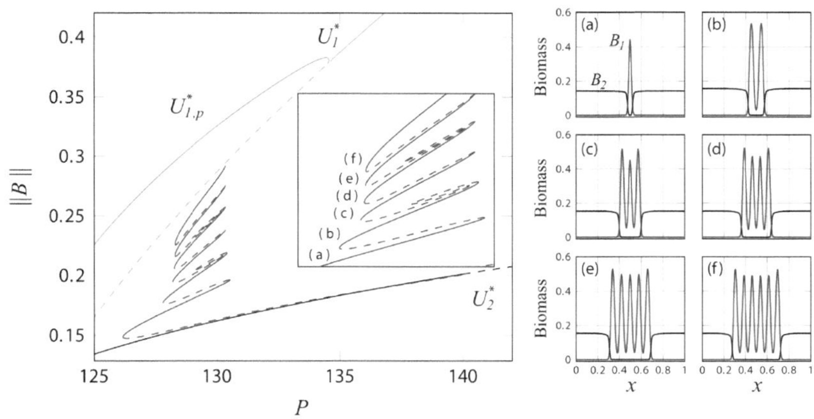

FIGURE 12.13: A bifurcation diagram that includes localized solution branches (left). The vertical axis is chosen here to be the L^2 norm of the biomass variables defined as $\|B\| = \int (B_1^2 + B_2^2)^{1/2} dx$. The inset shows a blow-up of the localized solution branches labeled in ascending order of localized-solution sizes as the spatial solution forms in panels (a-f) show. The blue (spiky) and black (flat) graphs denote B_1 and B_2, respectively. From [164].

$U_{1,p}^*$ pattern such solutions correspond to confined patterns of the species χ_1 that specializes in capturing light in an otherwise uniform distribution of the species χ_2 that specializes in capturing soil water. The bifurcation diagram in Figure 12.13 shows the solution branches associated with several localized solutions of increasing size and their stability properties. The localized solution branches snake up toward the periodic-pattern solution, acquiring an additional hump in each turn as the panels (a–f) show.

The multiplicity of stable localized solutions implies the possible coexistence of the two species in adjacent domains of the same landscape and under the same environmental conditions. This savanna-like form of species coexistence differs from facilitation by ecosystem engineers in that the species exclude each other from their locations. It results from the inability of either species to displace the other, or to the pinning of the front that is bi-asymptotic to U_2^* and $U_{1,p}^*$. The front pinning can be viewed as resulting from the interplay between competition for light and water; because of the different tradeoffs the two species make in capturing these resources, each species has an advantage and disadvantage in attempting to displace the other, which balance one another. The physical and ecological processes that maintain this balance over a precipitation range rather than at a single precipitation value (the Maxwell point) are not clear yet.

The localized structures occupy a limited subrange within the bistability precipitation range of the two pure-population states U_2^* and $U_{1,p}^*$. Outside this subrange, but still within the bistability range, one species displaces the other. At relatively high precipitation water is no longer a strong limiting factor and competition for light becomes more important. As a result, the $U_{1,p}^*$ state, which represents a periodic pattern of the better competitor for light, invades into areas occupied by the U_2^* state. At relatively low precipitation the vegetation is less dense and light is no longer a strong limiting factor. As a result the competition for water becomes more important, and the U_2^* state, which represents the better competitor for water, invades into areas occupied by the $U_{1,p}^*$ state.

12.4 Linking pattern formation and biodiversity

We described two mechanisms of species coexistence in water-limited systems associated with pattern formation: (i) local facilitation by an ecosystem engineer, both at the single-patch (localized structure) scale and at the landscape (periodic pattern) scale, (ii) spatial species segregation by front pinning that lead to savanna-like patterns (hybrid states). What are the implications of these mechanisms for biodiversity? While spatial heterogeneity has long been recognized as an important driver of biodiversity, the specific roles played by self-organized heterogeneity, i.e., heterogeneity induced by pattern formation, have hardly been studied (see Section 3.2.2). In this section we delineate possible research directions for identifying these roles. We first address the question how the model equations (12.1) or (12.9) can be used to derive community-level properties, such as functional diversity (Section 12.4.1), and then discuss possible applications to pattern-forming ecosystems (Section 12.4.2).

12.4.1 Derivation of community-level properties

Consider the simplified model equations, (12.9) and (12.10) with $p = 2$, for a community consisting of n functional groups uniformly distributed along the tradeoff axis χ (see Section 12.1.2). It will be useful to view the n biomass variables B_i, $i = 1, ..., n$ as discretized values of a continuous biomass function, $B = B(\mathbf{X}, \chi, T)$, with $B_i = B(\mathbf{X}, \chi_i, T)$. Figure 12.14 shows a typical asymptotic form of such a function obtained by solving (12.9) numerically for a non-pattern-forming community for which the state variables are space independent. The pulse-shape form contains information about functional diversity (pulse width), community abundance (pulse area) and community composition (pulse position). A variety of additional community level properties can be derived by studying pulse solutions of this kind along environmental gradients.

Figure 12.15 shows how biomass pulse solutions vary along rainfall and grazing gradients. According to this figure the pulse position along the χ axis (measured by the position of the pulse maximum) moves to lower χ values as

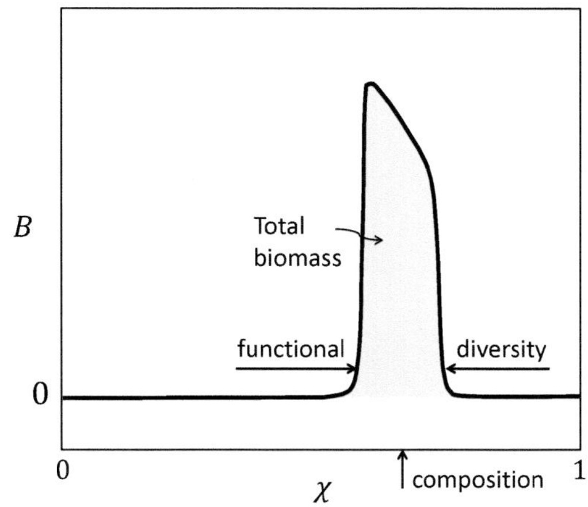

FIGURE 12.14: A typical pulse-shape biomass solution of (12.9). The biomass distribution along the tradeoff axis χ contains information about functional diversity (pulse width), composition (pulse position) and total biomass (pulse area).

the precipitation rate increases. This implies a compositional change toward a community that invests more in above-ground biomass at the expense of below-ground biomass. This trend, which holds for different grazing intensities, results from the better competitive abilities of taller functional groups in capturing light at higher precipitation.

The pulse shape changes too; both pulse height and pulse width increase with precipitation. The pulse height or abundance increases because, as there is more water, species can further approach their maximal potential size. The pulse width or diversity increases as the more abundant species reach their genetic size limitation, represented by K, slow down their water uptake and leave sufficient resources for less competitive species with higher K to appear. These pulse-shape changes reflect an increase in the total biomass (pulse area) of the community, which is a combined effect of population abundance changes and functional-diversity changes. Under grazing the pulses become shorter and wider. The increasing diversity associated with wider pulses is due to the fact that the grazing acts selectively on functional groups of the highest abundance and larger size. This reduces their competitive abilities with respect to light capture and allows the coexistence of inferior groups. This effect of the grazing stress, however, is limited to sufficiently low grazing intensities as discussed below.

A study of pulse solutions over a wide range of precipitation values, P, revealed the following behaviors [222]. The increase of the total biomass with precipitation begins once a precipitation threshold is exceeded and continues

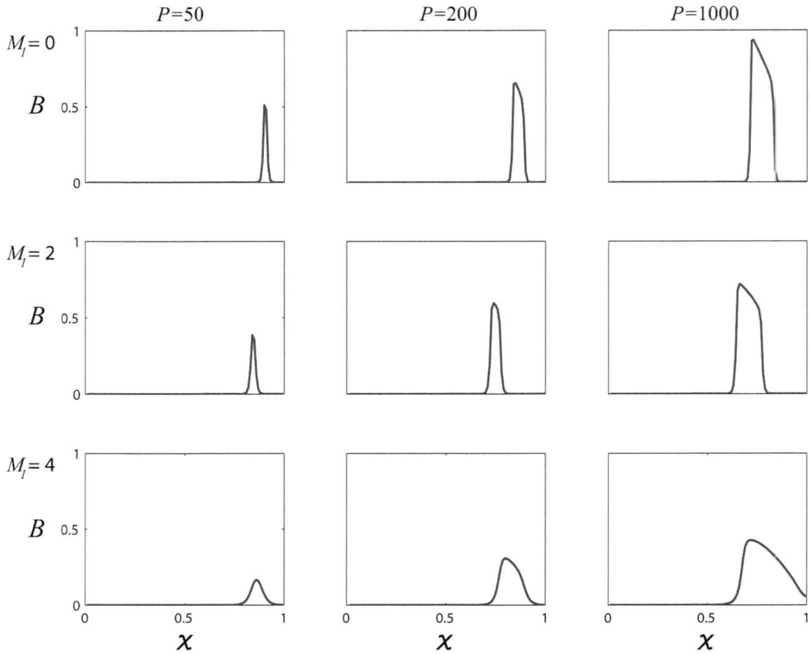

FIGURE 12.15: Asymptotic pulse solutions (biomass vs. the tradeoff parameter) of equations (12.9) for different precipitation rates P and grazing rates per unit biomass M_1 (assumed to be equal for all functional groups). Changes in pulse width, position and height provide information on how functional diversity, composition and abundance change along rainfall and grazing gradients. From [222].

monotonically, with a decelerating rate. The same qualitative behavior holds for the functional diversity. Adding a grazing stress, $M_1 > 0$, shifts the precipitation threshold upward but leads to a sharper increase of diversity with precipitation. A study of pulse solutions over a wide range of grazing intensities (M_1), revealed a non-monotonic dependence of functional diversity on grazing stress in productive ecosystems (relatively high rainfall). This result for functional diversity is inline with the *intermediate disturbance hypothesis* for species diversity [112, 49]. According to this hypothesis the highest species richness should be expected at intermediate disturbance levels; intermediate grazing can decrease competitive exclusion by biomass removal of tall plants, allowing coexistence of the tall more competitive species with short less competitive ones and thus lead to an overall increase of species richness. In less productive ecosystems the results are in line with the *grazing reversal hypothesis*, which states that that positive effects of grazing on plant richness should only be expected in high resource habitats with intense above ground competition [246].

While confirming known results and hypotheses[2] the model also provides new insights into the mechanisms and processes that affect various community-level properties along environmental gradients, such as community composition and functional diversity [222]. Most significant, however, are the tools that the spatially explicit model (12.1) (or (12.9)) provides for linking biodiversity to pattern formation, as discussed in the next section.

12.4.2 Prospects for future studies

12.4.2.1 Functional diversity in savanna-like landscapes

In Section 12.3.2 we studied front pinning as a mechanism of species coexistence. We considered a system consisting of a functional group $\chi_1 \ll 1$ that specializes in capturing light and a functional group $\chi_2 = 1$ that specializes in capturing soil water. Using the model equations (12.9) for two life forms we identified a bistability precipitation range of uniform and patterned states, and within this range a subrange of hybrid states that describe savanna-like coexistence of the two functional groups. This study is extendable to a large community of functional groups, χ_i, $i = 1, .., n$, spanning the interval $0 \leq \chi \leq 1$ and comprising a diverse pool of functional groups, part of which are pattern forming. Biomass solutions $B_i = B(\mathbf{X}, \chi_i, T)$ $(i = 1, ..., n)$ of the model equations (12.9) can then be used to derive community-level properties as in Section 12.4.1 by integrating the solutions over the physical space coordinates $\mathbf{X}$.

Several questions can be addressed using this approach. What hybrid states are possible in a given environment, taking into account the possible existence of a large sub-community of pattern-forming life forms? How does the diversity of hybrid states affect functional diversity? What spatial patterns are realizable along environmental gradients? How do transitions between different spatial patterns affect community-level properties such as functional diversity and its relations to bioproductivity and grazing intensity?

12.4.2.2 Impact of woody ecosystem engineers on functional diversity of herbaceous communities

Woody-herbaceous systems often consist of a few woody species and dozens of herbaceous species. Species richness in such systems is therefore determined by the community of herbaceous life forms, but the structure of this community may strongly depend on the presence of a woody ecosystem engineer and on the spatial patterns it forms. The impact of a woody engineer on herbaceous community can be studied by splitting equations (12.1) into an equation for a woody engineer, as was done in Section 12.2, and a set of equations describing a herbaceous community distributed along the tradeoff

---

[2] In fact, we could regard the model results as extending the earlier results obtained for species diversity to analogous results for functional diversity.

axis χ (or any other appropriate tradeoff axis). For simplicity, the parameters K_{min}, K_{max}, E_{min}, E_{max} can be chosen such that the functional groups that comprise the herbaceous community are not pattern forming.

Depending on the precipitation rate and the density of woody patches, various levels of competition and facilitation between the woody engineer and the herbaceous life forms can be realized. These varied woody-herbaceous interactions are bound to result in functional-diversity changes. As a concrete example of a possible change consider the functional diversity of the herbaceous community along the rainfall gradient. In the absence of a woody life form the diversity increases monotonically with precipitation, as discussed in Section 12.4.1. The presence of a woody engineer is likely to change this pattern because of the crossover from facilitative woody-herbaceous interactions to competitive interactions. The exclusion of herbaceous life forms by the woody engineer at high rainfall rates may result in a hump-shaped diversity-rainfall or diversity-productivity relationships.

Chapter summary

The relation between landscape heterogeneity and biodiversity has been studied in various contexts, but the impact of *self-organized* heterogeneity on community structure is still largely unexplored. Studying the manners by which pattern-formation mechanisms affect species coexistence and community-level properties is essential for understanding the relations between the abiotic environment and biodiversity and between biodiversity and ecosystem function (see Figure 1.2). Focusing on dryland plant communities we first extend the single model equations for a single life form (9.3) to equations describing a community of life forms. The extension includes the addition of interspecific competition for sunlight through biomass-dependent growth rates, and biomass-dependent grazing stress, representing selective stress towards tall plants with palatable leaf tissue.

Several model simplifications can be made to make the model more tractable. The first simplification is the elimination of one or two pattern-forming feedbacks as discussed in Section 9.2.3 for the single life form model. Two other simplifications involve the consideration of functional groups rather than species and tradeoffs between functional traits. Functional groups differ from one another by the values of selected functional traits. As a concrete example, we select the trait parameters K and E as representing functional traits that control the plant's investment in above-ground and below-ground biomass, respectively, and assume a tradeoff between the two. This leads to the characterization of the community by functional groups that are distributed along a single tradeoff axis $0 \leq \chi \leq 1$, where functional groups with low (high) χ values represent plants investing mostly in above (below) ground biomass.

As a first step toward understanding the link between pattern formation and biodiversity we consider the dynamics of two distinct life forms, looking for pattern-formation mechanisms for their coexistence. The first coexistence mechanism we study pertains to woody-herbaceous systems where the woody life form acts as an ecosystem engineer. By intercepting runoff a woody patch can increase the soil-water content beyond the level of bare soil and thereby facilitate the growth of the herbaceous life form. Facilitation is achieved when the infiltration feedback, which controls the local water increase, dominates the root-augmentation feedback, which controls the local water depletion by uptake. The relative strength of the two feedbacks depends on the trait parameters that control them, on the level of water stress and on the spatial patterns the woody engineer forms. Increased water stress does not affect much the infiltration rate at a woody patch but significantly decreases the water-uptake rate because of the smaller patch size (fewer individuals). As a result, water stress can induce facilitation of the herbaceous life form. Re-patterning of the woody engineer can also change the relative strength of the two feedbacks and lead to facilitation. One example is denser patterns of woody patches that are characterized by smaller patch sizes and lower water uptake. Another example is a transition from banded to spotted vegetation on a slope, which lengthens runoff pathways and results in increased infiltration.

The second mechanism for species coexistence does not involve facilitation; the two life forms, which make different compromises in capturing water and sunlight (characterized by low and high χ values), compete for these two resources. Coexistence can still be achieved by spatial segregation that leads to stable savanna-like patterns. A necessary condition for stable segregation is bistability of the two pure population states, where one of the two life forms is pattern forming. Under these conditions fronts that separate the two population states can be pinned in place and lead to stable hybrid states, i.e., patterns consisting of patterned domains of one life form in a landscape otherwise occupied by the other life form.

The next step in understanding the link between pattern formation and biodiversity is to consider a community consisting of many functional groups that are distributed along the χ axis and study the impact of the two coexistence mechanisms described above on community-level properties, such as functional diversity and its relations to rainfall and herbivory. Community-level properties can be extracted from plots of the community biomass function $B = B(\chi)$, where $B_i = B(\chi_i)$ is the biomass of the *ith* functional group (defined by a small χ increment around $\chi = \chi_i$). Solutions for $B = B(\chi)$ in homogeneous systems typically have pulse shapes from which three community-level properties can be deduced, functional diversity (pulse width), community composition (pulse location) and total biomass (pulse area). Studies of such pulse solutions in the context of the two species-coexistence mechanisms described above can clarify the roles these mechanisms play in shaping community-level properties. More generally, such studies can clarify how pattern formation processes affect the relationships between the abiotic environment, biodiversity and ecosystem function (see Figure 1.2).

Bibliography

[1] Z. Adeel, U. Safriel, D. Niemeijer, and R. White. Ecosystems and human well-being: Desertification synthesis. Technical Report of the Millennium Ecosystem Assessment, World Resources Institute, Washington, DC., 2005.

[2] A. Ana, I. Borthagaray, M. A. Fuentes, and P. A. Marquet. Vegetation pattern formation in a fog-dependent ecosystem. *Journal of Theoretical Biology*, 265:18–26, 2010.

[3] P. W. Anderson. More is different. *Science*, 177:393–396, 1972.

[4] I. Aranson, M. Assenheimer, V. Steinberg, and L. S. Tsimring. Large-scale flow and spiral core instability in Rayleigh–Bénard convection. *Physical Review E*, 55(5):R4877–4880, 1997.

[5] Igor Aranson, Lorenz Kramer, and Andreas Weber. Core instability and spatiotemporal intermittency of spiral waves in oscillatory media. *Phys. Rev. Lett.*, 72:2316–2319, Apr 1994.

[6] I. S. Aranson and L. Kramer. The world of the complex Ginzburg-Landau equation. *Review of Modern Physics*, 74(99), 2002.

[7] Cristina Armas, Susana Rodríguez-Echeverría, and Francisco I Pugnaire. A field test of the stress-gradient hypothesis along an aridity gradient. *Journal of Vegetation Science*, 22(5):818–827, 2011.

[8] M. Assenheimer and V. Steinberg. Observation of coexisting upflow and downflow hexagons in Boussinesq Rayleigh-Bénard convection. *Phys. Rev. Lett.*, 76:756–759, Jan 1996.

[9] N. K. Bajaj. *The Physics of Waves and Oscillations.* Tata Mc-Graw-Hill, 1984.

[10] P. Bak. Commensurate phases, incommensurate phases and the devil's staircase. *Rep. Prog. Phys.*, 45:587–629, 1982.

[11] I. V. Barashenkov and S. R. Woodford. Complexes of stationary domain walls in the resonantly forced Ginsburg-Landau equation. *Phys. Rev. E*, 71:026613, Feb 2005.

[12] G. I. Barenblatt. *Scaling, self-similarity, and intermediate asymptotics.* Cambridge University Press, 1996.

[13] A. Barlev. *Desert Poetry.* Me ve'Ma, 1994.

[14] M. Baudena, J. von Hardenberg, and A. Provenzale. Vegetation patterns and soil-atmosphere water fluxes in drylands. *Advances in Water Resources*, 53:131–138, 2013.

[15] Martin Baurmann, Thilo Gross, and Ulrike Feudel. Instabilities in spatially extended predator-prey systems: Spatio-temporal patterns in the neighborhood of Turing–Hopf bifurcations. *Journal of Theoretical Biology*, 245(2):220–229, 2007.

[16] Golan Bel, Aric Hagberg, and Ehud Meron. Gradual regime shifts in spatially extended ecosystems. *Theoretical Ecology*, 5:591–604, 2012.

[17] Andrew Belmonte, Qi Ouyang, and Jean-Marc Flesselles. Experimental survey of spiral dynamics in the Belousov–Zhabotinsky reaction. *J. Phys. II France*, 1997.

[18] J. Belnap and D. A. Gillette. Vulnerability of desert biological soil crusts to wind erosion: The influences of crust development, soil texture, and disturbance. *Journal of Arid Environments*, 39(2):133–142, 1998.

[19] E. Ben-Hur, O. Fragman-Sapir, R. Hadas, A. Singer, and R. Kadmon. Functional trade-offs increase species diversity in experimental plant communities. *Ecology Letters*, 15(11):1276–1282, 2012.

[20] E. Ben-Jacob, Nigel Goldenfeld, J. S. Langer, and Gerd Schön. Boundary-layer model of pattern formation in solidification. *Phys. Rev. A*, 29:330–340, Jan 1984.

[21] C. M. Bender and S. A. Orszag. *Advanced Mathematical Methods for Scientists and Engineers.* Springer, 1999.

[22] M. D. Bertness and R. Callaway. Positive interactions in communities. *Trends in Ecology and Evolution*, 9(5):191–193, 1994.

[23] A. J. Bloom, F. S. Chapin, and H. A. Mooney. Resource Limitation in Plants–An Economic Analogy. *Annual Review of Ecology and Systematics*, 16:363–392, 1985.

[24] E. Bodenschatz, W. Pesch, and G. Ahlers. Recent developments in Rayleigh-Bénard convection. *Annu. Rev. Fluid Mech.*, 32:709–778, 2000.

[25] Eberhard Bodenschatz, John R. de Bruyn, Guenter Ahlers, and David S. Cannell. Transitions between patterns in thermal convection. *Phys. Rev. Lett.*, 67:3078–3081, Nov 1991.

[26] Carl Boettiger, Noam Ross, and Alan Hastings. Early warning signals: the charted and uncharted territories. *Theoretical Ecology*, 6(3):255–264, 2013.

[27] T. Bohr, M. H. Jensen, G. Paladin, and A. Vulpiani. *Dynamical Systems Approach to Turbulence*. Cambridge University Press, Cambridge, 1998.

[28] J. A. Bonachela, M. A. Muoz, and S. A. Levin. Stochastic models of population extinction. *J. Stat. Phys.*, 148:723–739, 2012.

[29] F. Borgogno, P. D'Odorico, F. Laio, and L. Ridolfi. Mathematical models of vegetation pattern formation in ecohydrology. *Reviews of Geophysics*, 47:RG1005, 2009.

[30] D. D. Breshears, N. S. Cobb, P. M. Rich, K. P. Price, C. D. Allen, R. G. Balice, W. H. Romme, J. H. Kastens, M. L. Floyd, J. Belnap, et al. Regional vegetation die-off in response to global-change-type drought. *Proc. Natl. Acad. Sci. U. S. A.*, 102(42):15144–15148, 2005.

[31] J. M. Briggs, A. K. Knapp, J. M. Blair, J. L. Heisler, G. A. Hoch, M. S. Lett, and J. K. McCarron. Biological feedbacks in global desertification. *Bioscience*, 55:243–254, 2005.

[32] E. Buckingham. On physically similar systems: Illustrations of the use of dimensional equations. *Phys. Rev.*, 4(4):345–376, Oct 1914.

[33] F. H. Busse. Non-linear properties of thermal convection. *Reports on Progress in Physics*, 41:1929–1967, December 1978.

[34] Thomas Butler and Nigel Goldenfeld. Robust ecological pattern formation induced by demographic noise. *Phys. Rev. E.*, 80:030902, Sept 2009.

[35] R. M. Callaway. *Positive Interactions and Interdependence in Plant Communities*. Springer, 2007.

[36] R. M. Callaway, R. W. Brooker, P. Choler, Z. Kikvidze, C. J. Lortie, R. Michalet, L. Paolini, F. I. Pugnaire, B. Newingham, E. T. Aschehoug, C. Armas, D. Kikodze, and B. J. Cook. Positive interactions among alpine plants increase with stress. *Nature*, 417(6891):844–848, 2002.

[37] R. M. Callaway and L. R. Walker. Competition and facilitation: A synthetic approach to interactions in plant communities. *Ecology*, 78(7):1958–1965, 1997.

[38] S. E. Campbell. Soil stabilization by a prokaryotic desert crust: Implications for precambrian land biota. *Origins of Life*, 9:335–348, 1979.

[39] D. C. Canham, J. J. Cole, and W. K. Lauenroth, editors. *Models in Ecosystem Science*. Princeton University Press, 2003.

[40] R. S. Cantrell and C. Cosner. *Spatial Ecology via Reaction-Diffusion Equations (Wiley Series in Mathematical & Computational Biology)*. Wiley, 2003.

[41] J. Carr and R. L. Pego. Metastable patterns in solutions of ut = ϵ^2uxx – f(u). *Communications on Pure and Applied Mathematics*, 42(5):523–576, 1989.

[42] S. P. Carroll, A. P. Hendry, D. N. Reznick, and C. W. Fox. Evolution on ecological time-scales. *Functional Ecology*, 21:387–393, 2007.

[43] F. Carten, A. Marasco, G. Bonanomi, S. Mazzoleni, M. Rietkerk, and F. Giannino. Negative plant soil feedback explaining ring formation in clonal plants. *Journal of Theoretical Biology*, 313:153 –161, 2012.

[44] S. Chandrasekhar. *Hydrodynamic and Hydromagnetic Stability*. Clarendon Press, 1961.

[45] Hugues Chaté and Paul Manneville. Phase diagram of the two-dimensional complex Ginzburg–Landau equation. *Physica A: Statistical and Theoretical Physics*, 224(1-2):348–368, Feb 1996.

[46] Yiwei Cheng, Marc Stieglitz, Greg Turk, and Victor Engel. Effects of anisotropy on pattern formation in wetland ecosystems. *Geophysical Research Letters*, 38(4):n/a–n/a, 2011.

[47] M. G. Clerc, C. Falcon, and E. Tirapegui. Additive noise induces front propagation. *Phys. Rev. Lett.*, 94(14):148302, Apr 2005.

[48] J. D. Cole. On a quasi-linear parabolic equation occurring in aerodynamics. *Quart. Appl. Math.*, 9:225–236, 1951.

[49] J.H. Connell. Diversity in tropical rain forests and coral reefs. High diversity of trees and corals is maintained only in a non-equilibrium state. *Science*, pages 1302–1310, 1978.

[50] P. Coullet, C. Elphick, and D. Repaux. Nature of spatial chaos. *Phys. Rev. Lett.*, 58:431–434, 1987.

[51] P. Coullet and Emilsson. K. Strong resonances of spatially distributed oscillators: A laboratory to study patterns and defects. *Physica D: Nonlinear Phenomena*, 61(1-4):119–131, 1992.

[52] P. Coullet, J. Lega, B. Houchmanzadeh, and J. Lajzerowicz. Breaking chirality in nonequilibrium system. *Physical Review Letters*, 65:1352, 1990.

[53] C. M. Crain and M. D. Bertness. Ecosystem engineering across environmental gradients: Implications for conservation and management. *BioScience*, 56:211–218, 2006.

[54] Michael D. Cramer and Nichole N. Barger. Are Namibian fairy circles the consequence of self-organizing spatial vegetation patterning? *PloS one*, 8(8):e70876, 2013.

[55] M.C. Cross and H. Greenside. *Pattern Formation and Dynamics in Nonequilibrium Systems.* Cambridge University Press, 2009.

[56] M.C. Cross and P.C. Hohenberg. Pattern formation outside equilibrium. *Review of Modern Physics*, 65:851, 1993.

[57] K. Cuddington, J. Byers, A. Hastings, and W. Wilson, editors. *Ecosystem Engineers: Plants to Protists.* Academic Press, 2007.

[58] V. Dakos, M. Scheffer, E. H. van Nes, V. Brovkin, V. Petoukhov, and H. Held. Slowing down as an early warning signal for abrupt climate change. *Proceedings of the National Academy of Sciences*, 105(38):14308–14312, 2008.

[59] V. Dakos, E. H. van Nes, R. Donangelo, H. Fort, and M. Scheffer. Spatial correlation as leading indicator of catastrophic shifts. *Theo. Ecol.*, 3:163–174, 2010.

[60] Vasilis Dakos, Sonia Kéfi, Max Rietkerk, Egbert H Van Nes, and Marten Scheffer. Slowing down in spatially patterned ecosystems at the brink of collapse. *The American Naturalist*, pages E153–E166, 2011.

[61] A. De Wit, D. Lima, G. Dewel, and P. Borckmans. Spatiotemporal dynamics near a codimension-two point. *Phys. Rev. E*, 54:261–271, Jul 1996.

[62] D. L. DeAngelis and L. J. Gross, editors. *Individual-Based Models and Approaches on Ecology: Concepts and Models.* Chapman and Hall, 1992.

[63] V. Deblauwe, N. Barbier, P. Couteron, O. Lejeune, and J. Bogaert. The global biogeography of semi-arid periodic vegetation patterns. *Global Ecology and Biogeography*, 17:715–723, 2008.

[64] V. Deblauwe, P. Couteron, J. Bogaert, and N. Barbier. Determinants and dynamics of banded vegetation pattern migration in arid climates. *Ecological Monographs*, 82:3–21, 2012.

[65] S. C. Dekker, M. Rietkerk, and M. F. P. Bierkens. Coupling microscale vegetation-soil water and macroscale vegetation-precipitation feedbacks in semiarid ecosystems. *Global Change Biology*, 34:671–678, 2007.

[66] H. A. Dijkstra. Vegetation pattern formation in a semi-arid climate. *International Journal of Bifurcation and Chaos*, 21(12):3497–3509, 2011.

[67] E. Doedel. Auto: A program for the automatic bifurcation analysis of autonomous systems. *Congressus Numerantium*, 30:265–284, 1981.

[68] M. Dolnik, T. Bánsági, S. Ansari, I. Valent, and I.R. Epstein. Locking of Turing patterns in the chlorine dioxide-iodine-malonic acid reaction with one-dimensional spatial periodic forcing. *Phys. Chem. Chem. Phys.*, 13:12578–12583, 2011.

[69] M. Dolnik, I. Berenstein, A. M. Zhabotinsky, and I. R. Epstein. Spatial periodic forcing of Turing structures. *Physical Review Letters*, 87:238301, 2001.

[70] R. Donangelo, H. Fort, M. Scheffer, E. H. van Nes, and V. Dakos. Early warnings in ecosystens: Comparison between different indicators. *Int. J. Bifurcation Chaos*, 20:315–321, 2010.

[71] A. K. Duraiappah and S. Naeem. Ecosystems and human well-being: Biodiversity synthesis. Technical Report of the Millennium Ecosystem Assessment, World Resources Institute, Washington, DC., 2005.

[72] Sandra Díaz, Imanuel Noy-Meir, and Marcelo Cabido. Can grazing response of herbaceous plants be predicted from simple vegetative traits? *Journal of Applied Ecology*, 38(3):497–508, 2001.

[73] Ute Ebert and Wim van Saarloos. Front propagation into unstable states: universal algebraic convergence towards uniformly translating pulled fronts. *Physica D: Nonlinear Phenomena*, 146(14):1–99, 2000.

[74] R. E. Ecke, J. D. Farmer, and D. K. Umberger. Scaling of the Arnold tongues. *Nonlinearity*, 2(2):175, 1989.

[75] W. Eckhaus. *Studies in Nonlinear Stability Theory*. Springer, 1965.

[76] K. R. Elder, J. Viñals, and M. Grant. Dynamic scaling and quasiordered states in the two-dimensional Swift–Hohenberg equation. *Phys. Rev. A*, 46(12):7618–7629, Dec 1992.

[77] D. J. Eldridge, E. Zaady, and Shachak M. Infiltration through three contrasting biological soil crusts in patterned landscapes in the Negev, israel. *J. Stat. Phys.*, 148:723–739, 2012.

[78] S. P. Ellner and J. Guckenheimer. *Dynamic Models in Biology*. Princeton University Press, 2006.

[79] C. Elphick, A. Hagberg, Malomed B., and E. Meron. On the origin of traveling pulses in bistable systems. *Physics Letters A*, 230:33–37, 1997.

[80] C. Elphick, A. Hagberg, and E. Meron. Dynamical front transitions and vortex nucleation. *Physical Review E*, 51:1899–1915, 1995.

[81] C. Elphick, A. Hagberg, and E. Meron. Phase front instability in periodically forced oscillatory systems. *Physical Review Letters*, 80:5007–5010, 1998.

[82] C. Elphick, A. Hagberg, and E. Meron. Multi-phase patterns in periodically forced oscillatory systems. *Physical Review E*, 59:5285–5291, 1999.

[83] C. Elphick, E. Meron, and E. A. Spiegel. Spatiotemporal complexity in traveling wavetrains. *Phys. Rev. Lett.*, 61:496–499, 1988.

[84] Christian Elphick and Ehud Meron. Localized structures in surface waves. *Phys. Rev. A*, 40:3226–3229, Sep 1989.

[85] M. B. Eppinga, M. Rietkerk, M. J. Wassen, and P. C. De Ruiter. Linking habitat modification to catastrophic shifts and vegetation patterns in bogs. *Plant Ecology*, 200:53–68, 2009.

[86] Gerhard Ertl and Harm-Hinrich Rotermund. Spatiotemporal pattern formation in reactions at surfaces. *Current Opinion in Solid State and Materials Science*, 1(5):617–621, 1996.

[87] G. C. Evans. *The Quantitative Analysis of Plant Growth.* University of California Press, 1972.

[88] M. Evenari, L. Shanan, and N. H. Tadmor. *The Negev, the Challenge of a Desert.* Harvard University Press, 1971.

[89] C. Fernandez-Oto, M. Tlidi, D. Escaff, and M. G. Clerc. Strong interaction between plants induces circular barren patches: fairy circles. *Philos Trans A Math Phys Eng Sci*, 372(2027), 2014.

[90] R. J. Field and M. Burger. *Oscillations and Traveling Waves in Chemical Systems.* Wiley, 1985.

[91] R. J. Field, E. Koros, and R. M. Noyes. Oscillations in chemical cystems ii. thorough analysis of temporal oscillation in the bromate-cerium-malonic acid system. *J. Am. Chem. Soc.*, 94:8649–8664, 1972.

[92] R. J. Field and R. M. Noyes. Oscillations in chemical systems iv. Limit cycle behavior in a model of a real chemical reaction. *J. Chem. Phys.*, 60:1877–1884, 1974.

[93] R. FitzHugh. Impulses and physiological states in theoretical models of nerve membrane. *Biophysical J.*, 1:445–466, 1961.

[94] M. J. Follows, S. Dutkiewicz, S. Grant, and S. W. Chisholm. Emergent biogeography of microbial communities in a model ocean. *Science*, 315:1843–1846, 2007.

[95] P. Freitas and Rocha. C. Lyapunov functionals and stability for Fitzhugh–Nagumo systems. *Journal of Differential Equations*, 169:208–227, 2001.

[96] G. Freund, W. Pesch, and W. Zimmermann. Rayleigh–Bénard convection in the presence of spatial temperature modulations. *J. Fluid Mech.*, 673:318–348, 2011.

[97] Th. Friedrich, A. Lange, I. Rehberg, and R. Richter. A note on the magnetic spatial forcing of a ferrofluid layer. *Magnetohydrodynamics (0024-998X)*, 47(2):167–173, 2011.

[98] P. Gandhi, E. Knobloch, and C. Beaume. Localized states in periodically forced systems. *Phys. Rev. Lett.*, 114:034102, 2015.

[99] W. P. Ganley. Simple pendulum approximation. *Am. J. Phys.*, 53:73–76, 1985.

[100] K. J. Gaston. *Biodiversity - A Biology of Numbers and Differences.* Blackwell Science, 1996.

[101] S. Getzin, K. Wiegand, T. Wiegand, H. Yizhaq, J. von Hardenberg, and E. Meron. Adopting a spatially explicit perspective to study the mysterious fairy circles of Namibia. *Ecography*, 38:1–11, 2015.

[102] E. Gilad, M. Shachak, and E. Meron. Dynamics and spatial organization of plant communities in water limited systems. *Theoretical Population Biology*, 72:214–230, 2007.

[103] E. Gilad and J. von Hardenberg. A fast algorithm for convolution integrals with space and time variant kernels. *Journal of Computational Physics*, 216:326–336, 2006.

[104] E. Gilad, J. Von Hardenberg, A. Provenzale, M. Shachak, and E. Meron. Ecosystem engineers: From pattern formation to habitat creation. *Physical Review Letters*, 93(098105), 2004.

[105] E. Gilad, J. Von Hardenberg, A. Provenzale, M. Shachak, and E. Meron. A mathematical model for plants as ecosystem engineers. *Journal of Theoretical Biology*, 244:680, 2007.

[106] L. Glass. Synchronization and rhythmic processes in physiology. *Nature*, 410(6825):277–284, 2001.

[107] P. Glendinning. *Stability, Instability and Chaos: An Introduction to the Theory of Nonlinear Differential Equations.* Cambridge University Press, 1994.

[108] L. Goldberg and C. Tresser. Rotation orbits and the farey tree. *Ergod. Thy. & Dynam. Sys.*, 16:1011–1029, 1996.

[109] Raymond E. Goldstein, David J. Muraki, and Dean M. Petrich. Interface proliferation and the growth of labyrinths in a reaction-diffusion system. *Phys. Rev. E*, 53:3933–3957, Apr 1996.

[110] D. A. Golombek and R. E. Rosenstein. Physiology of circadian entrainment. *Physiol Rev*, 90:1063–1102, 2010.

[111] Karna Gowda, Hermann Riecke, and Mary Silber. Transitions between patterned states in vegetation models for semiarid ecosystems. *Phys. Rev. E*, 89:022701, Feb 2014.

[112] J. P. Grime. Competitive exclusion in herbaceous vegetation. *Nature*, (5396):344–347, 1973.

[113] V. Grimm and S. F. Railsback. *Individual-Based Modeling and Ecology.* Princeton University Press, 2005.

[114] J. Guckenheimer and P. Holmes. *Nonlinear Oscillations, Dynamical Systems, and Bifurcations of Vector Fields.* Applied Mathematical Sciences, No. 42. Springer, 1983.

[115] V. Guttal and C. Jayaprakash. Changing skewness: An early warning signal of regime shifts in ecosystems. *Ecology Letters*, 11:450–460, 2008.

[116] V. Guttal and C. Jayaprakash. Spatial variance and spatial skewness: Leading indicators of regime shifts in spatial ecological systems. *Theoretical Ecology*, 2:3–12, 2009.

[117] A. Hagberg and E. Meron. Complex patterns in reaction diffusion systems: A tale of two front instabilities. *Chaos*, 4:477–484, 1994.

[118] A. Hagberg and E. Meron. Pattern formation in non-gradient reaction diffusion systems: The effects of front bifurcations. *Nonlinearity*, 7:805–835, 1994.

[119] A. Hagberg and E. Meron. The dynamics of curved fronts: Beyond geometry. *Physical Review Letters*, 78:1166–1169, 1997.

[120] A. Hagberg, E. Meron, and T. Passot. Phase dynamics of nearly stationary patterns in activator-inhibitor systems. *Physical Review E*, 61:6471–6476, 2000.

[121] A. Hagberg and E. and Meron. Order parameter equations for front transitions: Nonuniformly curved fronts. *Physica D*, 123:460–473, 1998.

[122] D. Haim, G. Li, Q. Ouyang, W.D. McCormick, H.L. Swinney, A. Hagberg, and E. Meron. Breathing spots in a reaction diffusion system. *Physical Review Letters*, 77:190–193, 1996.

[123] Lev Haim, Yair Mau, and Ehud Meron. Spatial forcing of pattern-forming systems that lack inversion symmetry. *Phys. Rev. E*, 90:022904, Aug 2014.

[124] Vincent Hakim and Alain Karma. Theory of spiral wave dynamics in weakly excitable media: Asymptotic reduction to a kinematic model and applications. *Phys. Rev. E*, 60:5073–5105, Nov 1999.

[125] G. H. Hardy and E. M. Wright. *An Introduction to the Theory of Numbers.* Oxford University Press, 1979.

[126] A. Hastings and D. B. Wysham. Regime shifts in ecological systems can occur with no warning. *Ecology Letters*, 13(4):464–472, April 2010.

[127] M. Henriot, J. Burguete, and R. Ribotta. Entrainment of a spatially extended nonlinear structure under selective forcing. *Phys. Rev. Lett.*, 91(10):104501, 2003.

[128] R. HilleRisLambers, M. Rietkerk, F. Van den Bosch, H.H.T. Prins, and H. de Kroon. Vegetation pattern formation in semi-arid grazing systems. *Ecology*, 82:50–61, 2001.

[129] C. S. Holling. Resilience and stability of ecological systems. *Annual Review of Ecology and Systematics*, 4:1–23, 1973.

[130] C. S. Holling. Cross-scale morphology, geometry, and dynamics of ecosystems. *Ecological Monographs*, 62:447–502, 1992.

[131] E. E. Holmes, M. A. Lewis, J. E. Banks, and R. R. Veit. Partial differential equations in ecology: Spatial interactions and population dynamics. *Ecology*, 75(1):17–29, 1994.

[132] M. Holyoak, M. A. Leibold, and R. D. Holt, editors. *Metacommunities: Spatial Dynamics and Ecological Communities.* The University of Chicago Press, 2005.

[133] C. Holzapfel, K. Tielbörger, H. A. Parag, J. Kigel, and Sternberg M. Annual plant-shrub interactions along an aridity gradient. *Basic Appl. Ecol.*, 7:268–279, 2006.

[134] Eberhard Hopf. The partial differential equation ut + uux = uxx. *Communications on Pure and Applied Mathematics*, 3(3):201–230, 1950.

[135] M. Z. Hossain and J. M. Floryan. Instabilities of natural convection in a periodically heated layer. *Journal of Fluid Mechanics*, 733:33–67, 10 2013.

[136] R. B. Hoyle. *Pattern Formation: An Introduction to Methods.* Cambridge University Press, 2006.

[137] Jef Huisman, Nga, David M. Karl, and Ben Sommeijer. Reduced mixing generates oscillations and chaos in the oceanic deep chlorophyll maximum. *Nature*, 439:322–325, 2006.

[138] H. Ikeda, M. Mimura, and Y. Nishiura. Global bifurcation phenomena of traveling wave solutions for some bistable reaction-diffusion systems. *Non-Linear Anal.*, 13:507–526, April 1989.

[139] Bullock James M., Franklin Joe, Stevenson Mark J., Silvertown Jonathan, Coulson Sarah J., Gregory Steve J., and Tofts Richard. A plant trait analysis of responses to grazing in a long-term experiment. *Journal of Applied Ecology*, 38(2):253–267, 2001.

[140] J. Jiang, G. Daozhou, and D. L. DeAngelis. Towards a theory of ecotone resilience: Coastal vegetation on a salinity gradient. *Theoretical Population Biology*, 82:29–37, 2012.

[141] C. G. Jones, J. H. Lawton, and M. Shachak. Organisms as ecosystem engineers. *Oikos*, 69:373–386, 1994.

[142] C. G. Jones, J. H. Lawton, and M. Shachak. Positive and negative efects of organisms as ecosystem engineers. *Ecology*, 78:1946–1957, 1997.

[143] O. P. Judson. The rise of the individual-based model in ecology. *Trends in Ecology and Evolution*, 9:9–14, 1994.

[144] Norbert Juergens. The biological underpinnings of Namib desert fairy circles. *Science*, 339(6127):1618–1621, 2013.

[145] P. B. Kahn and Y. Zarmi. Weakly nonlinear oscillations: A perturbative approach. *Am. J. Phys.*, 72:538–552, 2004.

[146] P. B. Kahn and Y. Zarmi. Consistent application of the method of multiple-time scales to nonlinear systems. In *Nonlinear Phenomena Research Perspectives*, Nova Science Publishers, 2007.

[147] Prabha Kaira, Pablo S. Bodega, Christian Punckt, Harm Hinrich Rotermund, and Dagmar Krefting. Pattern formation in 4:1 resonance of the periodically forced co oxidation on pt(110). *Phys. Rev. E*, 77:046106, Apr 2008.

[148] Kyozi Kawasaki and Takao Ohta. Kink dynamics in one-dimensional nonlinear systems. *Physica A: Statistical Mechanics and its Applications*, 116(3):573–593, 1982.

[149] S. Kefi, M. Rietkerk, C. L. Alados, Y. Pueyo, V. P. Papanastasis, A. ElAich, and P. C. de Ruiter. Spatial vegetation patterns and imminent desertification in mediterranean arid ecosystems. *Nature*, 449:213–216, 2007.

[150] S. Kéfi, M. Rietkerk, M. Roy, A. Franc, P. C. de Ruiter, and M. Pascual. Robust scaling in ecosystems and the meltdown of patch size distributions before extinction. *Ecology Letters*, 14:29–35, 2011.

[151] M. A. Kessler and B. T. Werner. Self-Organization of Sorted Patterned Ground. *Science*, 299(5605):380–383, 2003.

[152] Shai Kinast, Yuval R. Zelnik, Golan Bel, and Ehud Meron. Interplay between Turing mechanisms can increase pattern diversity. *Phys. Rev. Lett.*, 112:078701, Feb 2014.

[153] Yuri S. Kivshar and Boris A. Malomed. Dynamics of solitons in nearly integrable systems. *Rev. Mod. Phys.*, 61:763–915, Oct 1989.

[154] C. A. Klausmeier and D. Tilman. Spatial models of competition. In *Competition and Coexistence*, Sommer and Worm Eds. Springer, 2002.

[155] C. A. Klausmeier. Regular and irregular patterns in semiarid vegetation. *Science*, 284:1826–1828, 1999.

[156] A. Y. Kletter, J. von Hardenberg, and E. Meron. Ostwald ripening in dryland vegetation. *Communications on Pure and Applied Analysis*, 11:261–273, 2012.

[157] E. Knobloch. Spatially localized structures in dissipative systems: Open problems. *Nonlinearity*, 21:T45, 2008.

[158] L. Korzinov, M.I. Rabinovich, and L.S. Tsimring. Symmetry breaking in nonequilibrium systems: Interaction of defects. *Physical Review A*, 46:7601, 1992.

[159] R. A. Kraenkel, J. G. Pereira, and M. A. Manna. The reductive perturbation method and the Korteweg-de Vries hierarchy. *Acta Appl. Math.*, 39:389–403, 1995.

[160] B. Krauskopf, H. M. Osinga, and J. Galán-Vioque, editors. *Numerical Continuation Methods for Dynamical Systems.* Wiley, 2007.

[161] Dagmar Krefting, Prabha Kaira, and Harm Hinrich Rotermund. Period doubling and spatiotemporal chaos in periodically forced co oxidation on pt(110). *Phys. Rev. Lett.*, 102:178301, Apr 2009.

[162] T. S. Kuhn. *The Structure of Scientific Revolutions.* University of Chicago Press, 1962.

[163] Yuri A. Kuznetsov. *Elements of Applied Bifurcation Theory*, volume 112 of *Applied Mathematcial Sciences.* Springer, third edition, 2004.

[164] P. Kyriazopoulos, N. Jonathan, and E. Meron. Species coexistence by front pinning. *Ecological Complexity*, 20C:271–281, 2014.

[165] Edelstein-Keshet. L. *Mathematical Models in Biology*, volume 46 of *SIAM Classics in Applied Mathematics.* SIAM, 2005.

[166] A. Lampert. Localized structures as spatial hosts for unstable modes. MSc thesis, Ben-Gurion University, 2007.

[167] A. Lampert and E. Meron. Localized structures as spatial hosts for unstable modes. *Europhysics Letters*, 78(1):1–5, 2007.

[168] K. J. Lee, W. D. McCormick, H. L. Swinney, and J. E. Pearson. Experimental observation of self-replicating spots in a reaction-diffusion system. *Nature*, 369:215–218, 1994.

[169] R. Lefever and O. Lejeune. On the origin of tiger bush. *Bull. Math. Biol.*, 59:263–294, 1997.

[170] O. Lejeune, Tlidi M., and R. Lefever. Vegetation spots and stripes: Dissipative structures in arid landscapes. *Int. J. Quantum Chem.*, 98:261–271, 2004.

[171] O. Lejeune, M. Tlidi, and P. Couteron. Localized vegetation patches: A self-organized response to resource scarcity. *Phys. Rev. E*, 66:010901, Jul 2002.

[172] S. A. Levin. The problem of pattern and scale in ecology. *Ecology*, 73:1943–11967, 1992.

[173] S. A. Levin and L. A. Segel. Pattern generation in space and aspect. *SIAM Review*, 27:45–67, 1985.

[174] J. Li, C. Y. Zhao, Y. J. Song, Y. Sheng, and H. Zhu. Spatial patterns of desert annuals in relation to shrub effects on soil moisture. *Journal of Vegetation Science*, 21:221–232, 2010.

[175] A. Lin, A. Hagberg, A. Ardelea, M. Bertram, H. L. Swinney, and E. Meron. Four-phase patterns in forced oscillatory systems. *Physical Review E*, 62:3790–3800, 2000.

[176] A. L. Lin, M. Bertram, K. Martinez, H. L. Swinney, A. Ardelea, and G. F. Carey. Resonant Phase Patterns in a Reaction-Diffusion System. *Phys. Rev. Lett.*, 84(18):4240–4243, 2000.

[177] Anna L. Lin, Aric Hagberg, Ehud Meron, and Harry L. Swinney. Resonance tongues and patterns in periodically forced reaction-diffusion systems. *Phys. Rev. E*, 69(6):066217, Jun 2004.

[178] Q.-X. Liu, A. Doelman, V. Rottschfer, M. de Jager, P. M. J. Herman, M. Rietkerk, and J. van de Koppel. Phase separation explains a new class of self-organized spatial patterns in ecological systems. *Proceedings of the National Academy of Sciences*, 2013.

[179] David J. B. Lloyd, Bjrn Sandstede, Daniele Avitabile, and Alan R. Champneys. Localized hexagon patterns of the planar Swift–Hohenberg equation. *SIAM J. Applied Dynamical Systems*, 7(3):1049–1100, 2008.

[180] M. Loreau. Linking biodiversity and ecosystems: Towards a unifying ecological theory. *Phil. Trans. R. Soc. B*, 265:49–60, 2010.

[181] M. Loreau, S. Naeem, P. Inchausti, J. Bengtsson, J.P. Grime, A. Hector, D.U. Hooper, M.A. Huston, D. Raffaelli, Schmid B., D. Tilman, and D.A. Wardle. Biodiversity and ecosystem functioning: Current knowledge and future challenges. *Science*, 294:804–808, 2001.

[182] Mary Lowe, J. P. Gollub, and T. C. Lubensky. Commensurate and incommensurate structures in a nonequilibrium system. *Phys. Rev. Lett.*, 51:786–789, Aug 1983.

[183] Tlidi M., R. Lefever, and A. Vladimirov. On vegetation clustering, localized bare soil spots and fairy circles. In *Dissipative Solutions: From Optics to Biology and Medicine*, Springer, 2008.

[184] D. Machiwal, Madan K. JHA, P. K. Singh, S. C. Mahnot, and A. Gupta. Planning and design of cost-effective water harvesting structures for efficient utilization of scarce water resources in semi-arid regions of rajasthan, india. *Water Resources Management*, 18:219–235, 2004.

[185] F. T. Maestre, F. Valladares, and J. F. Reynolds. Is the change of plant-plant interactions with abiotic stress predictable? A meta-analysis of field results in arid environments. *Journal of Ecology*, 93(4):748–757, 2005.

[186] Fernando T. Maestre, Ragan M. Callaway, Fernando Valladares, and Christopher J. Lortie. Refining the stress-gradient hypothesis for competition and facilitation in plant communities. *Journal of Ecology*, 97(2):199–205, 2009.

[187] A. E. Magurran. *Measuring Biological Diversity.* Blackwell Science, 2004.

[188] M. Mainguet. *Aridity: Droughts and Human Development.* Springer, 1999.

[189] H. Malchow, F. M. Hilker, R. R. Sarkar, and K. Brauer. Spatiotemporal patterns in an excitable plankton system with lysogenic viral infection. *Mathematical and Computer Modelling*, 42(9-10):1035–1048, 2005.

[190] H. Malchow, S. V. Petrovskii, and E. Venturino. *Spatiotemporal Patterns in Ecology And Epidemiology: Theory, Models, And Simulation.* Chapman & Hall/CRC Press, 2008.

[191] R. Maliva and T. Missimer. *Arid Lands Water Evaluation and Management.* Springer, 2012.

[192] Dan Malkinson and Katja Tielbrger. What does the stress-gradient hypothesis predict? Resolving the discrepancies. *Oikos*, 119(10):1546–1552, 2010.

[193] Boris A. Malomed, Alexander A. Nepomnyashchy, and Michael I. Tribelsky. Domain boundaries in convection patterns. *Phys. Rev. A*, 42:7244–7263, Dec 1990.

[194] A. Manor and N. Shnerb. Facilitation, competition, and vegetation patchiness: From scale free distributions to patterns. *J. Theoretical Biology*, 253:838–842, 2011.

[195] R. Manor, A. Hagberg, and E. Meron. Wavenumber locking in spatially forced pattern-forming systems. *Europhysics Letters*, 83(10005), 2008.

[196] R. Manor, A. Hagberg, and E. Meron. Wavenumber locking and pattern formation in spatially forced systems. *New Journal of Physics*, 11(063016), 2009.

[197] Marco Marani, Cristina Da Lio, and Andrea D'Alpaos. Vegetation engineers marsh morphology through multiple competing stable states. *Proceedings of the National Academy of Sciences*, 110(9):3259–3263, 2013.

[198] B. Marts, A. Hagberg, E. Meron, and A. L. Lin. Bloch-front turbulence in a periodically forced Belousov–Zhabotinsky reaction. *Physical Review Letters*, 93(108305):1–4, 2004.

[199] B. Marts, A. Hagberg, E. Meron, and A.L. Lin. Resonant and nonresonant patterns in forced oscillators. *Chaos*, 16(037113):1–8, 2006.

[200] Norman W. H. Mason and Francesco de Bello. Functional diversity: A tool for answering challenging ecological questions. *Journal of Vegetation Science*, 24(5):777–780, 2013.

[201] P. C. Matthews and S. M. Cox. Pattern formation with a conservation law. *Nonlinearity*, 13:1293–1320, 2000.

[202] Y. Mau. Pattern formation in spatially forced systems: Application to vegetation restoration. PhD thesis, Ben-Gurion University, 2013.

[203] Y. Mau, A. Hagberg, and E. Meron. Dual-mode spiral vortices. *Physical Review E*, 80:065203(R), 2009.

[204] Yair Mau, Aric Hagberg, and Ehud Meron. Spatial periodic forcing can displace patterns it is intended to control. *Phys. Rev. Lett.*, 109:034102, Jul 2012.

[205] Yair Mau, Lev Haim, Aric Hagberg, and Ehud Meron. Competing resonances in spatially forced pattern-forming systems. *Phys. Rev. E*, 88:032917, 2013.

[206] Yair Mau, Lev Haim, and Ehud Meron. Reversing desertification as a spatial resonance problem. *Phys. Rev. E*, 91:012903, 2015.

[207] R. M. May. Thresholds and breakpoints in ecosystems with a multiplicity of stable states. *Nature*, 296:471–477, 1977.

[208] K. D. M. McConnaughay and J. S. Coleman. Biomass allocation in plants: Ontogeny or optimality? A test along three resource gradients. *Ecology*, 80(8):2581–2593, 1999.

[209] A. Medvinsky, S. Petrovskii, I. Tikhonova, H. Malchow, and B. Li. Spatiotemporal complexity of plankton and fish dynamics. *SIAM Review*, 44(3):311–370, 2002.

[210] B. Meerson and P. V. Sasorov. Domain stability, competition, growth, and selection in globally constrained bistable systems. *Phys. Rev. E*, 53:3491–3494, 1996.

[211] E. Meron. Pattern formation in excitable media. *Physics Reports*, 218:1–66, 1992.

[212] E. Meron. Pattern-formation approach to modelling spatially extended ecosystems. *Ecological Modelling*, 234:70–82, 2012.

[213] E. Meron and E. Gilad. Dynamics of plant communities in drylands: A pattern formation approach. In *Complex Population Dynamics: Nonlinear Modeling in Ecology, Epidemiology and Genetics*, J. Kurths and L. Stone Eds. World-Scientific, 2007.

[214] E. Meron, E. Gilad, J. Von Hardenberg, M. Shachak, and Y. Zarmi. Vegetation patterns along a rainfall gradient. *Chaos Solitons and Fractals*, 19:367, 2004.

[215] E. Meron, H. Yizhaq, and E. Gilad. Localized structures in dryland vegetation: Forms and functions. *Chaos*, 17(037109), 2007.

[216] S. Métens, G. Dewel, P. Borckmans, and R. Engelhardt. Pattern selection in bistable systems. *Europhysics Letters*, 37:109–114, 1997.

[217] D. G. Míguez, E. M. Nicola, A. P. Muñuzuri, J. Casademunt, F. Sagués, and L. Kramer. Traveling-stripe forcing generates hexagonal patterns. *Phys. Rev. Lett.*, 93(4):048303, 2004.

[218] A. S. Mikhailov and K. Showalter. Control of waves, patterns and turbulence in chemical systems. *Physics Reports*, 425:79–194, 2006.

[219] H. C. Monger and B. T. Bestelmeyer. The soil-geomorphic template and biotic change in arid and semi-arid ecosystems. *Journal of Arid Environments*, 65(2):207–218, 2006.

[220] E. N. Mueller, J. Wainwright, A.J. Parsons, and L. Turnbull, editors. *Patterns of Land Degradation in Drylands*. Springer, 2014.

[221] J. Nagumo, S. Arimoto, and S. Yoshizawa. An active pulse transmission line simulating nerve axon. *Proc IRE*, 50:2061–2070, 1962.

[222] Jonathan Nathan, Yagil Osem, Moshe Shachak, and Ehud Meron. Linking functional diversity to environmental variability: An integrated mechanistic approach for water limited plant communities. *Submitted for publication*, 2014.

[223] R. Nathan and H. C. Muller-Landau. Spatial patterns of seed dispersal, their determinants and consequences for recruitment. *Trends in Ecology and Evolution*, 15(7):278–285, 2000.

[224] Ran Nathan. Long-Distance Dispersal of Plants. *Science*, 313(5788):786–788, 2006.

[225] A. H. Nayfeh. *Perturbation Methods*. Wiley, 2004.

[226] A. C. Newell and M. C. Cross. Convection patterns in large aspect ratio systems. *Physica D*, 10:299–328, 1984.

[227] A. C. Newell, T. Passot, and J. Lega. Order parameter equations for patterns. *Ann. Rev. Fluid Mech.*, 25:399–453, 1993.

[228] A. C. Newell and J. A. Whitehead. Finite bandwidth, finite amplitude convection. *Journal of Fluid Mechanics*, 38:279–303, 1969.

[229] A. C. Newell and J. V. Moloney. *Nonlinear Optics*. Addison-Wesley, 1992.

[230] Y. Nishiura and D. Ueyama. A skeleton structure of self-replicating dynamics. *Physica D: Nonlinear Phenomena*, 130:73–104, 1999.

[231] Amy Novick-Cohen and Lee A. Segel. Nonlinear aspects of the Cahn–Hilliard equation. *Physica D: Nonlinear Phenomena*, 10(3):277–298, 1984.

[232] Amy Novick-Cohen and Lee A. Segel. Nonlinear aspects of the Cahn-Hilliard equation. *Physica D: Nonlinear Phenomena*, 10(3):277–298, 1984.

[233] A. Novoplansky, D. Cohen, and T. Sachs. Ecological implications of correlative inhibition between plant shoots. *Physiologia Plantarum*, 77(1):136–140, 1989.

[234] R. V. O'Neill, D. L. Deangelis, J. B. Waide, and G. E. Allen. *A Hierarchical Concept of Ecosystems*. Monographs in Population Biology, No. 23. Princeton University Press, 1986.

[235] M. Or-Guil and M. Bode. Propagation of Turing–Hopf fronts. *Physica A: Statistical Mechanics and Its Applications*, 249(1-4):174–178, 1998.

[236] Michal Or-Guil, Ioannis G. Kevrekidis, and Markus Br. Stable bound states of pulses in an excitable medium. *Physica D: Nonlinear Phenomena*, 135(1-2):154–174, 2000.

[237] N. Oreskes, K. Shrader-Frechette, and K. Belitz. Verification, validation, and confirmation of numerical models in the earth sciences. *Science*, 263:641–646, 1994.

[238] Yagil Osem, Avi Perevolotsky, and Jaime Kigel. Site productivity and plant size explain the response of annual species to grazing exclusion in a mediterranean semi-arid rangeland. *Journal of Ecology*, 92(2):297–309, 2004.

[239] Qi Ouyang, Gemunu H. Gunaratne, and Harry L. Swinney. Rhombic patterns: Broken hexagonal symmetry. *Chaos: An Interdisciplinary Journal of Nonlinear Science*, 3(4):707–711, 1993.

[240] O. Ovaskainen and B. Meerson. Stochastic models of population extinction. *Trends Ecol Evol.*, 25:643–652, 2010.

[241] V. Petrov, Q. Ouyang, and H. L. Swinney. Resonant pattern formation in a chemical system. *Nature*, 388:655–657, 1997.

[242] S. T. A. Pickett and M. L. Cadenasso. The ecosystem as a multidimensional concept: Meaning, model, and metaphor. *Ecosystems*, 5:1–10, 2002.

[243] L. M. Pismen. *Patterns and Interfaces in Dissipative Dynamics*. Springer Series in Synergetics. Springer, 2006.

[244] Y. Pomeau. Front motion, metastability and subcritical bifurcations in hydrodynamics. *Physica D*, 23:3–11, 1986.

[245] William H. Press, Saul A. Teukolsky, William T. Vetterling, and Brian P. Flannery. *Numerical Recipes 3rd Edition: The Art of Scientific Computing*. Cambridge University Press, New York, NY, USA, 3 edition, 2007.

[246] M. Proulx and A. Mazumder. Reversal of grazing impact on plant species richness in nutrient-poor vs. nutrient-rich ecosystems. *Ecology*, page 2581–2592, 1998.

[247] F. I. Pugnaire and M. T. Luque. Changes in plant interactions along a gradient of environmental stress. *Oikos*, 93:42–49, 2001.

[248] J. Puigdefbregas. The role of vegetation patterns in structuring runoff and sediment uxes in drylands. *Earth Surface Processes and Landforms*, 30:133–147, 2003.

[249] M. I. Rabinovich, A. B. Ezersky, and P. D. Weidman. *The Dynamics of Patterns*. World Scientific, 2000.

[250] S. Ravi, P. D'Odorico, L. Wang, and S. Collins. Form and function of grass ring patterns in arid grasslands: The role of abiotic controls. *Oecologia*, 158:545–555, 2008.

[251] J. Realpe-Gomez, M. Baudena, T Galla, A. J. McKane, and M. Rietkerk. Demographic noise and resilience in a semi-arid ecosystem model. *Ecological Complexity*, 15:97–108, 2013.

[252] J. F. et al. Reynolds. Global desertification: Building a science for dryland development. *Science*, 316:847–851, 2007.

[253] L. Ridolfi, P. D'Odorico, and F. Laio. *Noise-Induced Phenomena in the Environmental Sciences*. Cambridge University Press, 2011.

[254] M. Rietkerk, M.C. Boerlijst, F. van Langevelde, R. HilleRisLambers, J. van de Koppel, L. Kumar, H. H. T. Prins, and A. M. De Roos. Self-organization of vegetation in arid ecosystems. *Am. Nat.*, 160:524–530, 2002.

[255] M. Rietkerk, S. C. Dekker, P. C. de Ruiter, and J. van de Koppel. Self-organized patchiness and catastrophic shifts in ecosystems. *Science*, 305:1926–1929, 2004.

[256] M. Rietkerk and J. van de Koppel. Regular pattern formation in real ecosystems. *Trends in Ecology and Evolution*, 23(3):169–175, 2008.

[257] M. Rietkerk, F. Van den Bosch, and J. van de Koppel. Site-specific properties and irreversible vegetation changes in semi-arid grazing systems. *Oikos*, 80:241–252, 1997.

[258] M. E. Ritchie. *Scale, Heterogeneity, and the Structure and Diversity of Ecological Communities*. Monographs in Population Biology, No. 45. Princeton University Press, 2010.

[259] P. Rohani, T. J. Lewis, D. Grnbaum, and G. D. Ruxton. Spatial self-organisation in ecology: Pretty patterns or robust reality? *Trends Ecol Evol.*, 12:70–74, 1997.

[260] Arkady B. Rovinsky and Michael Menzinger. Chemical instability induced by a differential flow. *Phys. Rev. Lett.*, 69:1193–1196, Aug 1992.

[261] A. B. Ryabov and B. Blasius. A graphical theory of competition on spatial resource gradients. *Ecology Letters*, 14:220–228, 2011.

[262] A. B. Ryabov, L. Rudolf, and B. Blasius. Vertical distribution and composition of phytoplankton under the influence of an upper mixed layer. *Journal of Theoretical Biology*, 263(1):120–133, 2010.

[263] E. J. Jr. Rykiel. Testing ecological models: The meaning of validation. *Ecological Modeling*, 90:229–2446, 1996.

[264] P. M. Saco, G. R. Willgoose, and G. R. Hancock. Eco-geomorphology of banded vegetation patterns in arid and semi-arid regions. *Hydrology and Earth System Sciences*, 11(6):1717–1730, 2007.

[265] O. Salazar and M. Casanova. Runoff water harvesting as a strategy for increasing agricultural production on hillslope areas in arid and semiarid zones. In *Water Recycling and Water Management*, Nova, 2011.

[266] M. Santillana and C. Dawson. A numerical approach to study the properties of solutions of the diffusive wave approximation of the shallow water equations. *Computational Geosciences*, 14:31–53, 2010.

[267] T. M. Scanlon, K. C. Kelly, S. A. Levin, and I. Rodriguez-Iturbe. Positive feedbacks promote power-law clustering of Kalahari vegetation. *Nature*, 449:209–212, 2007.

[268] Stefania Scarsoglio, Francesco Laio, Paolo DOdorico, and Luca Ridolfi. Spatial pattern formation induced by Gaussian white noise. *Mathematical Biosciences*, 229(2):174–184, 2011.

[269] M. Scheffer, J. Bascompte, W. A. Brock, V. Brovkin, S. R. Carpenter, V. Dakos, H. Held, E. H. van Nes, M. Rietkerk, and G. Sugihara. Early-warning signals for critical transitions. *Nature*, 461:387–393, 2009.

[270] M. Scheffer, S. Carpenter, J. A. Foley, C. Folke, and Walker B. Catastrophic shifts in ecosystems. *Nature*, 413:53–59, 2001.

[271] M. Scheffer, M. Holmgren, V. Brovkin, and M. Claussen. Synergy between small- and large-scale feedbacks of vegetation on the water cycle. *Global Change Biology*, 11(7):1003–1012, 2005.

[272] Marten Scheffer, Stephen R. Carpenter, Timothy M. Lenton, Jordi Bascompte, William Brock, Vasilis Dakos, Johan van de Koppel, Ingrid A. van de Leemput, Simon A. Levin, Egbert H. van Nes, Mercedes Pascual, and John Vandermeer. Anticipating critical transitions. *Science*, 338(6105):344–348, 2012.

[273] J. H. Schenk, S. Espino, C. M. Goedhart, M. Nordenstahl, H. I. Cabrera, and C. S. Jones. Hydraulic integration and shrub growth form linked across continental aridity gradients. *Proceedings of the National Academy of Sciences*, 105(32):11248–11253, 2008.

[274] W. H. Schlesinger, J. F. Reynolds, G. L. Cunningham, L. F. Huenneke, W. M. Jarrel, R. A. Virginia, and W. G. Whitford. Biological feedbacks in global desertification. *Science*, 247:1043–1048, 1990.

[275] L. A. Segel. Dissipative structure: an explanation and an ecological example. *J. Theor. Biol.*, 37:545–559, 1972.

[276] L. A. Segel. Distant side-walls cause slow amplitude modulation of cellular convection. *Journal of Fluid Mechanics*, 38:203–224, 1969.

[277] M. Segoli, E. D. Ungar, I. Giladi, A. Arnon, and M. Shachak. Untangling the positive and negative effects of shrubs on herbaceous vegetation in drylands. *Landscape Ecology*, 27:899–910, 2012.

[278] M. Segoli, E. D. Ungar, and M. Shachak. Fine-scale spatial heterogeneity of resource modulation in semi-arid islands of fertility. *Arid Land Research and Management*, 26:344–354, 2012.

[279] G. Seiden, S. Weiss, J.H. McCoy, W. Pesch, and E. Bodenschatz. Rayleigh-Bénard convection in the presence of spatial temperature modulations. *Phys. Rev. Lett.*, 101:214503, 2008.

[280] M. Shachak, J. R. Gosz, A. Perevolotsky, and S. T. A. Pickett. *Biodiversity in Drylands: Towards a Unified Framework*. Oxford University Press, 2005.

[281] M. Shachak and G. M. Lovett. Atmospheric deposition to a desert ecosystem and its implication for management. *Ecological Applications*, 8:455–463, 1998.

[282] M. Shachak, M. Sachs, and I. Moshe. Ecosystem management of desertified shrublands in Israel. *Ecosystems*, 1:475–483, 1998.

[283] E. Sheffer, J. von Hardenberg, H. Yizhaq, M. Shachak, and E. Meron. Emerged or imposed: A theory on the role of physical templates and self-organisation for vegetation patchiness. *Ecology Letters*, 16(2):127–139, 2013.

[284] E. Sheffer, J. von Hardenberg, H. Yizhaq, M. Shachak, and E. Meron. Self-organization of disordered vegetation patchiness. *In preparation*, 2014.

[285] E. Sheffer, H. Yizhaq, E. Gilad, M. Shachak, and E. Meron. Why do plants in resource deprived environments form rings? *Ecological Complexity*, 4:192–200, 2007.

[286] E. Sheffer, H. Yizhaq, E. Gilad, M. Shachak, and E. Meron. Mechanisms of vegetation-ring formation in water-limited systems. *J. Theoretical Biology*, 273:138–146, 2011.

[287] J. Sherratt, M. Smith, and J. Rademacher. Patterns of sources and sinks in the complex Ginzburg–Landau equation with zero linear dispersion. *SIAM Journal on Applied Dynamical Systems*, 9(3):883–918, 2010.

[288] Jonathan A. Sherratt and Alexios D. Synodinos. Vegetation patterns and desertification waves in semi-arid environments: Mathematical models based on local facilitation in plants. *Discrete and Continuous Dynamical Systems Series B*, 17:2815–2827, 2012.

[289] N. Shigesada and K. Kawasaki. *Biological Invasions: Theory and Practice.* Oxford University Press, UK, 1997.

[290] N. M. Shnerb, P. Sarah, H. Lavee, and S. Solomon. Reactive glass and vegetation patterns. *Physical Review Letters*, 90:0381011, 2003.

[291] Hiroto Shoji and Yoh Iwasa. Labyrinthine versus straight-striped patterns generated by two-dimensional Turing systems. *Journal of Theoretical Biology*, 237(1):104–116, 2005.

[292] Koen Siteur, Eric Siero, Maarten B. Eppinga, Jens D. M. Rademacher, Arjen Doelman, and Max Rietkerk. Beyond Turing: The response of patterned ecosystems to environmental change. *Ecological Complexity*, 20(0):81–96, 2014.

[293] R. V. Solé and J. Bascompte. *Self-Organization in Complex Ecosystems.* Princeton University Press, 2006.

[294] I. Stavi, H. Lavee, E. D. Ungar, and P. Sarah. Ecogeomorphic feedbacks in semiarid rangelands: A review. *Pedosphere*, 19(2):217–229, 2009.

[295] S. C. Stearns. Trade-offs in life-history evolution. *Functional Ecology*, 3:259–268, 1989.

[296] S. H. Strogatz. *Nonlinear Dynamics and Chaos: With Applications to Physics, Biology, Chemistry, and Engineering.* Westview Press, 2001.

[297] K. N. Suding, K. L. Gross, and G. R. Houseman. Alternative states and positive feedbacks in restoration ecology. *Trends in Ecology and Evolution*, 19:46–53, 2004.

[298] J. Swift and P.C. Hohenberg. Hydrodynamic fluctuations at the convective instability. *Physical Review A*, 15:319–328, 1977.

[299] W. Y. Tam, W. Horsthemke, Z. Noszticzius, and H. L. Swinney. Sustained spiral waves in a continuously fed unstirred chemical reactor. *The Journal of Chemical Physics*, 88(5):3395–3396, 1988.

[300] G. I. Taylor. The formation of a blast wave by a very intense explosion. i. Theoretical discussion. *Proc. Roy. Soc. A*, 201:159–174, 1950.

[301] G. I. Taylor. The formation of a blast wave by a very intense explosion. ii. The atomic explosion of 1945. *Proc. Roy. Soc. A*, 201:175–186, 1950.

[302] J. J. Tewksbury and J. D. Lloyd. Positive interactions under nurse-plants: Spatial scale, stress gradients and benefactor size. *Oecologia*, 127:425–434, 2001.

[303] J. Tews, U. Brose, V. Grimm, K. Tielbörger, M. C. Wichmann, M. Schwager, and F. Jeltsch. Animal species diversity driven by habitat heterogeneity/diversity: The importance of keystone structures. *Journal of Biogeography*, 31:79–92, 2004.

[304] S. Thompson and K. E. Daniels. A porous convection model for small-scale grass patterns, 2010.

[305] S. Thompson, G. Katul, and S. M. McMahon. Role of biomass spread in vegetation pattern formation within arid ecosystems. *Water Resources Research*, 44:W10421, 2008.

[306] R. E. Thorne. Charge-density-wave conductors. *Physics Today*, 49:42–47, 1996.

[307] D. Tilman and P. Kareiva, editors. *Spatial Ecology: The Role of Space in Population Dynamics and Interspecific Interactions.* Princeton University Press, 1997.

[308] David Tilman. *Resource Competition and Community Structure.* Monographs in Population Biology. Princeton University Press, Princeton, New Jersey, USA, 1982.

[309] D. J. Tongway, C. Valentin, J. Seghieri, and J.-C. Menaut, editors. *Banded Vegetation Patterning in Arid and Semiarid Environments: Ecological Processes and Consequences for Management.* Ecological Studies 149. Springer, 2001.

[310] D. J. Tritton, editor. *Physica Fluid Dynamics.* Oxford University Press, 2003.

[311] J. E. Truscott and J. Brindley. Ocean plankton populations as excitable media. *Bulletin of Mathematical Biology*, 56(5):981–998, 1994.

[312] Walter R. Tschinkel. The life cycle and life span of Namibian fairy circles. *PloS one*, 7(6):e38056, 2012.

[313] A. M. Turing. The chemical basis of morphogenesis. *Philosophical Transactions of the Royal Society of London. Series B, Biological Sciences*, 237:37–72, 1952.

[314] M. G. Turner. Landscape ecology: What is the state of the science? *Annu. Rev. Ecol. Evol. Syst.*, 36:319–344, 2005.

[315] John J. Tyson and James P. Keener. Singular perturbation theory of traveling waves in excitable media (a review). *Physica D: Nonlinear Phenomena*, 32(3):327–361, 1988.

[316] Scientific United Nations Educational and Cultural Organization (UNESCO). Map of the world distribution of arid regions: Map at scale 1:25,000,000 with explanatory note. Technical Report MAB Technical Notes 7, UNESCO, Paris, 1979.

[317] C. Valentine, J. M. d'Herbes, and J. Poesen. Soil and water components of banded vegetation patterns. *Catena*, 37:1–24, 1999.

[318] S. van der Stelt, A. Doelman, G. M. Hek, and J. Rademacher. Rise and fall of periodic patterns for a generalized Klausmeier-Gray-Scott model. *J. Nonlinear Sci.*, 23:39–95, 2013.

[319] A. G. van der Valk and B. G. Warner. The development of patterned mosaic landscapes: an overview. *Plant Ecology*, 200:1385–0237, 2009.

[320] W. van Saarloos and P. C. Hohenberg. Fronts, pulses, sources and sinks in generalized complex Ginzberg-Landau equations. *Phys. D*, 56(4):303–367, June 1992.

[321] V. K. Vanag and I. R. Epstein. Patterns and nanodroplets: The BZ-AOT microemulsion system. In *Self Organized Morphology in Nanostructured Materials*, K. Al-Shamery and J. Parisi, eds. Springer, 2008.

[322] J. Vandermeer and S. Yitbarek. Self-organized spatial pattern determines biodiversity in spatial competition. *Journal of Theoretical Biology*, 300:48–56, 2012.

[323] E. Verrecchia, A. Yair, G. J. Kidron, and K. Verrecchia. Physical properties of the psammophile cryptogamic crust and their consequences to the water regime of sandy softs, north-western Negev desert, Israel. *J Arid Environments*, 29:427–437, 1995.

[324] J. von Hardenberg, A. Y. Kletter, H. Yizhaq, J. Nathan, and E. Meron. Periodic vs. scale-free patterns in dryland vegetation. *Proceedings of the Royal Society B*, 277:1771–1776, 2010.

[325] J. von Hardenberg, E. Meron, M. Shachak, and Y. Zarmi. Diversity of vegetation patterns and desertification. *Physical Review Letters*, 87(198101), 2001.

[326] P. W. Voorhees. The theory of Ostwald ripening. *J. Stat. Phys.*, 38(1/2):231–252, 1985.

[327] B. H. Walker, D. Ludwig, C. S. Holling, and R. M. Peterman. Stability of semi-arid savanna grazing systems. *Journal of Ecology*, 69(2):pp. 473–498, 1981.

[328] Hongli Wang, Ke Zhang, and Qi Ouyang. Resonant-pattern formation induced by additive noise in periodically forced reaction-diffusion systems. *Phys. Rev. E*, 74:036210, Sep 2006.

[329] Haim Weissmann and Nadav M. Shnerb. Stochastic desertification. *EPL (Europhysics Letters)*, 106(2):28004, 2014.

[330] N. E. West. Structure and function of microphytic soil crusts in wildland ecosystems of arid to semi-arid regions. *Advances in Ecological Research*, 20:179–223, 1990.

[331] A. T. Winfree. Spiral waves of chemical activity. *Science*, 175:634–636, 1972.

[332] S. Yachi and M. Loreau. Biodiversity and ecosystem productivity in a fluctuating environment: The insurance hypothesis. *Proceedings of the National Academy of Sciences*, 96:1463–1468, 1999.

[333] A. Yair and M. Shachak. A case study of energy, water and soil flow chains in an arid ecosystem. *Oecologia*, 54:389–397, 1982.

[334] H. Yizhaq, E. Gilad, and E. Meron. Banded vegetation: Biological productivity and resilience. *Physica A*, 356:139, 2005.

[335] H. Yizhaq, S. Sela, T. Svoray, S. Assouline, and G. Bel. Effects of heterogeneous soil-water diffusivity on vegetation pattern formation. *Water Resources Research*, 339:5743–5758, 2014.

[336] A. Yochelis, A. Hagberg, C. Elphick, and E. Meron. Two-phase resonant patterns in forced oscillatory systems: Boundaries, mechanisms and forms. *Physica D*, 199:201–222, 2004.

[337] A. Yochelis, A. Hagberg, C. Elphick, and E. Meron. Frequency locking in extended systems: The impact of a Turing mode. *Europhysics Letters*, 69:170–176, 2005.

[338] A. Yochelis, A. Hagberg, E. Meron, A. Lin, and H.L. Swinney. Development of standing-wave labyrinthine patterns. *SIAM Journal on Applied Dynamical Systems*, 1:236–247, 2002.

[339] Y. R. Zelnik, S. Kinast, H. Yizhaq, G. Bel, and E. Meron. Regime shifts in models of dryland vegetation. *Philosophical Transactions R. Soc. A*, 371:20120358, 2013.

[340] Yuval R. Zelnik, Ehud Meron, and Golan Bel. Gradual regime shifts in fairy circles. *Submitted for publication*, 2014.

[341] Xiao-Chong Zhang, Gui-Quan Sun, and Zhen Jin. Spatial dynamics in a predator-prey model with Beddington–Deangelis functional response. *Phys. Rev. E*, 85:021924, 2012.

Index

T

U

V

W

For Product Safety Concerns and Information please contact our EU representative GPSR@taylorandfrancis.com Taylor & Francis Verlag GmbH, Kaufingerstraße 24, 80331 München, Germany

Batch number: 10722100

Printed by Printforce, the Netherlands